Dingyü Xue

Solving Optimization Problems with MATLAB®

Also of Interest

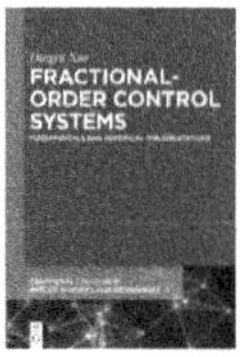

Fractional-Order Control Systems, Fundamentals and Numerical Implementations
Dingyü Xue, 2017
ISBN 978-3-11-049999-5, e-ISBN (PDF) 978-3-11-049797-7,
e-ISBN (EPUB) 978-3-11-049719-9

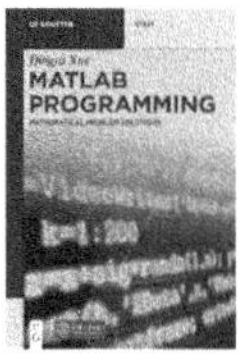

MATLAB® Programming, Mathematical Problem Solutions
Dingyü Xue, 2020
ISBN 978-3-11-066356-3, e-ISBN (PDF) 978-3-11-066695-3,
e-ISBN (EPUB) 978-3-11-066370-9

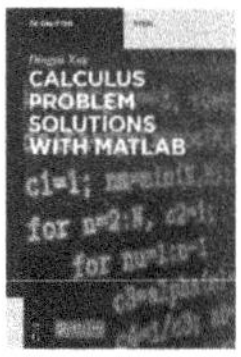

Calculus Problem Solutions with MATLAB®
Dingyü Xue, 2020
ISBN 978-3-11-066362-4, e-ISBN (PDF) 978-3-11-066697-7,
e-ISBN (EPUB) 978-3-11-066375-4

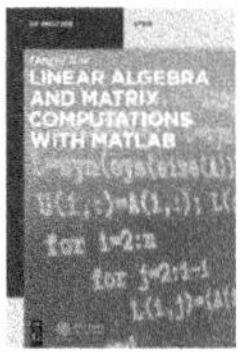

Linear Algebra and Matrix Computations with MATLAB®
Dingyü Xue, 2020
ISBN 978-3-11-066363-1, e-ISBN (PDF) 978-3-11-066699-1,
e-ISBN (EPUB) 978-3-11-066371-6

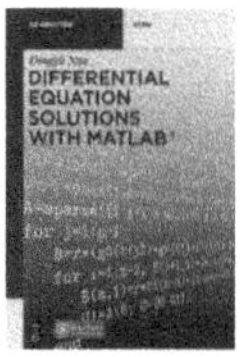

Differential Equation Solutions with MATLAB®
Dingyü Xue, 2020
ISBN 978-3-11-067524-5, e-ISBN (PDF) 978-3-11-067525-2,
e-ISBN (EPUB) 978-3-11-067531-3

Dingyü Xue

Solving Optimization Problems with MATLAB®

DE GRUYTER

清華大學出版社
TSINGHUA UNIVERSITY PRESS

Author
Prof. Dingyü Xue
School of Information Science and Engineering
Northeastern University
Wenhua Road 3rd Street
110819 Shenyang
China
xuedingyu@mail.neu.edu.cn

ISBN 978-3-11-066364-8
e-ISBN (PDF) 978-3-11-066701-1
e-ISBN (EPUB) 978-3-11-066369-3

Library of Congress Control Number: 2020931460

Bibliographic information published by the Deutsche Nationalbibliothek
The Deutsche Nationalbibliothek lists this publication in the Deutsche Nationalbibliografie; detailed bibliographic data are available on the Internet at http://dnb.dnb.de.

Cover image: Dingyü Xue
Typesetting: VTeX UAB, Lithuania
Printing and binding: CPI books GmbH, Leck

www.degruyter.com

Preface

Scientific computing is commonly and inevitably encountered in course learning, scientific research and engineering practice for each scientific and engineering student and researcher. For the students and researchers in the disciplines which are not pure mathematics, it is usually not a wise thing to learn thoroughly low-level details of related mathematical problems, and also it is not a simple thing to find solutions of complicated problems by hand. It is an effective way to tackle scientific problems, with high efficiency and in accurate and creative manner, with the most advanced computer tools. This method is especially useful in satisfying the needs for those in the area of science and engineering.

The author had made some effort towards this goal by addressing directly the solution methods for various branches in mathematics in a single book. Such a book, entitled "MATLAB based solutions to advanced applied mathematics", was published first in 2004 by Tsinghua University Press. Several new editions are published afterwards: in 2015, the second edition in English by CRC Press and in 2018, the fourth edition in Chinese were published. Based on the latest Chinese edition, a brand new MOOC project was released in 2018,[1] and received significant attention. The number of the registered students was around 14 000 in the first round of the MOOC course, and reached tens of thousands in later rounds. The textbook has been cited tens of thousands times by journal papers, books, and degree theses.

The author has over 30 years of extensive experience of using MATLAB in scientific research and education. Significant amount of materials and first-hand knowledge has been accumulated, which cannot be covered in a single book. A series entitled "Professor Xue Dingyü's Lecture Hall" of such works are scheduled with Tsinghua University Press, and the English editions are included in the DG STEM series with De Gruyter. These books are intended to provide systematic, extensive and deep explorations in scientific computing skills with the use of MATLAB and related tools. The author wants to express his sincere thanks to his supervisor, Professor Derek Atherton of Sussex University, who first brought him into the paradise of MATLAB.

The MATLAB series is not a simple revision of the existing books. With decades of experience and material accumulation, the idea of "revisiting" is adopted in authoring these books, in contrast to other mathematics and MATLAB-rich books. The viewpoint of an engineering professor is established and the focus is on solving various applied mathematical problems with tools. Many innovative skills and general-purpose solvers are provided to solve problems with MATLAB, which is not possible by any other existing solvers, so as to better illustrate the applications of computer tools in solving mathematical problems in every mathematics branch. It also helps

1 MOOC (in Chinese) address: https://www.icourse163.org/learn/NEU-1002660001

https://doi.org/10.1515/9783110667011-201

the readers broaden their viewpoints in scientific computing, and even finding innovative solutions by themselves to scientific computing which cannot be solved by any other existing methods.

The first title in the MATLAB series: “MATLAB Programming”, can be used as an entry-level textbook or reference book to MATLAB programming, so as to establish a solid foundation and deep understanding for the application of MATLAB in scientific computing. Each subsequent volumes tries to cover a branch or topic in mathematical courses. Bearing in mind the “computational thinking” in authoring the series, deep understanding and explorations are made for each mathematics branch involved. These MATLAB books are suitable for the readers who have already learnt the related mathematical courses, and want to revisit the courses to learn how to solve the problems by using computer tools. It can also be used as a companion in synchronizing the learning of related mathematics courses, and viewing the course from a different angle, so that the readers may expand their knowledge in learning the related courses, so as to better learn, understand and practice the materials in the courses.

This book is the fourth one in the MATLAB series. Two main topics – nonlinear equation solutions and optimization techniques – are covered in this book. We concentrate on solving the problems in these two fields. Analytical and numerical solutions of various nonlinear algebraic equations are discussed first, and solution approaches are provided to equations with multiple solutions. The subsequent chapters are devoted to introducing unconstrained optimization, linear and quadratic, constrained nonlinear, mixed-integer, multiobjective and dynamical programming. Practical attempts are made for finding global optimum solutions. Some intelligent optimization methods are introduced, and strict comparisons are made to assess the behaviors of conventional and intelligent solvers, and useful conclusions are reached.

At the time the books are published, the author wishes to express his sincere gratitude to his wife, Professor Yang Jun. Her love and selfless care over the decades provided the author immense power, which supports the authors’ academic research, teaching, and writing.

September 2019 Xue Dingyü

Contents

Preface — V

1 **An introduction to equations and optimization problems — 1**
1.1 Equations and their solutions — 1
1.2 Origins and development of optimization problems — 2
1.3 Structure of the book — 3
1.4 Exercises — 4

2 **Solutions of algebraic equations — 7**
2.1 Solutions of polynomial equations — 7
2.1.1 Polynomial equations of degrees 1 and 2 — 8
2.1.2 Analytical solutions of cubic equations — 9
2.1.3 Analytical solutions of quartic equation — 10
2.1.4 Higher-degree equations and Abel–Ruffini theorem — 13
2.2 Graphical methods for nonlinear equations — 13
2.2.1 Smooth graphics for implicit functions — 13
2.2.2 Univariate equations — 15
2.2.3 Equations with two unknowns — 17
2.2.4 Isolated equation solutions — 20
2.3 Numerical solutions of algebraic equations — 20
2.3.1 Newton–Raphson iterative algorithm — 20
2.3.2 Direct solution methods with MATLAB — 25
2.3.3 Accuracy specifications — 28
2.3.4 Complex domain solutions — 29
2.4 Accurate solutions of simultaneous equations — 31
2.4.1 Analytical solutions of low-degree polynomial equations — 32
2.4.2 Quasianalytical solutions of polynomial-type equations — 35
2.4.3 Quasianalytical solutions of polynomial matrix equations — 37
2.4.4 Quasianalytical solutions of nonlinear equations — 40
2.5 Nonlinear matrix equations with multiple solutions — 41
2.5.1 An equation solution idea and its implementation — 41
2.5.2 Pseudopolynomial equations — 46
2.5.3 A quasianalytical solver — 48
2.6 Underdetermined algebraic equations — 49
2.7 Exercises — 51

3 **Unconstrained optimization problems — 55**
3.1 Introduction to unconstrained optimization problems — 55
3.1.1 The mathematical model of unconstrained optimization problems — 55
3.1.2 Analytical solutions of unconstrained minimization problems — 56

3.1.3 Graphical solutions — 56
3.1.4 Local and global optimum solutions — 58
3.1.5 MATLAB implementation of optimization algorithms — 60
3.2 Direct solutions of unconstrained optimization problem with MATLAB — 62
3.2.1 Direct solution methods — 62
3.2.2 Control options in optimization — 65
3.2.3 Additional parameters — 69
3.2.4 Intermediate solution process — 70
3.2.5 Structured variable description of optimization problems — 72
3.2.6 Gradient information — 73
3.2.7 Optimization solutions from scattered data — 77
3.2.8 Parallel computation in optimization problems — 78
3.3 Towards global optimum solutions — 79
3.4 Optimization with decision variable bounds — 83
3.4.1 Univariate optimization problem — 83
3.4.2 Multivariate optimization problems — 85
3.4.3 Global optimum solutions — 87
3.5 Application examples of optimization problems — 87
3.5.1 Solutions of linear regression problems — 88
3.5.2 Least-squares curve fitting — 89
3.5.3 Shooting method in boundary value differential equations — 93
3.5.4 Converting algebraic equations into optimization problems — 96
3.6 Exercises — 97

4 Linear and quadratic programming — 103
4.1 An introduction to linear programming — 104
4.1.1 Mathematical model of linear programming problems — 104
4.1.2 Graphical solutions of linear programming problems — 105
4.1.3 Introduction to the simplex method — 106
4.2 Direct solutions of linear programming problems — 110
4.2.1 A linear programming problem solver — 110
4.2.2 Linear programming problems with multiple decision vectors — 116
4.2.3 Linear programming with double subscripts — 117
4.2.4 Transportation problem — 118
4.3 Problem-based description and solution of linear programming problems — 122
4.3.1 MPS file for linear programming problems — 122
4.3.2 Problem-based description of linear programming problems — 124
4.3.3 Conversions in linear programming problems — 129
4.4 Quadratic programming — 130
4.4.1 Mathematical quadratic programming models — 131

4.4.2 Direct solutions of quadratic programming problems — 131
4.4.3 Problem-based quadratic programming problem description — 132
4.4.4 Quadratic programming problem with double subscripts — 136
4.5 Linear matrix inequalities — 138
4.5.1 Description of linear matrix inequality problems — 138
4.5.2 Lyapunov inequalities — 139
4.5.3 Classifications of LMI problems — 141
4.5.4 MATLAB solutions of LMI problems — 142
4.5.5 Optimization solutions with YALMIP toolbox — 144
4.5.6 Trials on nonconvex problems — 146
4.5.7 Problems with quadratic constraints — 147
4.6 Exercises — 149

5 Nonlinear programming — 153
5.1 Introduction to nonlinear programming — 153
5.1.1 Mathematical models of nonlinear programming problems — 154
5.1.2 Feasible regions and graphical methods — 154
5.1.3 Examples of numerical methods — 157
5.2 Direct solutions of nonlinear programming problems — 159
5.2.1 Direct solution using MATLAB — 159
5.2.2 Handling of earlier termination phenomenon — 165
5.2.3 Gradient information — 166
5.2.4 Solving problems with multiple decision vectors — 168
5.2.5 Complicated nonlinear programming problems — 169
5.3 Trials with global nonlinear programming solver — 171
5.3.1 Trials on global optimum solutions — 171
5.3.2 Nonconvex quadratic programming problems — 174
5.3.3 Concave-cost transportation problem — 176
5.3.4 Testing of the global optimum problem solver — 178
5.3.5 Handling piecewise objective functions — 179
5.4 Bilevel programming problems — 181
5.4.1 Bilevel linear programming problems — 182
5.4.2 Bilevel quadratic programming problem — 183
5.4.3 Bilevel program solutions with YALMIP Toolbox — 184
5.5 Nonlinear programming applications — 185
5.5.1 Maximum inner polygon inside a circle — 185
5.5.2 Semiinfinite programming problems — 189
5.5.3 Pooling and blending problem — 193
5.5.4 Optimization design of heat exchange network — 196
5.5.5 Solving nonlinear equations with optimization techniques — 199
5.6 Exercises — 201

6 **Mixed integer programming — 207**
6.1 Introduction to integer programming — **207**
6.1.1 Integer and mixed-integer programming problems — **207**
6.1.2 Computational complexity of integer programming problems — **208**
6.2 Enumeration methods for integer programming — **209**
6.2.1 An introduction to the enumeration method — **209**
6.2.2 Discrete programming — **213**
6.2.3 0–1 programming — **214**
6.2.4 Trials on mixed-integer programming problems — **216**
6.3 Solutions of mixed-integer programming problems — **219**
6.3.1 Mixed-integer linear programming — **219**
6.3.2 Integer programming with YALMIP Toolbox — **222**
6.3.3 Mixed-integer nonlinear programming — **223**
6.3.4 A class of discrete programming problems — **226**
6.3.5 Solutions of ordinary discrete programming problems — **227**
6.4 Mixed 0–1 programming problems — **229**
6.4.1 0–1 linear programming problems — **229**
6.4.2 0–1 nonlinear programming problems — **233**
6.5 Mixed-integer programming applications — **235**
6.5.1 Optimal material usage — **235**
6.5.2 Assignment problem — **236**
6.5.3 Traveling salesman problem — **238**
6.5.4 Knapsack problems — **242**
6.5.5 Sudoku problems — **244**
6.6 Exercises — **247**

7 **Multiobjective programming — 253**
7.1 Introduction to multiobjective programming — **253**
7.1.1 Background introduction — **253**
7.1.2 Mathematical model of multiobjective programming — **254**
7.1.3 Graphical solution of multiobjective programming problems — **255**
7.2 Multiobjective programming conversions and solutions — **257**
7.2.1 Least-squares solutions of multiobjective programming problems — **258**
7.2.2 Linear weighting conversions — **260**
7.2.3 Best compromise solution of linear programs — **261**
7.2.4 Least-squares linear programming — **263**
7.3 Pareto optimal solutions — **264**
7.3.1 Nonuniqueness of multiobjective programming — **265**
7.3.2 Dominant solutions and Pareto frontiers — **265**
7.3.3 Computations of Pareto frontier — **267**
7.4 Minimax problems — **268**

7.5 Exercises — 275

8 Dynamic programming and shortest paths — 277
8.1 An introduction to dynamic programming — 277
8.1.1 Concept and mathematical models in dynamic programming — 277
8.1.2 Dynamic programming solutions of linear programming problems — 278
8.2 Shortest path problems in oriented graphs — 279
8.2.1 Examples of oriented graphs — 280
8.2.2 Manual solutions of shortest path problem — 281
8.2.3 Solution with dynamic programming formulation — 282
8.2.4 Matrix representation of graphs — 283
8.2.5 Finding the shortest path — 284
8.2.6 Dijkstra algorithm implementation — 288
8.3 Optimal paths for undigraphs — 290
8.3.1 Matrix description — 290
8.3.2 Route planning for cities with absolute coordinates — 292
8.4 Exercises — 293

9 Introduction to intelligent optimization methods — 297
9.1 Intelligent optimization algorithms — 297
9.1.1 Genetic algorithms — 297
9.1.2 Particle swarm optimization methods — 299
9.2 MATLAB Global Optimization Toolbox — 299
9.3 Examples and comparative studies of intelligent optimization methods — 301
9.3.1 Unconstrained optimization problems — 302
9.3.2 Constrained optimization problems — 305
9.3.3 Mixed-integer programming — 312
9.3.4 Discrete programming problems with the genetic algorithm — 315
9.4 Exercises — 317

Bibliography — 319

MATLAB function index — 321

Index — 325

1 An introduction to equations and optimization problems

Optimization techniques are the most important mathematical tools in science and engineering. It is also a very effective means in handling scientific and engineering problems.

It is not an exaggeration to say that, if the ideas and solution methods of optimization problems are mastered, the actual capabilities of scientific research may be boosted to an upper level. In the past, one was very happy if a solution to a problem could be found. Equipped with the optimization ideas and solution methods, it is natural that one may pursuit the best solutions of the problems.

There is a very close relationship between optimization and equation solutions. Here a brief historic review of equation solutions is presented first, then the developments in optimization and mathematical programming are proposed. Finally, the outline of the book is presented.

1.1 Equations and their solutions

Equations are ubiquitous mathematical models. In science, engineering and daily life, such mathematical models can be found everywhere.

Definition 1.1. Equations are equalities containing one or more variables. The variables are referred to as unknown variables. The values of the unknown variables satisfying the equalities are referred to as the equation solutions.

Definition 1.2. If several equations are given together, and the equations have certain variables such that the equations are satisfied at the same time, the equations are referred to as simultaneous equations.

In modern mathematics, expressions and equality signs are used to represent equations. There are expressions on both sides of the equality sign. In other words, the equality sign can be used to join two expressions together. Welch physicist and mathematician Robert Recorde (c1512–1558) invented in 1557 the equality sign, and described equations in mathematical symbols.

Equations are often classified into algebraic and differential equations. An algebraic equation describes a static relationship between variables. That is, the solutions of such equations are constants. In differential equations, there are dynamic relationships between the variables. Differential equations will be discussed fully in Volume V, and they are not treated in this book.

Algebraic equations are roughly classified into linear, polynomial, and nonlinear equations. Besides, there are parametric and implicit equations, and so on. Linear

https://doi.org/10.1515/9783110667011-001

equations were studied fully in Volume III. Other types of equations are studied in-depth in this book.

In fact, long before Recorde, who was using equality sign to express equations, studies of various equations were carried out. For instance, in about 2000 BCE, ancient Babylonians studied quadratic equations with one unknown. In 628 CE, Indian mathematician Brahmagupta (c598–c668 CE) described the solution method verbally rather than in mathematical formulas. Chinese mathematicians Liu Hui (c225–295) and Wang Xiaotong (580–640) studied solutions of many special cubic equations. In 1554, Italian mathematician Gerolamo Cardano (1501–1576) published a mathematics book, presenting the cubic equation solution formula, an extension to the formula of Italian mathematician Scipione del Ferro (1465–1526), and also the quartic equation solution formula by one of his students, Italian mathematician Lodovico de Ferrari (1522–1565). Cardano was the first mathematician who used negative numbers. Norwegian mathematician Niels Henrik Abel (1802–1829) showed in 1824 that polynomial equations of degree 5 or higher have no general solution formulas.

There is another category of special equations, they hold for any unknowns. These equations are also known as identities, for instance,

$$x^2 - y^2 = (x - y)(x + y), \quad \sin^2\theta + \cos^2\theta = 1. \tag{1.1.1}$$

These equations can be directly proved through symbolic computation in MATLAB. Identities will not be mentioned any more in this book.

1.2 Origins and development of optimization problems

The concept of optimization is originated in the study of calculus. French mathematicians Pierre de Fermat (1607–1665) and Joseph-Louis Lagrange (1736–1813) proposed methods for finding minimum and maximum values using calculus formulas. Apart from simple function optimization problems, other optimization problems are also known as mathematical programming problems. For a related historic development review, we refer to [18].

The Soviet Union scholar Leonid Vitaliyevich Kantorovich (1912–1986) laid out a solid foundation in the field of optimization, especially in the field of linear programming. American mathematician George Bernard Dantzig (1914–2005) proposed the simplex method[18] to solve linear programming problems. There is an interesting story about Dantzig and his simplex method.[15, 16] In 1939, as a first-year doctorate student at the University of California, Berkeley, Dantzig was late to attend the class delivered by Professor Jerzy Neyman (1894–1981). He mistakenly thought that the two world-class unsolved problems on the blackboard were the homework, and he gave the solutions of the problems. An effective solution method for linear programming problems was then proposed. American mathematician and computer scientist John von

Neumann (1903–1957) presented the dual simplex theory and computation method, to further improve the efficiency in solving linear programming problems.

American applied scientist Richard Ernest Bellman (1920–1984) initiated a new field in optimization – dynamic programming. Multistage decision making problems are handled using this method.

The idea and technology of optimization established the mathematical foundation of many science and engineering fields. The word "optimal" can be used to join the nouns of many fields so that new and perspective fields are initiated. For instance, if the key word "optimal" is used in a search engine, many related fields such as optimal control, optimal design, optimal resource allocation, optimal system, optimal stopping theory, optimal capital structure, and so on, can be found. These fields are closely related to optimization techniques.

1.3 Structure of the book

Optimization problems are closely related to equation solutions. In Chapter 2, in-depth discussions on various solution methods are made for algebraic equations, including analytical solutions of polynomial equations and graphical and numerical solutions of complicated nonlinear simultaneous equations. Especially, the numerical and quasianalytical solutions of nonlinear matrix equations with multiple solutions are explored with MATLAB solvers, which in theory can be used to find all possible solutions of nonlinear equations of any complexity.

In Chapter 3, the simplest category of optimization problems – the unconstrained optimization problems – are studied. The solutions are demonstrated with graphical and simple numerical algorithms. We concentrate on how to use the powerful tools provided in MATLAB to solve directly the unconstrained optimization problems. The concepts of local and global optimum solutions are introduced, and an attempt is made to provide a general-purpose MATLAB tool aiming at finding global minimum solutions. The applications in unconstrained optimization techniques are also demonstrated.

Normal mathematical programming problems are classified into linear, nonlinear, mixed, multiobjective, and dynamic programming problems. The subsequent chapters are arranged accordingly to present the solutions of various programming problems.

In Chapter 4, the focus is on convex optimization problems – linear and quadratic programming problems. The simplex method is demonstrated through a simple example. The existing tools in MATLAB are mainly used to demonstrate the direct solutions of the problems. The problem-specific expressions supported in recent versions of MATLAB are addressed such that linear and quadratic programming problems can be described and solved in a direct and straightforward manner. Besides, linear matrix inequality problem and tools are introduced.

In Chapter 5, nonlinear programming problems are mainly studied. The graphical method is illustrated first, then the nonlinear programming solver provided in Optimization Toolbox is illustrated, and direct solutions of complicated problems are presented. In particular, a global optimum solution function is written for nonconvex nonlinear optimization problems. Applications of nonlinear programming solutions such as semiinfinite programming problem, pooling and blending problems, and heat exchanger design problems are presented in this chapter.

In Chapter 6, mixed-integer programming problems are mainly presented. For small-scale integer programming problems, enumeration methods are demonstrated for a variety of examples. MATLAB-based solvers for mixed linear and nonlinear programming problems are presented to demonstrate the solutions of complicated mixed-integer programming problems. Mixed 0–1 programming problems are also explored in this chapter. Several application examples such as assignment, knapsack, traveling salesman, and sudoku problems are studied.

In Chapter 7, we concentrate on presenting the essentials of multiobjective programming problems. The graphical and numerical methods for simple multiobjective programming are illustrated. Then, we focus on how to convert multiobjective programming problems into single objective ones, and the methods studied in other chapters can be used to solve the converted problems directly. The concepts and extraction of Pareto frontiers are provided. Finally, the modeling and solving of minimax problems are addressed.

In Chapter 8, the concepts and solution methods of dynamic programming problems are introduced. Then, the idea of matrix representation of oriented graphs is proposed. Shortest path problems, which can be solved by the existing tools or by Dijkstra algorithm, are illustrated. Finally, the shortest path solutions for undigraphs are also studied.

In the solutions using search-based optimization solvers introduced in the first few chapters, we mainly complained that when the initial search points are inappropriately chosen, local optimum solutions can be found. In Chapter 9, some of the commonly used intelligent optimization solvers in MATLAB are briefly introduced. The genetic algorithm, pattern search method, particle swarm optimization method, as well as simulated annealing method, are discussed. Comparative studies of the intelligent optimization solvers and best ones studied in the earlier chapters are made. Useful results can be demonstrated through examples.

1.4 Exercises

1.1 See whether you can solve the following equations with MATLAB:

(1) $$\begin{cases} x_1 + x_2 = 35, \\ 2x_1 + 4x_2 = 94, \end{cases}$$

(2) $\begin{cases} x^2 e^{-xy^2/2} + e^{-x/2}\sin(xy) = 0, \\ y^2\cos(x+y^2) + x^2 e^{x+y} = 0. \end{cases}$

1.2 Find the minimum of the function within the interval $-2 \leqslant x \leqslant 11$[23] for

$$f(x) = x^6 - \frac{52}{25}x^5 + \frac{39}{80}x^4 + \frac{71}{10}x^3 - \frac{79}{20}x^2 - x + \frac{1}{10}.$$

1.3 Draw the surface of the following functions. Rotate the surfaces to see at which values of x and y you can find the valleys of the surfaces if

(1) $f(x,y) = -(y+47)\sin\sqrt{\left|\frac{x}{2} + (y+47)\right|} - x\sin\sqrt{|x-(y+47)|}$,

(2) $f(x,y) = 20 + \left(\frac{x}{30} - 1\right)^2 + \left(\frac{y}{20} - 1\right)^2 - 10\left[\cos\left(\frac{x}{30} - 1\right)\pi + \cos\left(\frac{y}{20} - 1\right)\pi\right]$.

2 Solutions of algebraic equations

Equations are often encountered as mathematical models in scientific research and engineering practice. The so-called equations are equalities containing unknown variables. Equations are often classified into algebraic and differential equations, and so on. In this chapter, solution methods are mainly explored for algebraic equations, where analytical and numerical solutions are both considered. Attempts are made to find all possible solutions of equations with multiple solutions.

In the third volume of the series, linear equations have been considered. The cases of unique solutions, infinitely many solutions, and least-squares solutions for $\boldsymbol{AX} = \boldsymbol{B}$ type equations have been investigated. Also solutions of linear equations like $\boldsymbol{XA} = \boldsymbol{B}$, $\boldsymbol{AXB} = \boldsymbol{C}$, and multiterm linear algebraic equations have been considered. Besides, solutions of Sylvester equations and multiterm Sylvester equations have also been explored. The numerical and analytical solutions of the above equations can be obtained with the powerful facilities provided in MATLAB. For the solution methods of linear equations, the readers may refer to the materials in Volume III.

In this chapter, we concentrate on the solutions of polynomial and ordinary nonlinear equations. In Section 2.1, solution formulas for lower-degree polynomial equations are discussed, with low-level MATLAB implementations. It can be seen from the numerical viewpoint that the MATLAB solvers are more suitable for lower-degree polynomial equations with multiple solutions. In Section 2.2, graphical solution methods are introduced for equations with one or two unknowns, and practical methods for finding the solutions are discussed. In Section 2.3, numerical solutions of ordinary nonlinear equations are proposed. The classical Newton–Raphson iterative method and its MATLAB implementation are presented first. Then, the nonlinear equation solver provided in MATLAB is presented for nonlinear algebraic and matrix equations. In Section 2.4, analytical solutions for low-degree algebraic equations are studied with symbolic computation methods, and for high-degree algebraic equations and nonlinear matrix equations, quasianalytical solution methods are presented. In Section 2.5, solution methods for matrix equations and pseudopolynomial equations are explored, with high-precision methods. In Section 2.6, solutions for some underdetermined equations and studied.

2.1 Solutions of polynomial equations

Polynomial equations are often encountered in real applications. The mathematical form of polynomial equations is presented first in this section, followed by the solutions methods of polynomial equations.

Definition 2.1. The general mathematical form of polynomial equations is

$$x^n + a_1x^{n-1} + a_2x^{n-2} + \cdots + a_{n-1}x + a_n = 0. \tag{2.1.1}$$

https://doi.org/10.1515/9783110667011-002

The roots and coefficients of the polynomial equation satisfy the following Viéte theorem, also known as Viéte formula. The theorem was proposed by French mathematician François Viéte (1540–1603).

Theorem 2.1 (Viéte theorem). *Assume that the roots of the polynomial equation are* x_1, $x_2, \dots, x_n$, *then*

$$\begin{cases} x_1 + x_2 + \cdots + x_n = -a_1, \\ (x_1x_2 + x_1x_3 + \cdots + x_1x_n) + (x_2x_1 + x_2x_3 + \cdots + x_2x_n) + \cdots = a_2, \\ \vdots \\ x_1x_2 \cdots x_n = (-1)^n a_n. \end{cases} \tag{2.1.2}$$

In this section, we concentrate on the formulas for low-degree polynomial equations, with MATLAB implementations. The well-known Abel–Ruffini theorem for high-degree equations is then proposed.

2.1.1 Polynomial equations of degrees 1 and 2

For polynomial equations of degrees 1 and 2, there are simple solution formulas. Here examples are explored to demonstrate the direct solutions of polynomial equations of degrees 1 and 2.

Example 2.1. What is the solution of polynomial equation $x + c = 0$?

Solutions. It is obvious that the solution of the polynomial equation of degree 1 is $x = -c$, no matter what the value of c is.

Example 2.2. In the Chinese mathematics book entitled "*Sunzi Suanjing*", also known as "*The Mathematical Classic of Master Sun*", published between the fourth to fifth century CE, a well-known "chick–rabbit cage" problem was proposed: "In a cage there are chicks and rabbits. If counted by heads, there are 35 heads altogether, while when counted by legs, there are 94 legs. How many chicks and rabbits are there?"

Solutions. Various methods are available in classical mathematics books to solve this problem. If algebraic equations are applied, by denoting the numbers of chicks and rabbits by x_1 and x_2, respectively, it is easy to write down the following linear equation set with two unknowns:

$$\begin{cases} x_1 + x_2 = 35, \\ 2x_1 + 4x_2 = 94. \end{cases}$$

It can be seen from the first equation that, by letting $x_1 = 35 - x_2$ and then substituting it into the second one, we have

$$2(35 - x_2) + 4x_2 = 70 - 2x_2 + 4x_2 = 70 + 2x_2 = 94.$$

It is immediately found that $x_2 = 12$, and after substitution, we have $x_1 = 35 - x_2 = 23$. When the solution is substituted back into the two equations, it can be seen that the errors are zero, indicating that the solution is correct. Of course, there are other methods of solving this equation, and they are not discussed here, for the time being.

Example 2.3. Find the solution formula for the quadratic equation $ax^2 + bx + c = 0$.

Solutions. Dating back to 1800 BCE, the ancient Babylonians started studying this problem. Later many mathematicians were studying the solution methods. It was not until 1615, when French mathematician François Viéte published a book, and the relationship between polynomial equation coefficients and solutions was proposed, known as Viéte theorem stated earlier.

Factoring out coefficient a, we have $a(x^2+bx/a+c/a) = 0$. If a is not zero, we need only to explore the equation $x^2+bx/a+c/a = 0$. We can use the "method of completing the square" to derive the solutions of quadratic equations:

$$\begin{aligned} x^2 + \frac{b}{a}x + \frac{c}{a} &= x^2 + \frac{b}{a}x + \left(\frac{b}{2a}\right)^2 - \left(\frac{b}{2a}\right)^2 + \frac{c}{a} \\ &= \left(x + \frac{b}{2a}\right)^2 - \frac{b^2 - 4ac}{4a^2} = 0 \end{aligned}$$

It can be seen from the last equation that

$$\left(x + \frac{b}{2a}\right)^2 = \frac{b^2 - 4ac}{4a^2}.$$

Computing the square roots for both sides, and after easy manipulations, we can find the two solutions:

$$x_{1,2} = \frac{-b \pm \sqrt{b^2 - 4ac}}{2a}.$$

2.1.2 Analytical solutions of cubic equations

The solutions of cubic and higher-degree polynomial equations are not so simple. Ancient Babylonians studied cubic equations. The well-known mathematicians in China such as Liu Hui (c225–295 CE) published "*Jiuzhang Suanshu Zhu*" (also known as "Notes on the Nine Chapters on the Mathematical Art") in 265 CE, where cubic equation problems were studied. Wang Xiaotong (580–640) in Tang Dynasty solved 25 special cubic equations in his book "*Jigu Suanjing*" (meaning "Continuation of Ancient Mathematics").

Definition 2.2. The standard form of a cubic equation is $x^3 + c_2x^2 + c_3x + c_4 = 0$.

For simplicity, monic manipulation was assumed. Letting $x = t - c_2/3$, the cubic equation can be converted into the form of $t^3 + pt + q = 0$, where

$$p = -\frac{1}{3}c_2^2 + c_3,$$
$$q = \frac{2}{27}c_3^2 - \frac{1}{3}c_2c_3 + c_4. \tag{2.1.3}$$

Example 2.4. Validate the formula (2.1.3) with MATLAB.

Solutions. With MATLAB, variable substitution can be made with the following statements:

```
>> syms x c2 c3 c4; f=x^3+c2*x^2+c3*x+c4;
   syms t; f1=subs(f,x,t-c2/3); f2=collect(f1,t)
```

Therefore, the following results are immediately found, such that the conclusions in (2.1.3) are validated:

$$f_2 = t^3 + \left(-\frac{1}{3}c_2^2 + c_3\right)t + \frac{2}{27}c_2^3 - \frac{1}{3}c_3c_2 + c_4 = 0.$$

Theorem 2.2. *The closed-form solution formula for t of the cubic equation is*

$$t = \xi^k u + \xi^{2k} v, \quad k = 0, 1, 2, \tag{2.1.4}$$

where $\xi = (-1 + \sqrt{3}\mathrm{j})/2$, *and*

$$u = \sqrt[3]{-\frac{q}{2} + \sqrt{\frac{q^2}{4} + \frac{p^3}{27}}}, \quad v = \sqrt[3]{-\frac{q}{2} - \sqrt{\frac{q^2}{4} + \frac{p^3}{27}}}. \tag{2.1.5}$$

While in real computation with MATLAB, the independent computation of u and v may lead to errors, since the cubic root computation is not unique. To better compute the values of u and v, the value of u can be found first, then, with Viéte theorem, the value of v can be found from $v = -p/(3u)$.

Of course, there is a special case to consider, namely $u = v = 0$, then the equation has a triple solution $t = 0$, which in turn yields $x = -c_2/3$.

Substituting the solution of t back to $x = t - c_2/3$, the three solutions of the equation can be found. The closed-form solution formula was proposed by Italian mathematician Gerolamo Cardano (1501–1576) in 1545.

2.1.3 Analytical solutions of quartic equation

An ordinary quartic equation can be converted into a special form of quartic equation via variable substitution method, and finally, the closed-form solution formula for the quartic equations can be obtained once a cubic equation is solved.

Definition 2.3. A quartic equation is of the form $x^4 + c_2x^3 + c_3x^2 + c_4x + c_5 = 0$.

Example 2.5. For an ordinary quartic equation, if we let $x = y - c_2/4$, what will be the form of the converted original equation?

Solutions. With the following commands to accomplish variable substitution:

```
>> syms x c2 c3 c4 c5 y; f=x^4+c2*x^3+c3*x^2+c4*x+c5;
   f1=subs(f,x,y-c2/4); f2=collect(f1,y)
```

the original equation can be transformed into the form of quartic equation $y^4 + py^2 + qy + s = 0$, where

$$\begin{aligned} p &= c_3 - \frac{3}{8}c_2^2, \\ q &= \frac{1}{8}c_2^3 - \frac{1}{2}c_3c_2 + c_4, \\ s &= -\frac{3}{256}c_2^4 + \frac{1}{16}c_3c_2^2 - \frac{1}{4}c_4c_2 + c_5. \end{aligned} \tag{2.1.6}$$

From the special form, the equation of y can be converted into the following form:

$$\left(y^2 + \frac{p}{2} + m - \sqrt{2m}y + \frac{q}{2\sqrt{2m}}\right)\left(y^2 + \frac{p}{2} + m + \sqrt{2m}y - \frac{q}{2\sqrt{2m}}\right) = 0, \tag{2.1.7}$$

where m is the solution of a cubic equation $8m^3 + 8pm^2 + (2p^2 - 8s)m - q^2 = 0$. The algorithm was proposed by Italian mathematician Lodovico de Ferrari (1522–1565). Of course, the condition is that $m \neq 0$. If $m = 0$, it means that $q = 0$ and the equation of y can be written as $y^4 + py^2 + s = 0$, from which it is easy to find y^2 and then y. With the solutions y, the corresponding solutions of x can be obtained with substitution formula.

In MATLAB, function `roots()` is proposed to compute the roots of polynomial equations using matrix eigenvalue methods. The syntax of the function is ***r***=`roots(`***p***`)`, where ***p*** is the coefficient vector of the polynomial in the descending order. The numerical solution ***r*** is returned in a column vector.

This function is often complained about for equations with repeated roots, since the error obtained may be large. Considering the above mentioned algorithms for low-degree equations, a replacement for function `roots()` is written, which may find accurate numerical solutions for low-degree polynomial equations. If the degree is higher than 4, the original MATLAB function `roots()` is embedded to find numerical solutions.

```
function r=roots1(c)
i=find(c~=0); c=c(i(1):end); n=length(c)-1; c=c/c(1);
switch n
   case 1, r=-c(2);
```

```
    case 2
      d=sqrt(c(2)^2-4*c(3)); r=[-c(2)+d; -c(2)-d]/2;
    case 3
      p=c(3)-c(2)^2/3; q=2*c(2)^3/27-c(2)*c(3)/3+c(4); v=0;
      u=(-q/2+sqrt(q^2/4+p^3/27))^(1/3); if u~=0, v=-p/3/u; end
      xi=(-1+sqrt(3)*1i)/2; w=xi.^[0,1,2]'; r=w*u+w.^2*v-c(2)/3;
    case 4
      p=-3*c(2)*c(2)/8+c(3); q=c(2)^3/8-0.5*c(2)*c(3)+c(4);
      s=-3*c(2)^4/256+c(2)*c(2)*c(3)/16-0.25*c(2)*c(4)+c(5);
      rr=roots1([8,8*p,2*p^2-8*s,-q^2]); m=rr(1); d=sqrt(2*m);
      if q==0, r1=roots1([1,p,s]); r=[sqrt(r1); -sqrt(r1)];
      else
        r=[roots1([1,-d,p/2+m+q/2/d]); roots1([1,d,p/2+m-q/2/d])];
      end, r=r-c(2)/4;
    otherwise, r=roots(c);
end
```

Example 2.6. Solve the equation $(s-5)^4 = 0$.

Solutions. Of course, if the equation is given in this form, we know that the quadruple root is 5. In real applications, the expanded form of the equation may be given, and the factorized form is not known. Instead, the expanded one $x^4 + 4x^3 + 6x^2 + 4x + 1 = 0$ is given. How can we solve the problem? Function `roots()` provided in MATLAB can be tried, also the new function `roots1()` can be called:

```
>> P=conv([1,-5],conv([1,-5],conv([1,-5],[1,-5])))
   d1=roots(P), d2=roots1(P), d3=roots(sym(P))
   e1=norm(polyval(P,d1)), e2=norm(polyval(P,d2))
```

It can be seen that function `roots1()` can be used to find the expected quadruple root $s = 5$, while with `roots()` function, the solutions obtained are 5.0010, $5 \pm 0.001\mathrm{j}$, and 4.9990. It can be seen that the errors in the MATLAB function may be large when there are repeated roots. Under the double precision framework, other numerical solution algorithms have the same problem. In real applications, this may lead to trouble. When the roots are substituted back to the equation, the errors are $e_1 = 2.1690 \times 10^{-12}$ and $e_2 = 0$. It can be seen through the above demonstration that if symbolic structures are used, accurate solutions can always be found.

Example 2.7. Solve the following polynomial equation with complex coefficients under the double precision framework, and evaluate the precision:

$$f(s) = s^4 + (5+3\mathrm{j})s^3 + (6+12\mathrm{j})s^2 - (2-14\mathrm{j})s - (4-4\mathrm{j}) = 0.$$

Solutions. So far two solvers, `roots()` and `roots1()`, have been discussed for polynomial equations. In fact, when constructing the original equation, it is assumed that the equation has a triple solution of $-1 + 1j$, and a real root of -2. The following statements can be used to directly solve the following equation. Compared with the results with the given analytical solution, it can be seen that `roots()` yields a solution with the error norm of 3.5296×10^{-5}, where `roots1()` function yields the error norm of 0, indicating the solution is accurate. Function `roots1()` is also suitable for dealing with polynomial equations with complex coefficients:

```
>> p=[1,5+3i,6+12i,-2+14i,-4+4i];
   r1=roots(p), err1=norm(r1-[-2; -1-1i; -1-1i; -1-1i])
   r2=roots1(p), err2=norm(r2-[-1-1i; -1-1i; -1-1i; -2])
```

2.1.4 Higher-degree equations and Abel–Ruffini theorem

Definition 2.4. An algebraic algorithm means that the closed-form formula is composed of finitely many addition, subtraction, multiplication, division, exponentiation and root operations.

Theorem 2.3 (Abel–Ruffini theorem). *A polynomial equation with degree 5 or over has no algebraic algorithm. The theorem is also known as Abel impossibility theorem.*

Italian mathematician Paolo Ruffini (1765–1822) provided an incomplete proof of the theorem in 1799. Norwegian mathematician Niels Henrik Abel (1802–1929) gave the proof in 1824.

For ordinary higher-degree polynomial equations, finding numerical solutions is the only way to solve them.

2.2 Graphical methods for nonlinear equations

If equations have one or two unknowns, graphical methods can be used to solve them. If there are too many unknowns, it is not suitable to use graphical methods. Other methods should be tried. In this section, graphical solution methods for equations with one or two unknowns are proposed, and advantages and disadvantages of graphical methods are summarized.

2.2.1 Smooth graphics for implicit functions

A MATLAB function `fimplicit()` for implicit function graphics is provided. Normally, smooth curves for implicit functions can be drawn, while in some specific cases, manual adjustment of control parameters is needed, so as to draw smooth curves. An example is given below to draw smooth curves for implicit functions.

Example 2.8. Draw the smooth curve for the implicit function

$$y^2 \cos(x + y^2) + x^2 e^{x+y} = 0, \quad x, y \in (-2\pi, 2\pi).$$

Solutions. The implicit function can be described as a symbolic expression, or by an anonymous function with dot operations. With the following MATLAB commands, the curves of the implicit function can be drawn, as shown in Figure 2.1. It can be seen that the curves of the implicit function can be drawn automatically, however, there are burrs in the curves. For instance, at the upper-left and lower-right corners, there are burrs.

```
>> syms x y; p=2*pi;
   f=y^2*cos(x+y^2)+x^2*exp(x+y); fimplicit(f,[-p,p])
```

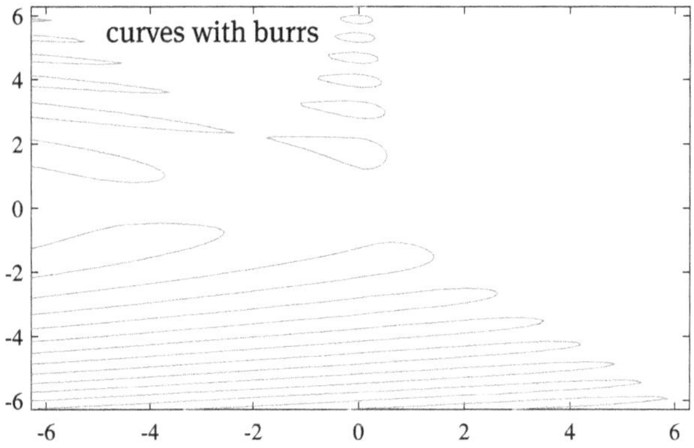

Figure 2.1: Curves under default setting.

The smoothness of the implicit function curves is governed by the control option `'Meshdensity'`, with a default value of 151. If the curves are found unsmooth, a large value, for instance, 500, can be set, and the new implicit curves are shown in Figure 2.2. It can be seen that the new curves are satisfactory. Even though we zoom the curves, the curves are smooth.

```
>> fimplicit(f,[-p,p],'Meshdensity',500)
```

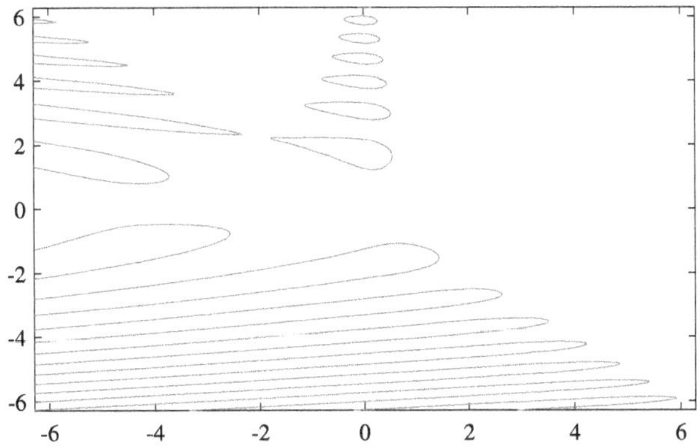

Figure 2.2: Smooth curves for the implicit function.

2.2.2 Univariate equations

Definition 2.5. The mathematical form of the equation with one unknown is

$$f(x) = 0. \tag{2.2.1}$$

For arbitrary univariate equation $f(x) = 0$, symbolic expression can be used to describe the equation. Then, with function `fplot()`, the curve of the function can be drawn. The intersection of the curves with horizontal axis can be located. They are the solutions to the equations.

Example 2.9. Analytical solutions of square root equations are usually complicated. Normally, it is hard to find the analytical solutions. Use the graphical method to solve the following root equation:

$$\sqrt{2x^2+3} + \sqrt{x^2+3x+2} - \sqrt{2x^2-3x+5} - \sqrt{x^2-5x+2} = 0.$$

Solutions. With a symbolic expression to represent the left side of the equation, function `fplot()` can be called, then, superimposed with horizontal axis, the curves in Figure 2.3 can be obtained. It can be seen from the result that there is only one intersection of the equation with the horizontal axis. The intersection is the solution of the equation.

```
>> syms x                          % declare symbolic variables
   f=sqrt(2*x^2+3)+sqrt(x^2+3*x+2)-sqrt(2*x^2-3*x+5)-sqrt(x^2-5*x+2);
   fplot(f), line([-5 5],[0,0]) % draw the curve and the horizontal axis
```

If one wants to get the solution, the icon in the toolbar of the graphics window can be clicked, to zoom the area around the intersection. The user may repeatedly

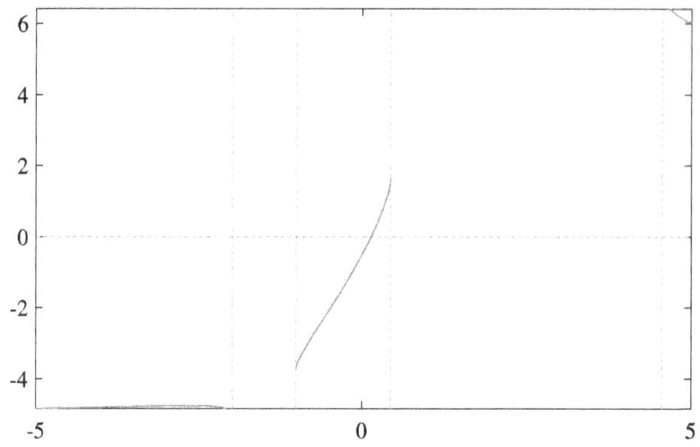

Figure 2.3: Curves and solutions of the univariate equation.

use this facility such that the scale of the x axis is almost the same. Then, the value can be regarded as the solution of the equation. For this particular equation, through zooming, the solution obtained is $x = 0.13809878$, as shown in Figure 2.4. When substituted back to the equation, the error is 6.018×10^{-9}.

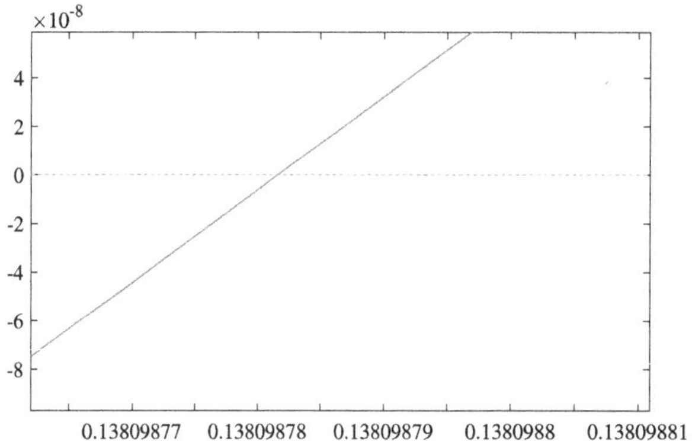

Figure 2.4: Zoomed results and equation solutions.

Example 2.10. Solve the transcendental univariate equation

$$e^{-0.2x} \sin(3x + 2) + \cos x = 0, \quad x \in (0, 20).$$

Solutions. There is no analytical solution for this transcendental equation. Numerical solution method must be used. The graphical method is a practical way to solve the equation. In the previous example, a symbolic expression was used to describe the original equation. Here an anonymous function is used to describe the equation, and the two methods are equivalent. With function `fplot()`, the plots of the equation can be drawn directly, superimposed with horizontal axis, as shown in Figure 2.5. Therefore, the intersections of the curve and the horizontal axis are the solutions of the equation.

```
>> f=@(x)exp(-0.2*x).*sin(3*x+2)-cos(x); %dot operation should be used
   fplot(f,[0,20]), line([0,20],[0,0])    %superimpose horizontal axis
```

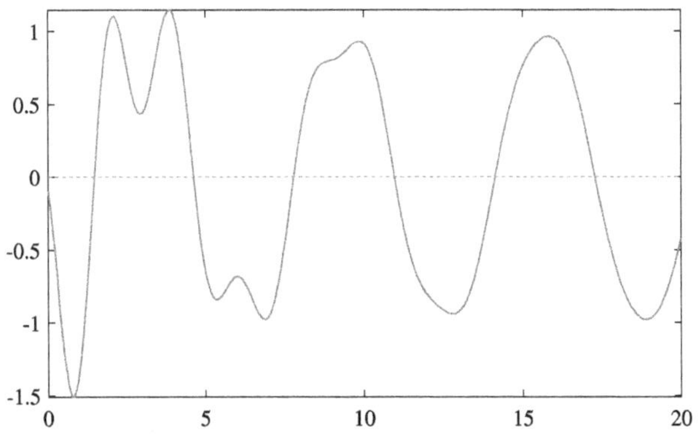

Figure 2.5: Graphical solution of transcendental equation.

It can be seen that in the given region, there are 6 intersections. These values of x at the intersections are all the solutions of the equation. Similar to the case in the previous example, local zooming can be made to find individually the solutions. However, the process may be complicated. For instance, one of the solutions can be found at $x = 10.9601289577$. When substituted back to the original equation, the error is -3.4739×10^{-11}. Better methods will be explored later.

2.2.3 Equations with two unknowns

The mathematical form of the simultaneous equations with two unknowns is presented below. In this section, we shall introduce graphical methods to solve simultaneous equations with two unknowns, and point out problems in using graphical methods.

Definition 2.6. The mathematical form of the simultaneous equations with two unknowns are

$$\begin{cases} f(x,y) = 0, \\ g(x,y) = 0. \end{cases} \tag{2.2.2}$$

It can be seen that in the mathematical form, $f(x,y) = 0$ can be regarded as an implicit function of two independent variables x and y. Therefore, with function `fimplicit()`, the curve of the implicit function can be drawn. All the points on the curve satisfy the equation. Similarly $g(x,y) = 0$ is also an implicit function. Function `fimplicit()` can also be used to draw the equation. When the curves of the two implicit functions are drawn together, the intersections are the solutions of the simultaneous equations.

Example 2.11. Find the solutions of the following equations:

$$\begin{cases} x^2 e^{-xy^2/2} + e^{-x/2}\sin(xy) = 0, \\ y^2\cos(x+y^2) + x^2 e^{x+y} = 0, \end{cases}$$

where $-2\pi \leqslant x, y \leqslant 2\pi$.

Solutions. In order to solve the simultaneous equations, one should declare the symbolic variables x and y. Then, the two equations can be expressed with symbolic expressions. Next, function `fimplicit()` can be used to draw the curves of the two equations, as shown in Figure 2.6. The intersections of the curves are the solutions of the simultaneous equations. It can be seen that there are quite a few solutions in the region of interest. It may also be seen that the smoothness of the curves may not affect the solution process of the equations. Therefore, the curves can be drawn under the default settings.

```
>> syms x y
   f1=x^2*exp(-x*y^2/2)+exp(-x/2)*sin(x*y);
   f2=y^2*cos(x+y^2)+x^2*exp(x+y);
   fimplicit([f1 f2],[-2*pi,2*pi])
```

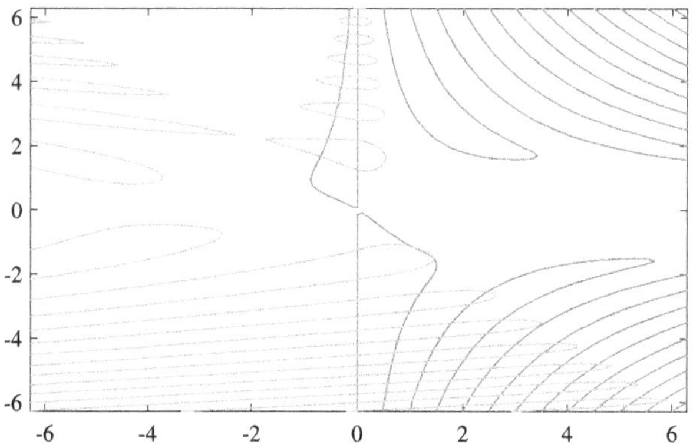

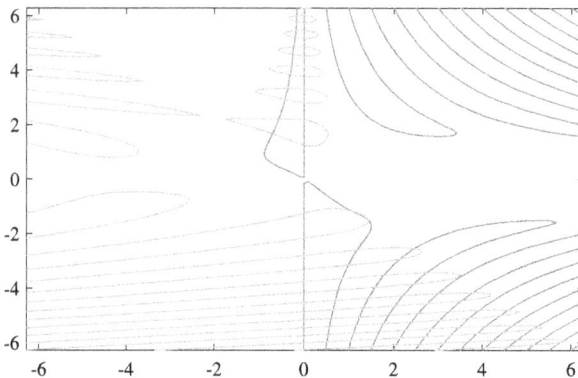

Figure 2.6: Graphical method for the equations.

If we want to find the information at a certain intersection, local zooming should be employed to find roughly the values of x and y. Normally, it is seen that the method may not be able to find accurate solutions. Besides, since there are too many intersections in the simultaneous equations, the local zooming method to find the solutions one by one is not a good choice. Better methods should be introduced to find all the intersections.

Example 2.12. Solve the following equations with the graphical method:

$$\begin{cases} x^2 + y^2 = 5, \\ x + 4y^3 + 3y^2 = 2. \end{cases}$$

Solutions. Symbolic expressions can be used to express the two equations. Then, the implicit functions can be drawn as shown in Figure 2.7. It can be seen that there are two intersections. Can we conclude that the equations have two solutions?

```
>> syms x y
   f1=x^2+y^2-5; f2=x+4*y^3+3*y^2-2;
   fimplicit([f1,f2],[-pi,pi])
```

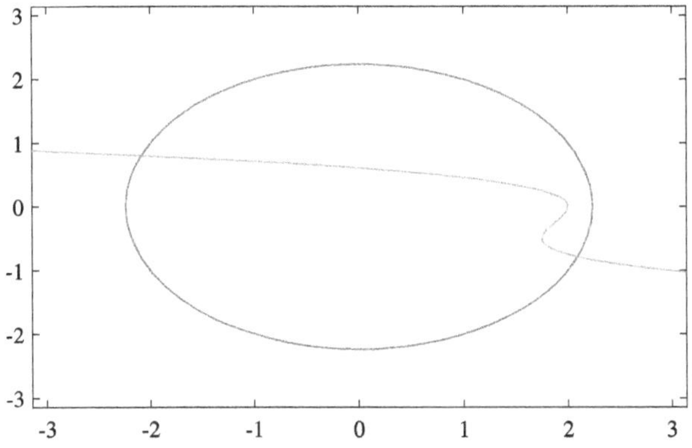

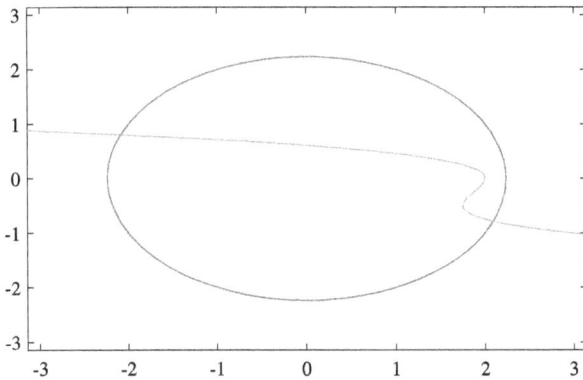

Figure 2.7: Illustration of graphical solutions.

Slightly changing the format of the second expression, we have $x = -4y^3 - 3y^2 + 2$. Substituting it into the first equation, we have

$$16y^6 + 24y^5 + 9y^4 - 16y^3 - 11y^2 - 1 = 0.$$

It can be seen that this is a polynomial of y of degree 6, and there should be six solutions, not the two solutions shown in Figure 2.7. Why there are only two solutions found in the figure? Since these solutions are real, the other four should be two pairs of complex conjugate solutions. With graphical methods, only real solutions can be found. Complex roots cannot be displayed graphically.

2.2.4 Isolated equation solutions

Observing the equations in Example 2.11, it is not hard to find that the point $x = y = 0$ is a solution of the equations, since the two equations are satisfied. It can be seen that the curves seem to avoid this particular point in the first equation. The curves of the second equation do not include this point. This point is not evolved from any other point. Therefore, this solution is an isolated one.

There is no any existing method to find isolated solutions. The users should use their experience to judge whether a point is an isolated solution of the equations. For instance, through observation the point $x = y = 0$ is an isolated solution of the equations, since it can be validated when substituted into the two equations.

2.3 Numerical solutions of algebraic equations

The graphical method discussed so far is only one of the methods in finding numerical solutions of nonlinear equations. There are advantages and disadvantages of graphical methods. They can only be used to solve equations with one or two unknowns, for finding real solutions. Besides, for equations with more than two unknowns, the equations cannot be solved with graphical methods. In this section, further solution ideas and implementations are discussed.

2.3.1 Newton–Raphson iterative algorithm

For simplicity, solutions for equations with one unknown are studied first. The Newton–Raphson iterative method is named after British scientist Isaac Newton and British mathematician Joseph Raphson (c1648–c1715). It is an iterative method for solving algebraic equations.

Assume that the equation with one unknown can be described by $f(x) = 0$, and at point $x = x_0$ the function value $f(x_0)$ is known. A tangent line at point $(x_0, f(x_0))$ can be drawn, as shown in Figure 2.8. The intersection of the tangent line with the horizontal axis is x_1, which can be regarded as the first approximate solution of the equation. It can be seen from Figure 2.8 that, since the slope of the tangent line is $f'(x_0)$, the position x_1 can be computed from

$$x_1 = x_0 - \frac{f(x_0)}{f'(x_0)}, \tag{2.3.1}$$

where $f'(x_0)$ is the value of the derivative of $f(x)$ with respect to x, at point x_0. Drawing another tangent line at x_1, a new approximation x_2 can be found. From x_2, point x_3 can be found, etc. Assuming that the approximation x_k is found, the next approximation

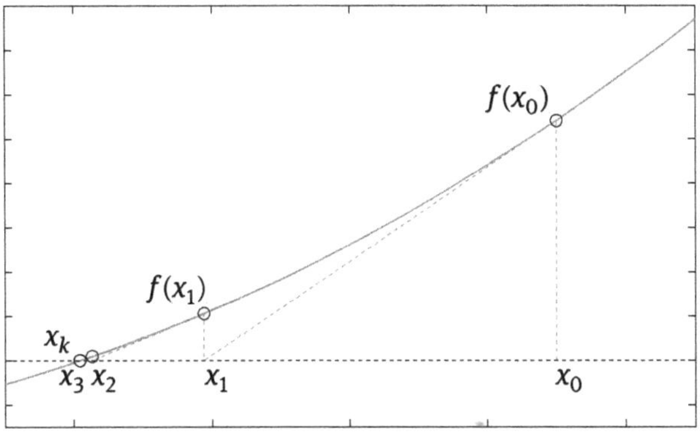

Figure 2.8: Illustration of Newton–Raphson iterative method.

can be found from

$$x_{k+1} = x_k - \frac{f(x_k)}{f'(x_k)}, \quad k = 0, 1, 2, \ldots \tag{2.3.2}$$

If the conditions $|x_{k+1} - x_k| \leqslant \varepsilon_1$ or $|f(x_{k+1})| \leqslant \varepsilon_2$ are satisfied, where ε_1 and ε_2 are preselected error tolerances, then x_k is a solution of the original equation.

Definition 2.7. For a multivariate function $\boldsymbol{f}(\boldsymbol{x})$, its Jacobian matrix is defined as

$$\boldsymbol{J}[\boldsymbol{f}(\boldsymbol{x})] = \begin{bmatrix} \partial f_1(\boldsymbol{x})/\partial x_1 & \partial f_1(\boldsymbol{x})/\partial x_2 & \ldots & \partial f_1(\boldsymbol{x})/\partial x_n \\ \partial f_2(\boldsymbol{x})/\partial x_1 & \partial f_2(\boldsymbol{x})/\partial x_2 & \ldots & \partial f_2(\boldsymbol{x})/\partial x_n \\ \vdots & \vdots & \ddots & \vdots \\ \partial f_m(\boldsymbol{x})/\partial x_1 & \partial f_m(\boldsymbol{x})/\partial x_2 & \ldots & \partial f_m(\boldsymbol{x})/\partial x_n \end{bmatrix}. \tag{2.3.3}$$

Theorem 2.4. *For a multivariate equation $\boldsymbol{f}(\boldsymbol{x}) = \boldsymbol{0}$, where $\boldsymbol{x}$ is a vector or a matrix and the nonlinear function $\boldsymbol{f}(\boldsymbol{x})$ is of the same size, the following formula can be used to iteratively solve the equation:*

$$\boldsymbol{x}_{k+1} = \boldsymbol{x}_k - \boldsymbol{J}^{-1}[\boldsymbol{f}(\boldsymbol{x}_k)]\boldsymbol{f}(\boldsymbol{x}_k). \tag{2.3.4}$$

If $\|\boldsymbol{x}_{k+1} - \boldsymbol{x}_k\| \leqslant \varepsilon_1$ or $\|\boldsymbol{f}(\boldsymbol{x}_{k+1})\| \leqslant \varepsilon_2$, then $\boldsymbol{x}_k$ is considered the solution of the equation. In the algorithm, the Jacobian matrix is involved.

Based on the above algorithm, the following MATLAB function is written

```
function x=nr_sols(f,df,x0,epsilon,key)
if nargin<=4, key=0; end, x1=x0;
if nargin==3, epsilon=eps*10; end
while (1), x=x0-df(x0)\f(x0);
   if norm(x-x0)<epsilon || norm(f(x))<epsilon, break;
   else, x0=x; x1=[x1,x]; end
```

```
end
if key==1, x=x1; end
```

The function handles f and d, which are respectively the handles of the equation and Jacobian matrix, should be provided, together with the initial search point $\boldsymbol{x}_0$. Also the error tolerance ε should be supplied. If the option `key` is assigned to 1, the intermediate search results are returned in matrix $\boldsymbol{x}$.

For univariate equations, the availability of the derivative function is itself a hard question. An alternative secant method can be adopted, to compute the approximate derivatives, so as to find the equation solutions. The solution formula in (2.3.2) can be rewritten as

$$x_{k+1} = x_k - f(x_k)\frac{x_k - x_{k-1}}{f(x_k) - f(x_{k-1})}, \quad k = 0, 1, 2, \ldots \tag{2.3.5}$$

If dot operations are used, the secant method can also be used in finding the solutions of multivariate equations.

```
function x=sec_sols(f,x1,x0,epsilon,key)
if nargin<=4, key=0; end, xm=[x0 x1];
if nargin==3, epsilon=eps*10; end
while (1), x=x0-f(x1).*(x0-x1)./(f(x0)-f(x1));
   if norm(x-x0)<epsilon || norm(f(x))<epsilon, break;
   else, x1=x0; x0=x; xm=[xm,x]; end
end
if key==1, x=xm; end
```

Example 2.13. Selecting initial value of $x_0 = 10$, find one root for the univariate transcendental equation in Example 2.10.

Solutions. Symbolic computation can be used in deriving the first-order derivative of the given function

```
>> syms x; f=exp(-0.2*x)*sin(3*x+2)+cos(x), diff(f)
```

and it is found that the derivative function is

$$f'(x) = 3\mathrm{e}^{-x/5}\cos(3x+2) - \sin x - \frac{1}{5}\mathrm{e}^{-x/5}\sin(3x+2).$$

With the function and its derivative, anonymous functions can be used to describe them, and then assigning the initial value at $x_0 = 10$, the Newton–Raphson solver can be used to solve the equation. The intermediate points found are $[10, 10.8809, 11.0700, 11.0593, 11.0593]^\mathrm{T}$, and the solution is $x = 11.0593$. Substituting it back to the equation, the error obtained is -5.1348×10^{-16}. It can be seen that the precision is rather high. The solution process is shown in Figure 2.9. It can be seen that with a few iteration steps, the solution of the equation is found.

```
>> f=@(x)exp(-0.2*x)*sin(3*x+2)+cos(x);
   d=@(x)3*exp(-x/5)*cos(3*x+2)-sin(x)-(exp(-x/5)*sin(3*x+2))/5;
   x=nr_sols(f,d,10,1e-15,1), x1=x(end), f(x1)
   fplot(f,[9,12]), hold on, plot(x,zeros(size(x)),'o'), hold off
```

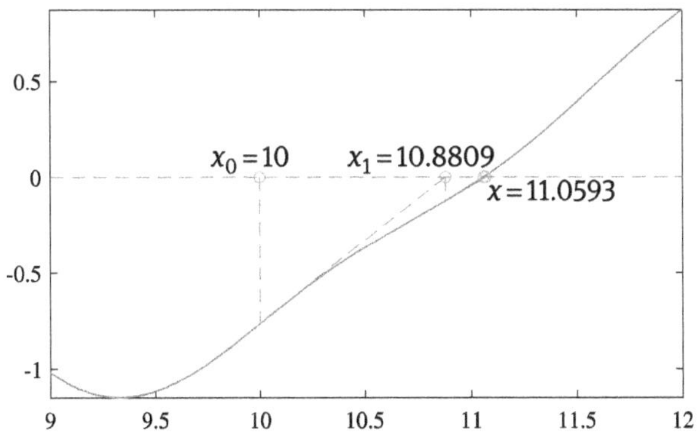

Figure 2.9: Intermediate process when solving the given equation.

Example 2.14. Use the secant method to find a solution to the equation in Example 2.13.

Solutions. We can select two initial points $x_0 = 9$ and $x_1 = 10$, then, by calling the solver, the intermediate results are obtained as:

$$\boldsymbol{x} = [9, 10, 12.9816, 11.3635, 10.7250, 11.0222, 11.0633, 11.0592, 11.0593, 11.0593],$$

and the solution process is illustrated in Figure 2.10. It can be seen that the efficiency of the solver is not as good as that of the Newton–Raphson algorithm, but since there is no need for the user to provide information of the derivatives, the algorithm and implementation are meaningful.

```
>> f=@(x)exp(-0.2*x)*sin(3*x+2)+cos(x);
   x=sec_sols(f,10,9,1e-15,1), x1=x(end), f(x1)
```

Example 2.15. Solve the simultaneous equations with two unknowns in Example 2.11. Find one solution starting from the initial point (1, 1).

Solutions. Since there are two independent variables x and y, the equations cannot be solved directly. They should be rewritten as equations for a vector $\boldsymbol{x}$. The simplest

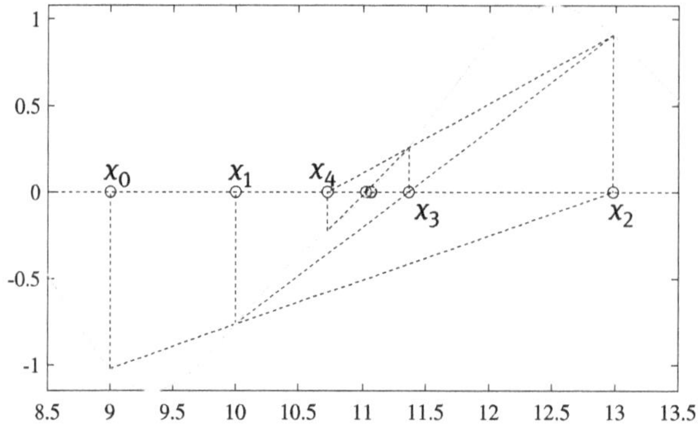

Figure 2.10: Intermediate solutions.

method is to select $x_1 = x$ and $x_2 = y$, such that the original equation can be rewritten as

$$\boldsymbol{f}(x_1, x_2) = \begin{bmatrix} x_1^2 e^{-x_1 x_2^2/2} + e^{-x_1/2}\sin(x_1 x_2) \\ x_2^2 \cos(x_1 + x_2^2) + x_1^2 e^{x_1+x_2} \end{bmatrix}.$$

Jacobian matrices cannot be easily found manually. Symbolic computation can be employed in finding their analytical expressions. The original equations can be expressed symbolically, and with function `jacobian()` the matrix can be found:

```
>> syms x1 x2
   f=[x1^2*exp(-x1*x2^2/2)+exp(-x1/2)*sin(x1*x2);
      x2^2*cos(x1+x2^2)+x1^2*exp(x1+x2)];
   J=jacobian(f,[x1,x2])
```

The Jacobian matrix of the equations can be found as

$$\boldsymbol{J} = \begin{bmatrix} 2x_1 e^{-x_1 x_2^2/2} - e^{-x_1/2}\sin(x_1 x_2)/2 - x_1^2 x_2^2 e^{-x_1 x_2^2/2}/2 + x_2 e^{-x_1/2}\cos(x_1 x_2) \\ 2x_1 e^{x_1+x_2} - x_2^2 \sin(x_2^2 + x_1) + {x_1}^2 e^{x_1+x_2} \end{bmatrix.$$

$$\begin{matrix} x_1 e^{-x_1/2}\cos(x_1 x_2) - {x_1}^3 x_2 e^{-x_1 x_2^2/2} \\ 2x_2 \cos(x_2^2 + x_1) - 2x_2^3 \sin(x_2^2 + x_1) + {x_1}^2 e^{x_1+x_2} \end{matrix}\Bigg].$$

With the original functions and Jacobian matrix, the two anonymous functions can be written manually, and then the Newton–Raphson solver can be used to solve the equation:

```
>> f=@(x)[x(1)^2*exp(-x(1)*x(2)^2/2)+exp(-x(1)/2)*sin(x(1)*x(2));
          x(2)^2*cos(x(1)+x(2)^2)+x(1)^2*exp(x(1)+x(2))];
   J=@(x)[2*x(1)*exp(-x(1)*x(2)^2/2)...
```

```
        -exp(-x(1)/2)*sin(x(1)*x(2))/2-...
        x(1)^2*x(2)^2*exp(-x(1)*x(2)^2/2)/2+...
        x(2)*exp(-x(1)/2)*cos(x(1)*x(2)),...
      x(1)*exp(-x(1)/2)*cos(x(1)*x(2))-...
        x(1)^3*x(2)*exp(-x(1)*x(2)^2)/2;
      2*x(1)*exp(x(1)+x(2))-x(2)^2*sin(x(2)^2+x(1))+...
        x(1)^2*exp(x(1)+x(2)),...
      2*x(2)*cos(x(2)^2+x(1))-2*x(2)^3*sin(x(2)^2+x(1))+...
        x(1)^2*exp(x(1)+x(2))];
   x=nr_sols(f,J,[1; 1],1e-15,1), length(x), norm(f(x(:,end)))
```

From the initial search point, the solution found is $\boldsymbol{x} = [5.1236, -12.2632]^{\mathrm{T}}$, with 18 intermediate points. Substituting the solution back to the original equations, the norm of the error is 3.9323×10^{-12}. It can be seen that a solution can be found. In the solution process, the Jacobian matrix is needed and should be expressed as an anonymous function. The process is error prone.

If the secant method is used, the following statements are needed, and the solution found is $\boldsymbol{x} = [1.0825, -1.1737]^{\mathrm{T}}$. The error norm is 2.2841×10^{-14}.

```
>> x=sec_sols(f,[1;1],[1.5; -3],1e-14,1)
   length(x), norm(f(x(:,end)))
```

Although the Jacobian matrix is not needed in the function call, the number of iterations is 265. Therefore, the efficiency is relatively low. For algebraic equations, better methods are needed.

2.3.2 Direct solution methods with MATLAB

From the presentation so far, it seems that the equation solution process is rather complicated. The needed information is sometimes difficult to acquire. Therefore, better solution methods are needed.

In MATLAB, a more practical solver, `fsolve()`, is provided. There is no need for the user to provide function handles of the Jacobian matrix. Only the function handle and initial search point are needed. The solver can be used to solve nonlinear simultaneous equations of any complexity. From the given initial search point, a solution can be found with the following syntax:

$\boldsymbol{x}$=`fsolve(`f,$\boldsymbol{x}_0$`)`

where f is the function handle of the equation functions, and $\boldsymbol{x}_0$ is the initial vector or matrix. The size of function f is exactly the same as that of the argument $\boldsymbol{x}_0$. Normally, the returned $\boldsymbol{x}$ is a numerical solution of the equation.

The function may return at most four arguments. The complete syntax of the solver is as follows:

`[x,F,flag,out]=fsolve(f,x0,opts)`

where $\boldsymbol{x}$ returns an equation solution, $\boldsymbol{F}$ contains the values of the equation functions at $\boldsymbol{x}$. If `flag` is positive, the solution process is successful. The argument `out` returns some intermediate information. The user may specify some control options in `opts`, to indicate the algorithm or error tolerance, and so on. These will be demonstrated through examples later.

Example 2.16. Solve the equations in Example 2.15 again with the new solver.

Solutions. An anonymous function is still needed to describe the original equations. There is no need to provide the Jacobian matrix function handle. With the direct call to the solver, the solution $\boldsymbol{x}_1 = [0, 2.1708]^{\mathrm{T}}$ is found. When substituted into the original equations, the norm of the errors is 3.2618×10^{-14}. Although the solution is not the same as that obtained in Example 2.15, it is still a solution of the equations. Besides, it can be seen that the number of iterations is 14, and the number of function f evaluations is 38. The result is similar to that obtained in Example 2.15, while the benefit here is that there is no need to provide the derivative functions. Therefore, the function is more suitable in real applications, and it is recommended to use directly in equation solution tasks.

```
>> f=@(x)[x(1)^2*exp(-x(1)*x(2)^2/2)+exp(-x(1)/2)*sin(x(1)*x(2));
          x(2)^2*cos(x(1)+x(2)^2)+x(1)^2*exp(x(1)+x(2))];
   x0=[1; 1]; [x1,f1,key,cc]=fsolve(f,x0)
```

If the initial search point is changed to $\boldsymbol{x}_0 = [2, 1]^{\mathrm{T}}$, the solution found is $\boldsymbol{x}_1 = [-0.7038, 1.6617]^{\mathrm{T}}$, with the norm of the errors $f_1 = 2.0242 \times 10^{-12}$.

```
>> x0=[2; 1]; x1=fsolve(f,x0), f1=norm(f(x1))
```

Compared with the Newton–Raphson method studied earlier, the solution process is significantly reduced. Since the descriptions of equations themselves, rather than Jacobian matrices, are needed, the solution process becomes very simple. This method can be used in practical equation solutions.

In the previous examples, the unknowns $\boldsymbol{x}$ and equation functions $\boldsymbol{f}(\boldsymbol{x})$ are vectors of the same size, and anonymous functions can be written to describe the equations and solve them. If in the equation $\boldsymbol{f}(\boldsymbol{x}) = \boldsymbol{0}$, $\boldsymbol{x}$ and $\boldsymbol{f}$ are matrices of the same size, function `fsolve()` can still be used in finding the numerical solutions of the equations. Let us see an example where Riccati equations are solved.

Definition 2.8. The mathematical form of an algebraic Riccati equation is

$$\boldsymbol{A}^{\mathrm{T}}\boldsymbol{X} + \boldsymbol{X}\boldsymbol{A} - \boldsymbol{X}\boldsymbol{B}\boldsymbol{X} + \boldsymbol{C} = \boldsymbol{0}, \tag{2.3.6}$$

where all the matrices are of size $n \times n$.

Riccati equations are named after Italian mathematician Jacopo Francesco Riccati (1676–1754). Originally, they correspond to a class of special first-order differential equations. We require that $\boldsymbol{B}$ be a positive-definite matrix, and $\boldsymbol{C}$ symmetric. Since differential equations are hard to solve, they are simplified as the above algebraic Riccati equations.

From the mathematical point of view, the matrices can be of any form. In the control science community, the function `are()` provided in the Control System Toolbox can be used to find one of the numerical solutions of Riccati equations. If one wants to find all the solutions, function `vpasolve()` can be called directly. The solutions of Riccati equations and other polynomial matrix equations will be demonstrated next.

Example 2.17. Solve the algebraic Riccati equation with the following matrices:

$$\boldsymbol{A}=\begin{bmatrix}-1 & 1 & 1\\ 1 & 0 & 2\\ -1 & -1 & -3\end{bmatrix},\quad \boldsymbol{B}=\begin{bmatrix}2 & 1 & 1\\ -1 & 1 & -1\\ -1 & -1 & 0\end{bmatrix},\quad \boldsymbol{C}=\begin{bmatrix}0 & -2 & -3\\ 1 & 3 & 3\\ -2 & -2 & -1\end{bmatrix}.$$

Solutions. The matrices should be entered first into MATLAB environment. Then, the `are()` function in the Control System Toolbox can be used to solve the algebraic Riccati equation, and compute the norm of the error matrix:

```
>> A=[-1,1,1; 1,0,2; -1,-1,-3]; B=[2,1,1; -1,1,-1; -1,-1,0];
   C=[0,-2,-3; 1,3,3; -2,-2,-1]; X=are(A,B,C) % solve equations
   norm(A'*X+X*A-X*B*X+C)     % find the norm of error matrix
```

Only one of the solutions can be found with the `are()` function, shown below. The error is 1.3980×10^{-14}, meaning that the solver is rather accurate:

$$\boldsymbol{X}=\begin{bmatrix}0.21411546 & -0.30404517 & -0.57431474\\ 0.83601813 & 1.60743230 & 1.397651600\\ -0.004386346 & 0.20982900 & 0.24656718\end{bmatrix}.$$

Since the equation is quadratic, it is natural that the equation may have multiple solutions. How can we find the other solutions? It is obvious to select another initial matrix, for instance, a matrix of ones, to search for another solution.

```
>> f=@(x)A.'*x+x*A-x*B*x+C; x0=ones(3); % with another initial point
   [X1,f1,flag,cc]=fsolve(f,x0), norm(f1)
```

The solution obtained is shown as follows, with an error of 6.3513×10^{-9}:

$$\boldsymbol{X}_1=\begin{bmatrix}2.1509892 & 2.9806867 & 2.4175971\\ -0.9005114 & -1.3375371 & -1.2847861\\ 0.95594506 & 1.8384489 & 1.7300025\end{bmatrix}.$$

It is obvious that this solution is different from that obtained with `are()` function. The returned $\boldsymbol{f}_1$ matrix is the value of equation matrix at the solution, i. e., $\boldsymbol{f}(\boldsymbol{X}_1)$. Besides, the argument `flag` returned is 1, meaning the solution process is successful, since it is a positive one. The returned argument `cc` contains the following information, indicating the number of iteration is 11, and the number of function calls is 102, meaning the solution efficiency is rather high:

```
     iterations: 11
      funcCount: 102
      algorithm: 'trust-region-dogleg'
  firstorderopt: 4.9891e-08
        message: Equation solved.
```

2.3.3 Accuracy specifications

It can be seen from the above example that the error may sometimes be rather large. Is there a way to control the size of the error?

Earlier in Theorem 2.4, in the convergence conditions of the iteration process, two error tolerances ε_1 and ε_2 were described. They can be assigned independently. A command `opts=optimset` can be used to set the control options `opts` in the solution process. The argument `opts` is a structured variable, many of its components can be modified by the users. For instance, `AbsTol` is the absolute error tolerance, which is in accordance with the constant ε_1. The more commonly used relative error tolerance can be assigned in `RelTol`. Besides, the error tolerance of the function, ε_2, can be assigned by setting `FunTol`. Therefore, these error tolerances can be used to control the solution precision.

Apart from these two options, the maximum numbers of iterations `MaxIter` and function calls `MaxFunEvals` can be set, otherwise warning messages may be shown to indicate that the numbers exceeded the maximum allowed ones. In this case, one may assign appropriate parameters to be large numbers. On the other hand, the solution obtained can be used as the initial value, to call `fsolve()` again in the solution process, until a convergent solution is found. This method can be implemented with loop structures.

Let us see the following example to demonstrate the specifications of accuracy and other settings.

Example 2.18. Select the matrix of ones to solve again the algebraic Riccati equation in Example 2.17.

Solutions. If the control options are expressed in a structured variable `ff`, it can be assigned directly in the function call, to find more accurate solutions. The new norm of error matrix of the solution is 2.5020×10^{-15}, significantly better compared with that under default specifications.

```
>> A=[-1,1,1; 1,0,2; -1,-1,-3]; B=[2,1,1; -1,1,-1; -1,-1,0];
   C=[0,-2,-3; 1,3,3; -2,-2,-1]; f=@(x)A.'*x+x*A-x*B*x+C;
   ff=optimset; ff.TolX=eps; ff.TolFun=1e-20;
   x0=ones(3); X1=fsolve(f,x0,ff), norm(f(X1))
```

Example 2.19. Solve the algebraic Riccati equation with

$$\boldsymbol{A}=\begin{bmatrix}-2 & 1 & -3\\ -1 & 0 & -2\\ 0 & -1 & -2\end{bmatrix},\quad \boldsymbol{B}=\begin{bmatrix}2 & 1 & -1\\ 1 & 2 & 0\\ -1 & 0 & -4\end{bmatrix},\quad \boldsymbol{C}=\begin{bmatrix}5 & -4 & 4\\ 1 & 0 & 4\\ 1 & -1 & 5\end{bmatrix}.$$

Solutions. The following commands can be tried to solve the Riccati equation:

```
>> A=[-2,1,-3; -1,0,-2; 0,-1,-2]; B=[2,1,-1; 1,2,0; -1,0,-4];
   C=[5 -4 4; 1 0 4; 1 -1 5]; X=are(A,B,C)
```

and we will be prompted with the error message "No solution: (***A***,***B***) may be uncontrollable or no solution exists". Although function `are()` may fail, the numerical solution of the original equation may still exist. For instance, the following command can be used:

```
>> f=@(x)A.'*x+x*A-x*B*x+C; x0=-ones(3);
   [X1,f1,flag,cc]=fsolve(f,x0), norm(f(X1))
```

and a solution can be found below, with an error norm of 1.2515×10^{-14}:

$$\boldsymbol{X}_1=\begin{bmatrix}5.5119052 & -4.2335285 & 0.18011933\\ 3.8739991 & -3.6090612 & -0.36369470\\ 6.6867568 & -4.9302000 & 1.07355930\end{bmatrix}.$$

2.3.4 Complex domain solutions

Another benefit of `fsolve()` function is that, if complex values are selected as the initial search point, real or complex solutions may be found. Besides, the solver can be used to handle equations with complex coefficients.

Example 2.20. Find the complex solutions of the equation in Example 2.17.

Solutions. If a complex initial matrix is used, sometimes complex solutions may be found. It can be validated that the complex conjugate of the solution also satisfies the original equation.

```
>> A=[-1,1,1; 1,0,2; -1,-1,-3]; B=[2,1,1; -1,1,-1; -1,-1,0];
   C=[0,-2,-3; 1,3,3; -2,-2,-1]; f=@(x)A.'*x+x*A-x*B*x+C;
   ff=optimset; ff.TolX=1e-15; ff.TolFun=1e-20;
   x0=eye(3)+ones(3)*1i; X1=fsolve(f,x0,ff), norm(f(X1))
   X2=conj(X1), norm(f(X2))
```

The solution found from the initial matrix is given below, and the error is about 1.1928×10^{-14}. The complex conjugate of the matrix also satisfies the original equation:

$$\boldsymbol{X}_1=\begin{bmatrix} 1.0979+2.6874\mathrm{j} & 1.1947+4.5576\mathrm{j} & 0.7909+4.1513\mathrm{j} \\ -3.5784+1.3112\mathrm{j} & -5.8789+2.2236\mathrm{j} & -5.4213+2.0254\mathrm{j} \\ -4.7771-0.4365\mathrm{j} & -7.8841-0.7403\mathrm{j} & -7.1258-0.6743\mathrm{j} \end{bmatrix}.$$

For this specific problem, if the real part of the initial matrix is assigned to a matrix of ones, and the imaginary part is set to an identity matrix, the solution obtained is a real matrix, indicating that a complex initial point may yield real solutions as well. It is worth mentioning that the result may come with tiny imaginary parts, and the error in this example is 6.4366×10^{-19}. The real solution can be extracted with `real()` function.

```
>> x0=ones(3)+eye(3)*1i; X1=fsolve(f,x0,ff), norm(f(X1))
   norm(imag(X1)), X2=real(X1), norm(f(X2))
```

Example 2.21. If matrix $\boldsymbol{A}$ in the algebraic Riccati equation is changed into a complex matrix, solve the equation again when

$$\boldsymbol{A}=\begin{bmatrix} -1+8\mathrm{j} & 1+\mathrm{j} & 1+6\mathrm{j} \\ 1+3\mathrm{j} & 5\mathrm{j} & 2+7\mathrm{j} \\ -1+4\mathrm{j} & -1+9\mathrm{j} & -3+2\mathrm{j} \end{bmatrix}.$$

Solutions. The matrices can be entered first into MATLAB workspace, then an anonymous function can be used to describe the original equation. The direct transpose $\boldsymbol{A}^\mathrm{T}$ of matrix $\boldsymbol{A}$, rather than the Hermitian transpose $\boldsymbol{A}^\mathrm{H}$, is involved. Therefore, in the anonymous function, `A.'` rather than `A'` is used, otherwise the description of matrix equation is wrong, and the solution is useless. The following commands can be used to find and validate the solution. The complex conjugate of the solution is also tested.

```
>> A=[-1+8i,1+1i,1+6i; 1+3i,5i,2+7i; -1+4i,-1+9i,-3+2i];
   B=[2,1,1; -1,1,-1; -1,-1,0]; C=[0,-2,-3; 1,3,3; -2,-2,-1];
   f=@(x)A.'*x+x*A-x*B*x+C; X0=ones(3);
   X1=fsolve(f,x0), norm(f(X1)), X2=conj(X1), norm(f(X2))
```

The solution found is shown below

$$X_1 = \begin{bmatrix} 0.0727 + 0.1015\mathrm{j} & 0.2811 - 0.1621\mathrm{j} & -0.3475 - 0.0273\mathrm{j} \\ -0.0103 + 0.0259\mathrm{j} & -0.2078 + 0.2136\mathrm{j} & 0.2940 + 0.2208\mathrm{j} \\ -0.0853 - 0.1432\mathrm{j} & -0.0877 - 0.0997\mathrm{j} & -0.0161 - 0.1826\mathrm{j} \end{bmatrix}.$$

If it is substituted back into the equation, the norm of the error is 5.4565×10^{-13}. For this particular example, even though a real initial matrix is used, complex solutions could still be found. Besides, although the solution can be found directly, the complex conjugate matrix appears to not satisfy the original equation. This is different from the case in real matrix equations, where if a solution matrix is found, its complex conjugate also satisfies the equation.

If the `are()` function in the Control System Toolbox is used as

```
>> X=are(A,B,C), norm(f(X))
```

the solution matrix obtained is given below, and there is no warning or error messages displayed. However, if the solution is substituted back to the original equation, the error is 10.5210, which is far too large, indicating that the solution does not at all satisfy the original equation:

$$X = \begin{bmatrix} 0.512140 & -0.165860 & 0.020736 \\ -0.390440 & 0.322260 & -0.051222 \\ -0.093655 & -0.075522 & 0.110550 \end{bmatrix}.$$

2.4 Accurate solutions of simultaneous equations

It has been pointed out that graphical methods can only be used in finding real solutions of given equations. Complex solutions cannot be found at all. A specific example was given in Example 2.12. Besides, if simultaneous equations have more solutions, zooming methods can be used to find the positions of the solutions individually, which is a rather complicated process. The graphical method can be used to find one solution at a time. The use of the method may sometimes be very inconvenient.

An analytic solver `solve()` is provided in the MATLAB Symbolic Math Toolbox, which can be used to solve the algebraic equations with analytical approaches. When the analytical solutions do not exist, an alternative solver `vpasolve()` can be used to find the high-precision solutions, the error of the solutions may be as small as 10^{-30} or even smaller, much more accurate than the numerical ones obtained in the double precision framework. In this book, the solutions are referred to as quasianalytical, in contrast to the analytical and numerical solutions in the double precision framework.

2.4.1 Analytical solutions of low-degree polynomial equations

Equations of polynomial equations of degrees 1 and 2 can be solved directly with `solve()` function. The solver can also be used in solving equations with other parameters. However, when other parameters exist, function `solve()` cannot be used to solve polynomial equations of degrees 3 and 4, unless specific control options are provided. Examples are used to demonstrate the solution processes.

The syntaxes of the `solve()` function are

S=`solve(`eqn_1`,`eqn_2`,...,`eqn_n`)`, %solve equation

S=`solve(`eqn_1`,`eqn_2`,...,`eqn_n`,`x_1`,...,`x_n`)`, %indicate the unknowns

where the equations eqn_i to be solved should be expressed with symbolic expressions, and the independent variables x_i should be symbolic variables. The obtained solution S is a structured variable, and the solutions can be extracted from $S.x_i$ commands. In the syntaxes, eqn_i can be a single equation, or a vector or matrix of equations. All the equations can also be described by a single vector or matrix symbolic expression eqn_1, so that the equations can be solved directly.

Of course, the following syntax can also be used to find the equation solutions directly:

`[`x_1`,...,`x_n`]=solve(...)`

It can be seen from the function call that the syntax with direct arguments is more practical than that with structured variables. Therefore, this syntax is used throughout the book where possible.

Example 2.22. Solve directly the chick–rabbit cage problem in Example 2.2.

Solutions. The symbolic variables should be declared first, then the equations can be rewritten with symbolic expressions. The function `solve()` can be used to directly solve the given equations. Note that double equality signs should be used in the equation descriptions.

```
>> syms x1 x2
   [x1,x2]=solve(x1+x2==35, 2*x1+4*x2==94)
```

The solutions found are $x_1 = 23$ and $x_2 = 12$. The solutions can be assigned to other argument names. Besides, if the right-hand sides of the equations are zeros, they can be omitted. For instance, the above solution command can be modified as follows:

```
>> [x0,y0]=solve(x1+x2-35, 2*x1+4*x2-94)
```

Example 2.23. Solve the following simultaneous equations, and find the conditions for real solutions:

$$\begin{cases} ax + cy = 2, \\ bx^2 + cx + ay^2 - 4xy = -3. \end{cases}$$

Solutions. It can be seen that through a simple transformation, the equation can be mapped into a polynomial equation of x or y, of degree 2, such that the corresponding formula can be used to find all the analytical solutions. However, it can be seen that the manual solution process is not so simple for this example. Computers should be used instead to replace the tedious tasks.

The necessary symbolic variables should be declared first, and the two equations can then be described by symbolic expressions. Then, the following commands can be used to solve the equations directly:

```
>> syms x y a b c
   f1=x*a+c*y-2; f2=b*x^2+c*x+a*y^2-4*x*y+3;
   [x0,y0]=solve(f1,f2)
```

A pair of solutions can be found as

$$x_0 = -\frac{c\left(8a + 4bc + ac^2 \pm a\Delta\right)}{2a\left(a^3 + 4ac + bc^2\right)} - \frac{2}{a},$$

$$y_0 = \frac{8a + 4bc + ac^2 \pm a\Delta}{2\left(a^3 + 4ac + bc^2\right)},$$

where $\Delta = \sqrt{c^4 - 48ac - 8a^2c - 12bc^2 - 12a^3 - 16c^2 - 16ab + 64}$.

It can be seen that when the expression under the square root sign is greater than or equal to zero, the equations have real solutions:

$$c^4 - 48ac - 8a^2c - 12bc^2 - 12a^3 - 16c^2 - 16ab + 64 \geqslant 0.$$

Example 2.24. Solve the polynomial equation with complex coefficients described in Example 2.7 in the symbolic computation framework.

Solutions. Since the analytical solutions of the equation exist, function `solve()` can be called to solve the equations. Finally, the analytical solutions can be obtained as $[-2, -1-\mathrm{j}, -1-\mathrm{j}, -1-\mathrm{j}]^{\mathrm{T}}$.

```
>> syms s; f=s^4+(5+3i)*s^3+(6+12i)*s^2-(2-14i)*s-(4-4i);
   s0=solve(f)
```

Example 2.25. Solve analytically the following quartic equation:

$$7s^4 + 119s^3 + 756s^2 + 2128s + 2240 = 0.$$

Solutions. Since there are algebraic formulas for finding the solutions of a quartic equation, the equation can be expressed in MATLAB environment, and then we can use `solve()` function to solve the equation:

```
>> syms s; f=7*s^4+119*s^3+756*s^2+2128*s+2240;
   x=solve(f)
```

The solutions obtained are $x = -5, -4, -4, -4$. In fact, although the algebraic solution formulas exist, the solution obtained may be irrational numbers. In other words, the genuine analytical solutions may not be easily found. For instance, if we add 1 to the left-hand side of the equation, the equation is still a quartic one, but the analytical solutions may not be found easily with the solver.

```
>> f=f+1; x=solve(f)
```

The equation, in fact, cannot be solved directly with the solver `solve()`. The results obtained are as follows:

```
root(z^4 + 17*z^3 + 108*z^2 + 304*z + 2241/7, z, 1)
root(z^4 + 17*z^3 + 108*z^2 + 304*z + 2241/7, z, 2)
root(z^4 + 17*z^3 + 108*z^2 + 304*z + 2241/7, z, 3)
root(z^4 + 17*z^3 + 108*z^2 + 304*z + 2241/7, z, 4)
```

indicating that the polynomial equation $z^4 + 17z^3 + 108z^2 + 304z + 2\,241/7 = 0$ has four solutions. If one wants to find its solutions, function `vpa()` can be used, or, alternatively, the solver `vpasolve()` can be called, and the norm of the error is 4.7877×10^{-36}:

```
>> x1=vpa(x), norm(subs(f,s,x1))
```

In fact, if we do need the analytical solutions of the cubic or quartic equations, the option `MaxDegree` should be assigned. The solution obtained would be rather lengthy. An example is shown below to demonstrate the process in finding the analytical solutions.

Example 2.26. Find the analytical solution of the modified quartic equation in Example 2.25.

Solutions. The option `'MaxDegree'` can be set to 4 to solve again the quartic equation

```
>> syms s; f=7*s^4+119*s^3+756*s^2+2128*s+2241;
   x=solve(f,'MaxDegree',4), x(1)
```

The solutions obtained are tedious to write. Through careful manual variable substitutions, the first solution can be simplified as

$$x_1 = -\frac{r_2^2}{6r} - \frac{\sqrt{9r^2r_2^2/2 - 9r^4r_2^2 - 12r_2^2/7 + 27r^3/4}}{6rr_2} - \frac{17}{4},$$

where

$$r = \sqrt[6]{\frac{1}{14} + \frac{\sqrt{3}\sqrt{7}\sqrt{67}\mathrm{i}}{882}}, \quad r_2 = \sqrt[4]{\frac{9r^2}{4} + 9r^4 + \frac{12}{7}}.$$

With such a formula, the solution of any precision can be obtained. For instance, the following commands may yield an error as small as 7.9004×10^{-104}:

```
>> norm(vpa(subs(f,s,x),100))
```

2.4.2 Quasianalytical solutions of polynomial-type equations

For an ordinary polynomial algebraic equation, function `vpasolve()` can be used in finding all the possible quasianalytical solutions. For a nonlinear equation, only one quasianalytical solution can be found at a time. In this section, quasianalytical solutions of some classes of equations are explored.

Example 2.27. Solve again the simultaneous equations in Example 2.12.

Solutions. The graphical method cannot be used to effectively solve the simultaneous equations in the example. Since it contains complex as well as real roots, symbolic expressions should be used to describe the equations, and then, with function `vpasolve()`, their solutions can be found directly

```
>> syms x y;
   [x0,y0]=vpasolve(x^2+y^2==5,x+4*y^3+3*y^2==2)
```

The solutions obtained are:

$$\boldsymbol{x}_0 = \begin{bmatrix} -2.0844946216518881587217059536735 \\ 2.2517408643882313559856 \mp 0.0056679616287447182117\mathrm{j} \\ -2.2559104938211695667 \pm 0.29064185911298833754 35\mathrm{j} \\ 2.0928338805177645801934398246011 \end{bmatrix},$$

$$\boldsymbol{y}_0 = \begin{bmatrix} 0.80924790534443253378482794183867 \\ 0.047383446412235832 56 \pm 0.26935104521931155183\mathrm{j} \\ -0.80829228997732274456 \mp 0.81116945942300081 79\mathrm{j} \\ -0.78743021821425870977966291830 05 \end{bmatrix}.$$

The solutions are returned in the vectors $\boldsymbol{x}_0$ and $\boldsymbol{y}_0$. It is not a simple thing to validate the solutions, since the equation expressions should be entered and evaluated. An alternative way is to express the two equations as symbolic expressions, then, with the variable substitution method, the norm of the error vector is 1.0196×10^{-38}:

```
>> f1=x^2+y^2-5; f2=x+4*y^3+3*y^2-2;
   [x0,y0]=vpasolve(f1,f2)
   norm([subs(f1,{x,y},{x0,y0}),subs(f2,{x,y},{x0,y0})])
```

It can be seen from the example that, for the user, the solution process for complicated high-degree polynomial equations is as simple as the chick–rabbit cage problem, since what is required is writing the equations as symbolic expressions, and then using an appropriate command and waiting for the solutions.

Simply put, a vector form can be used to express the two equations, then function `vpasolve()` can be called to solve the equations again. The results obtained are exactly the same as those obtained earlier:

```
>> F=[x^2+y^2-5; x+4*y^3+3*y^2-2]; [x0,y0]=vpasolve(F)
   norm(subs(F,{x,y},{x0,y0}))
```

Example 2.28. Solve the more complicated algebraic equations:

$$\begin{cases} \dfrac{1}{2}x^2+x+\dfrac{3}{2}+\dfrac{2}{y}+\dfrac{5}{2y^2}+\dfrac{3}{x^3}=0 \\ \dfrac{y}{2}+\dfrac{3}{2x}+\dfrac{1}{x^4}+5y^4=0. \end{cases}$$

Solutions. These equations are different from those in Example 2.12, since in Example 2.12 they could be converted manually into a higher-degree polynomial equation of a single variable. For these equations, the conversions are not possible by hand. Not only finding the solutions of the equations themselves, but also answering the question of how many solutions there are is difficult without the use of computers.

It is not necessary to worry about these low-level problems if computers are used. What the user needs to do is to code the equations into a computer in a standard way, such that the quasianalytical solutions can be found.

```
>> syms x y;
   f1(x,y)=x^2/2+x+3/2+2/y+5/(2*y^2)+3/x^3;
   f2(x,y)=y/2+3/(2*x)+1/x^4+5*y^4;
   [x0,y0]=vpasolve(f1,f2), size(x0)         % solve and count the solutions
   e1=norm(f1(x0,y0)), e2=norm(f1(x0,y0)) % evaluate the errors
```

If all the roots are substituted back to the original equations, a small error of size 10^{-33} can be found, indicating all the solutions are very accurate. For such a complicated equation, the solution process is again as simple as that for the chick–rabbit cage problem.

Even though, as indicated by the well-known Abel–Ruffini theorem, high-degree polynomial equations have no analytical solutions, quasianalytical methods can still be used to find the solutions with very high accuracy.

2.4.3 Quasianalytical solutions of polynomial matrix equations

In Definition 2.8, an algebraic Riccati equation is described. It is an equation involving a quadratic form of $\boldsymbol{X}$, so it is a polynomial equation. Describing the equation in symbolic form is a crucial step in the solution process. Here function `vpasolve()` is demonstrated with examples to show the solutions of the equations.

Example 2.29. Find all the roots of the algebraic Riccati equation in Example 2.17.

Solutions. If all the roots of the equation are expected, function `vpasolve()` should be explored. The unknown matrix $\boldsymbol{X}$ should be declared as a symbolic variable, and based on it, the Riccati equation should be coded as a symbolic expression, then function `vpasolve()` can be called to solve the equation:

```
>> A=[-1,1,1; 1,0,2; -1,-1,-3]; B=[2,1,1; -1,1,-1; -1,-1,0];
   C=[0,-2,-3; 1,3,3; -2,-2,-1];
   X=sym('x%d%d',3); F=A.'*X+X*A-X*B*X+C
   tic, X0=vpasolve(F), toc
```

With symbolic manipulation, the following Riccati simultaneous equations can be derived:

$$\begin{cases} x_{12}-x_{13}-x_{31}(1+x_{11}-x_{12})+x_{11}(x_{12}-2x_{11}+x_{13}-2)-x_{21}(1+x_{11}+x_{12}-x_{13})=0, \\ x_{11}-x_{13}-x_{32}(1+x_{11}-x_{12})+x_{12}(x_{12}-2x_{11}+x_{13}-1)-x_{22}(x_{11}+x_{12}-x_{13}-1)=2, \\ x_{11}+2x_{12}-x_{33}(1+x_{11}-x_{12})+x_{13}(x_{12}-2x_{11}+x_{13}-4)-x_{23}(x_{11}+x_{12}-x_{13}-1)=3, \\ x_{22}-x_{23}-x_{31}(1+x_{21}-x_{22})+x_{11}(1+x_{22}-2x_{21}+x_{23})-x_{21}(1+x_{21}+x_{22}-x_{23})=-1, \\ x_{21}-x_{23}-x_{32}(1+x_{21}-x_{22})+x_{12}(x_{22}-2x_{21}+x_{23})-x_{22}(x_{21}+x_{22}-x_{23})=-3, \\ x_{21}+2x_{22}-x_{33}(1+x_{21}-x_{22})+x_{13}(1+x_{22}-2x_{21}+x_{23})-x_{23}(3+x_{21}+x_{22}-x_{23})=-3, \\ x_{32}-x_{33}-x_{31}(4+x_{31}-x_{32})+x_{11}(1+x_{32}-2x_{31}+x_{33})-x_{21}(x_{31}+x_{32}-x_{33}-2)=2, \\ x_{31}-x_{33}-x_{32}(3+x_{31}-x_{32})+x_{12}(1+x_{32}-2x_{31}+x_{33})-x_{22}(x_{31}+x_{32}-x_{33}-2)=2, \\ x_{31}+2x_{32}-x_{33}(6+x_{31}-x_{32})+x_{13}(1+x_{32}-2x_{31}+x_{33})-x_{23}(x_{31}+x_{32}-x_{33}-2)=1. \end{cases}$$

After 23.04 seconds of waiting, all the 20 solutions of the equation can be found, and the 5th, 10th, 15th, and 18th solutions are real. The rest of them are complex conjugate ones. The four real solution matrices are respectively:

$$X_5 = \begin{bmatrix} 1.9062670985148 & 2.6695228037028 & 4.1090269897993 \\ -4.3719461205750 & -3.2277001659457 & -5.7232367559600 \\ -8.1493167800300 & -2.8535676463847 & -12.90505951546 \end{bmatrix},$$

$$X_{10} = \begin{bmatrix} 8.69508738924215 & -8.369677652147 & 15.08579547886 \\ -15.265261644743 & 14.485759397031 & -23.336518219659 \\ -20.7167885668890 & 17.582212700878 & -33.22526573531 \end{bmatrix},$$

$$X_{15} = \begin{bmatrix} 0.21411545933325 & -0.3040451651414 & -0.5743147449581 \\ 0.83601813100313 & 1.60743227422054 & 1.397651628726543 \\ -0.0043863464229 & 0.209828998159396 & 0.246567175609337 \end{bmatrix},$$

$$X_{18} = \begin{bmatrix} 2.1509892346834 & 2.980686747114 & 2.4175971297531 \\ -0.900511398366 & -1.3375371059663 & -1.284786114934488 \\ 0.95594505835086 & 1.83844891740592 & 1.7300024774690319 \end{bmatrix}.$$

The real solutions of the equation in Example 2.17 is, in fact, the 15th solution, which can be extracted and validated, and it can be seen that the norm of the error matrix is 1.8441×10^{-39}:

```
>> k=15;    % extracting the 15th solution
   X1=[X0.x11(k) X0.x12(k) X0.x13(k);
       X0.x21(k) X0.x22(k) X0.x23(k);
       X0.x31(k) X0.x32(k) X0.x33(k)];
   norm(A.'*X1+X1*A-X1*B*X1+C)
```

Example 2.30. Find all the roots of the equation in Example 2.19 with the quasianalytical method.

Solutions. The following commands can be used to find all the 20 solutions of the equation, with 8 being real and the rest being complex conjugates. The total elapsed time is 36.75 seconds.

```
>> A=[-2,1,-3; -1,0,-2; 0,-1,-2]; B=[2,1,-1; 1,2,0; -1,0,-4];
   C=[5 -4 4; 1 0 4; 1 -1 5];
   X=sym('x%d%d',3); F=A.'*X+X*A-X*B*X+C
   tic, X0=vpasolve(F), toc
```

If the A^T term in the algebraic Riccati equation is substituted by a free matrix D, a generalized Riccati equation is obtained.

Definition 2.9. The mathematical form of the generalized Riccati equation is

$$DX + XA - XBX + C = 0, \tag{2.4.1}$$

where all the matrices are of dimension $n \times n$.

There is no direct solution function available in MATLAB to solve generalized Riccati equations. Function `vpasolve()` can still be tried to find all the solutions to the equations.

Example 2.31. With the given matrices, solve the generalized Riccati equations

$$\boldsymbol{A}=\begin{bmatrix}-1&1&1\\1&0&2\\-1&-1&-3\end{bmatrix},\quad \boldsymbol{B}=\begin{bmatrix}2&1&1\\-1&1&-1\\-1&-1&0\end{bmatrix},\quad \boldsymbol{C}=\begin{bmatrix}0&-2&-3\\1&3&3\\-2&-2&-1\end{bmatrix},\quad \boldsymbol{D}=\begin{bmatrix}2&-1&-1\\1&1&-1\\1&-1&0\end{bmatrix}.$$

Solutions. As before, the matrices should be entered into MATLAB first. Then, symbolic expressions can be used to describe the equations, and the solver can be called to solve the equations directly:

```
>> A=[-1,1,1; 1,0,2; -1,-1,-3]; B=[2,1,1; -1,1,-1; -1,-1,0];
   C=[0,-2,-3; 1,3,3; -2,-2,-1]; D=[2,-1,-1; 1,1,-1; 1,-1,0];
   X=sym('x%d%d',3); F=D*X+X*A-X*B*X+C
   tic, X0=vpasolve(F), toc
```

After about 25 seconds of waiting, 20 solutions can be found with the above statements, where 8 of them are real, and the rest are complex conjugate.

Example 2.32. In Example 5.42 in Volume III, a higher-degree matrix equation has been explored, namely

$$\boldsymbol{A}\boldsymbol{X}^3+\boldsymbol{X}^4\boldsymbol{D}-\boldsymbol{X}^2\boldsymbol{B}\boldsymbol{X}+\boldsymbol{C}\boldsymbol{X}-\boldsymbol{I}=\boldsymbol{0},$$

where the given matrices are:

$$\boldsymbol{A}=\begin{bmatrix}2&1&9\\9&7&9\\6&5&3\end{bmatrix},\quad \boldsymbol{B}=\begin{bmatrix}0&3&6\\8&2&0\\8&2&8\end{bmatrix},\quad \boldsymbol{C}=\begin{bmatrix}7&0&3\\5&6&4\\1&4&4\end{bmatrix},\quad \boldsymbol{D}=\begin{bmatrix}3&9&5\\1&2&9\\3&3&0\end{bmatrix}.$$

Use the method studied here to solve the equation.

Solutions. It is natural to write the following commands. Unfortunately, after 1 843.8 seconds of waiting, only one solution of the equation is found, followed by the warning message “Solutions might be lost”, indicating that the method cannot find all the solutions.

```
>> A=[2 1 9; 9 7 9; 6 5 3]; B=[0 3 6; 8 2 0; 8 2 8];
   C=[7 0 3; 5 6 4; 1 4 4]; D=[3 9 5; 1 2 9; 3 3 0];
   f=A*X^3+X^4*D-X^2*B*X+C*X-eye(3);
   tic, X0=vpasolve(f), toc
```

Example 2.33. Find the quasianalytical solutions of the Riccati equation with complex matrices.

Solutions. It is natural to solve the Riccati equation with complex matrices employing the following statements. Since complex solution matrices are involved, the solution process is rather time consuming. With the same procedures and statements, about 379 seconds are needed to find the solutions, while in the real matrix case, 23.04 seconds are needed. All the 20 solutions can be found, where all of them are complex.

```
>> A=[-1+8i,1+1i,1+6i; 1+3i,5i,2+7i; -1+4i,-1+9i,-3+2i];
   B=[2,1,1; -1,1,-1; -1,-1,0]; C=[0,-2,-3; 1,3,3; -2,-2,-1];
   X=sym('x%d%d',3); F=A.'*X+X*A-X*B*X+C
   tic, X0=vpasolve(F), toc
```

2.4.4 Quasianalytical solutions of nonlinear equations

If symbolic expressions are used to describe the nonlinear simultaneous equations, function `vpasolve()` can be used to find their solutions. In contrast to polynomial-type equations, only one of the solutions can be found. If the original equations have more solutions, the user may select an initial search point, and find the quasianalytical solution from a given point of $\boldsymbol{x}_0$. The corresponding syntax of the function is

$\boldsymbol{x}$=`vpasolve`(eqn$_1$,...,eqn$_n$,x_1,...,x_n,$\boldsymbol{x}_0$)

where $\boldsymbol{x}_0$ is the initial search point of the unknown solution. If the equations are simultaneous equations of polynomial-type, there is no impact of the initial point $\boldsymbol{x}_0$ selection. An alternative syntax of the function is

$\boldsymbol{x}$=`vpasolve`(eqn$_1$,...,eqn$_n$,x_1,...,x_n,$\boldsymbol{x}_{\rm m}$,$\boldsymbol{x}_{\rm M}$)

where $\boldsymbol{x}_{\rm m}$ and $\boldsymbol{x}_{\rm M}$ are the lower and upper bounds of the unknown vector $\boldsymbol{x}$.

Example 2.34. Solve the nonlinear algebraic equation in Example 2.11 from an initial starting point of (2, 2).

Solutions. A natural way to solve the nonlinear equation directly is using

```
>> syms x1 x2
   f=[x1^2*exp(-x1*x2^2/2)+exp(-x1/2)*sin(x1*x2);
      x2^2*cos(x1+x2^2)+x1^2*exp(x1+x2)];
   [x0,y0]=vpasolve(f,[2 ; 2]), norm(subs(f,{x1,x2},{x0,y0}))
```

The solution obtained is as follows. When it is substituted into the equation, the error norm is as small as 2.0755×10^{-37}. It can be seen that the accuracy of the solution is much higher than for the numerical methods under the double precision framework:

$$\begin{aligned} x_0 &= 3.0029898235869693047992458712192, \\ y_0 &= -6.2769296697194948789764344182923. \end{aligned}$$

It can be seen from the example that, although the solutions with very high-precision are found, there is still one problem to be solved, that is, how to find all the intersections shown in Figure 2.6 in the region of interest. This is the problem to be solved in the next section.

2.5 Nonlinear matrix equations with multiple solutions

Although the solver vpasolve() can be used to find all the solutions of certain equation types, for matrix equations, especially for nonlinear matrix equations, the problem of finding all the solutions remains unsolved. There is no other existing MATLAB function for finding solutions of arbitrary nonlinear matrix equations.

Several algorithms have been studied earlier to find equation solutions from given initial points. However, with these methods it may be difficult to find all the possible solutions one at a time. Therefore, a simpler solver is expected to fulfill this task. In this section, an idea is conceived, and based on it, a general-purpose solver in MATLAB is written, trying to find all the possible solutions. Further, based on a similar method, we attempt to find all the quasianalytical solutions with an alternative solver.

2.5.1 An equation solution idea and its implementation

It can be seen from the previous solver that, if a random initial value is selected, a solution of the equation can be found. From the initial search point, a loop structure can be used. In the loop, fsolve() can be used to find one solution. If there is an earlier obtained solution, the precision of the new solution is compared with the recorded one, and the less accurate is discarded. If the solution is new, then it is recorded.

If an infinite loop structure is used, then it is likely to find all the solutions. Based on such an idea, a general-purpose solver can be written. In fact, we have released several versions of the solver. In this version, some special treatments are introduced. For instance, the zero matrix is tried to see whether it is an isolated solution. If a more accurate solution is found, it is used to replace the less accurate one. If a complex solution is found, its complex conjugate is tested to see whether it satisfies the equation. Based on these considerations, the following solver can be written. There is a benefit in the solver that the longer the solver is executed, the more likely it is that accurate solutions can be found.

```
function more_sols(f,X0,varargin)
[A,tol,tlim,ff]=default_vals({1000,eps,30,optimset},varargin{:});
if length(A)==1, a=-0.5*A; b=0.5*A; else, a=A(1); b=A(2); end
ar=real(a); br=real(b); ai=imag(a); bi=imag(b);
ff.Display='off'; ff.TolX=tol; ff.TolFun=1e-20;
```

```
[n,m,i]=size(X0); X=X0; tic
if i==0, X0=zeros(n,m); % check whether it is an isolated solution
   if norm(f(X0))<tol, i=1; X(:,:,i)=X0; end
end
while (1) % infinite loop, press Ctrl+C to terminate
   x0=ar+(br-ar)*rand(n,m); % generate a random initial search point
   if abs(imag(A))>1e-5, x0=x0+(ai+(bi-ai)*rand(n,m))*1i; end
   [x,aa,key]=fsolve(f,x0,ff); t=toc; if t>tlim, break; end
   if key>0, N=size(X,3); % find the number of roots
      for j=1:N, if norm(X(:,:,j)-x)<1e-4; key=0; break; end, end
      if key==0            % if a more accurate solution is found
         if norm(f(x))<norm(f(X(:,:,j))), X(:,:,j)=x; end
      elseif key>0, X(:,:,i+1)=x; % record the solution
         if norm(imag(x))>1e-5 && norm(f(conj(x)))<1e-8
            i=i+1; X(:,:,i+1)=conj(x);  % test the complex conjugate
         end
         assignin('base','X',X); i=i+1, tic  % update the information
      end, assignin('base','X',X);
end, end
```

A low-level common function `default_vals()` is written, which is also shown in other volumes. For convenience, this function is listed below

```
function varargout=default_vals(vals,varargin) % read default arguments
if nargout=length(vals), error('number of arguments mismatch');
else, nn=length(varargin)+1; % with look to assign default arguments
   varargout=varargin;
   for i=nn:nargout, varargout{i}=vals{i};
end, end, end
```

The syntax of `more_sols()` solver is as follows:

`more_sols(`f`,`$\boldsymbol{X}_0$`,`a`,`ϵ`,`t_{lim}`,opts)`

where f is the function handle of the equations, which can be in an anonymous function or in a MATLAB function; $\boldsymbol{X}_0$ is a three-dimensional array, used to describe the size of the solution, or initially found solutions. If the solver is called for the first time, the argument should be assigned to `zeros(`n`,`m`,0)`, that is, a blank three-dimensional array is assigned, where n and m indicate the size of the solution matrix. The solutions of the equation are stored automatically in MATLAB workspace in the variable $\boldsymbol{X}$. If one wants to continuously solve the equations, $\boldsymbol{X}_0$ should be assigned to $\boldsymbol{X}$. The default value of a is 1 000, meaning the search for the solutions within the interval $[-500, 500]$. The argument ϵ is `eps`. The argument t_{lim} is assigned to 30, meaning that if within

30 seconds, a new solution is not found, the solver is completed automatically. The users may also assign the control options `opts`, with the default setting assigned as `optimset`. The argument a can also be selected as a complex number, indicating that complex solutions are also needed. Besides, a can also be assigned to the solution interval $[a, b]$.

Example 2.35. Solve again the univariate equation in Example 2.10.

Solutions. It can be seen from Figure 2.5 that there are 6 intersections in the region of interest. With the solver `more_sols()`, all 6 solutions can be found directly, as labeled on Figure 2.11.

```
>> f=@(x)exp(-0.2*x).*sin(3*x+2)-cos(x);
   more_sols(f,zeros(1,1,0),[0,20])
   x0=X(:); x1=x0(x0>=0 & x0<=20), fplot(f,[0,20]), hold on
   plot(x1,f(x1),'o',[0,20],[0,0],'--'), hold off
```

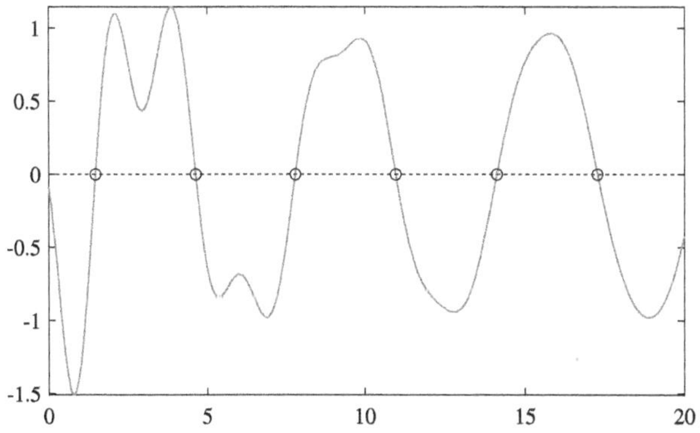

Figure 2.11: All real solutions of a transcendental equation.

The solutions obtained are

$$x_1 = [1.4720, 4.6349, 7.7990, 10.9601, 14.1159, 17.2666].$$

Example 2.36. Find all solutions of the equation in Example 2.11 within the region of interest, which is $-2\pi \leqslant x_1, x_2 \leqslant 2\pi$.

Solutions. It is natural to use the following form to solve the nonlinear algebraic equations directly. An anonymous function should be used to describe the simultaneous equations. Then, the solver `more_sols()` can be called to find all solutions of the equations in the region of interest. Here A is set to 13, slightly larger than 4π. The number of solutions obtained is slightly larger than of those existing in the region of interest.

```
>> f=@(x)[x(1)^2*exp(-x(1)*x(2)^2/2)+exp(-x(1)/2)*sin(x(1)*x(2));
          x(2)^2*cos(x(1)+x(2)^2)+x(1)^2*exp(x(1)+x(2))];
   A=13; more_sols(f,zeros(2,1,0),A)
```

All the solutions in the region of interest can be extracted, and it can be seen that there are altogether 110 solutions, all superimposed on the curves, as shown in Figure 2.12.

```
>> ii=find(abs(X(1,1,:))<=2*pi && abs(X(2,1,:))<=2*pi);
   X1=X(:,:,ii); size(ii) %extract the solutions in the region
   x=X1(1,1,:); x=x(:); y=X1(2,1,:); y=y(:); plot(x,y,'o')
   syms x y; f1=x^2*exp(-x*y^2/2)+exp(-x/2)*sin(x*y);
   f2=y^2*cos(x+y^2)+x^2*exp(x+y); hold on;
   fimplicit([f1,f2],[-2*pi,2*pi],'Meshdensity',800)
   hold off, axis(2*pi*[-1,1,-1,1])
```

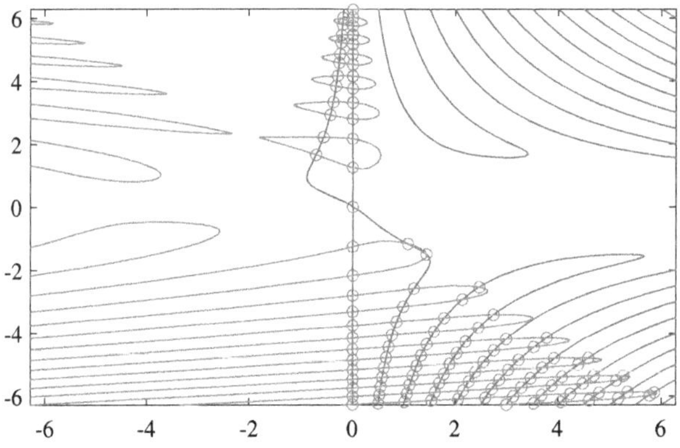

Figure 2.12: Graphical display of the solutions.

Example 2.37. Solve the generalized Riccati equation in Example 2.31 with numerical methods.

Solutions. As before, the matrices should be entered first. Next, an anonymous function can be written to describe the equations. Then, the following commands can be used in solving the equation directly, where 8 real solutions are obtained:

```
>> A=[-1,1,1; 1,0,2; -1,-1,-3]; B=[2,1,1; -1,1,-1; -1,-1,0];
   C=[0,-2,-3; 1,3,3; -2,-2,-1]; D=[2,-1,-1; 1,1,-1; 1,-1,0];
   F=@(X)D*X+X*A-X*B*X+C, more_sols(F,zeros(3,3,0))
```

Continuing the solution process with the following statements, by assigning A to a complex quantity, all 20 complex solutions are obtained, and they are the same as those in Example 2.31.

```
>> more_sols(F,X,1000+1000i)
```

Example 2.38. If the generalized Riccati equation in Example 2.31 is converted into the form of $\boldsymbol{DX} + \boldsymbol{XA} - \boldsymbol{BX}^2 + \boldsymbol{C} = \boldsymbol{0}$, solve the equation with different methods.

Solutions. If the solver `more_sols()` is used directly, altogether 19 real solutions can be found.

```
>> A=[-1,1,1; 1,0,2; -1,-1,-3]; B=[2,1,1; -1,1,-1; -1,-1,0];
   C=[0,-2,-3; 1,3,3; -2,-2,-1]; D=[2,-1,-1; 1,1,-1; 1,-1,0];
   F=@(X)D*X+X*A-B*X^2+C; more_sols(F,zeros(3,3,0))
```

If one searches for solutions in the complex domain, 57 solutions can be found. Now let us try `vpasolve()` to find all the quasianalytical solutions. After 5 873.8 seconds of waiting, 60 solutions were found.

```
>> X=sym('x%d%d',[3,3]);
   F=D*X+X*A-B*X^2+C, tic, X0=vpasolve(F), toc
```

The numbers of solutions with the two methods are similar. If we compare the 11th solution, it is found that x_{11} for the latter is located far away from the origin, with a magnitude larger than 10^{10}, while the same for the former is located reasonably close. The values of x_{11} obtained by the two methods are compared in Table 2.1. Be-

Table 2.1: Comparisons of the two solutions in x_{11} term.

quasianalytical	numerical	quasianalytical	numerical
$-3.006466262658 \times 10^{63}$	–	–	−431.5070709
–	−71.6679956	−3.4447376962310	−3.444737696
−1.5146484375	–	–	−1.450204069
−0.393316966940371	−0.3933169669	0.1563285970559477	0.1563285971
0.19217475249284	0.1921747525	0.925385242867366	0.9253852429
1.10128812201771	1.101288122	1.712893235210277	1.712893235
1.83357133024232	1.83357133	3.121670783661601	3.121670784
3.1272528908906671	3.127252891	5.803147834137380	5.803147834
7.0244296789169	7.02442972	7.387972365540733	7.387972366
–	15.92032583	38.0	–
–	52.986452	–	577.0832185
79 369 526 682 545 291 264.0	–	169 534 813 891 083 632 640.0	–
$5.7961230691484403 \times 10^{59}$	–	$4.69257498760911 9\times 10^{67}$	–

sides, it is seen that the solution $x_{11} = 38$ is the 21th one, and, when substituted back to the equation, the error norm is 18 981 334.363, meaning the solution is not valid. The solution $x_{11} = -1.5146$ is the 15th solution, which does not satisfy the original equation either. Some of the double precision solutions found by `more_sols()`, such as −71.668, cannot be found with `vpasolve()` function. Therefore, although function `vpasolve()` may handle polynomial matrix equations, the solutions are not reliable.

```
>> length(find(abs(X0.x11)>1e10)), k=21;
   x0=[X0.x11(k) X0.x12(k) X0.x13(k);
       X0.x21(k) X0.x22(k) X0.x23(k);
       X0.x31(k) X0.x32(k) X0.x33(k)]
   norm(F(x0))
```

Example 2.39. Solve the following nonlinear matrix equation:

$$e^{\boldsymbol{AX}} \sin \boldsymbol{BX} - \boldsymbol{CX} + \boldsymbol{D} = \boldsymbol{0},$$

where matrices $\boldsymbol{A}$, $\boldsymbol{B}$, $\boldsymbol{C}$, and $\boldsymbol{D}$ are given in Example 2.31. Find all its real solutions.

Solutions. The following statements can be used to solve the complicated nonlinear matrix equations directly. So far, 92 real solutions are found, and stored in the file data36c.mat. The readers may try for themselves and see whether new solutions can be found.

```
>> A=[2 1 9; 9 7 9; 6 5 3]; B=[0 3 6; 8 2 0; 8 2 8];
   C=[7 0 3; 5 6 4; 1 4 4]; D=[3 9 5; 1 2 9; 3 3 0];
   f=@(X)expm(A*X)*funm(B*X,@sin)-C*X+D;
   more_sols(f,zeros(3,3,0),5); X
```

2.5.2 Pseudopolynomial equations

A pseudopolynomial equation is a special case of a nonlinear matrix equation, where the unknown matrix is a scalar. In this section the definition and solution methods of pseudopolynomial equations are presented.

Definition 2.10. The mathematical form of a pseudopolynomial is given by

$$p(s) = c_1 s^{\alpha_1} + c_2 s^{\alpha_2} + \cdots + c_{n-1} s^{\alpha_{n-1}} + c_n s^{\alpha_n} = 0 \tag{2.5.1}$$

where α_i are real numbers.

It can be seen that a pseudopolynomial equation is an extension of an ordinary polynomial equation, since the unknown matrix is a scalar. Here we explore the solution process.

Example 2.40. Solve the pseudopolynomial equation[52]

$$x^{2.3} + 5x^{1.6} + 6x^{1.3} - 5x^{0.4} + 7 = 0.$$

Solutions. A natural method is to introduce a new variable $z = x^{0.1}$, such that the original equation can be mapped into a polynomial equation of z:

$$z^{23} + 5z^{16} + 6z^{13} - 5z^{4} + 7 = 0.$$

It can be seen that there are 23 solutions. With the mapping $x = z^{10}$, all the solutions to the original pseudopolynomial equation solution can be found. The idea can be implemented in the following MATLAB statements:

```
>> syms x z; f1=z^23+5*z^16+6*z^13-5*z^4+7;
   p=sym2poly(f1); r=roots(p);  f=x^2.3+5*x^1.6+6*x^1.3-5*x^0.4+7;
   r1=r.^10, double(subs(f,x,r1))
```

However, if one tries to substitute the solutions back to the original equation, it is found that most of them do not satisfy the original pseudopolynomial equation. How many genuine solutions are there? Only two of the solutions in x satisfy the original equation, namely, $x = -0.1076 \pm \mathrm{j}0.5562$, the remaining 21 are extraneous roots. With the following statements, the same two solutions can be found:

```
>> f=@(x)x.^2.3+5*x.^1.6+6*x.^1.3-5*x.^0.4+7;
   more_sols(f,zeros(1,1,0),100+100i), x0=X(:)
```

From the mathematical viewpoint, the genuine solutions are located on the first Riemann sheet, and other “solutions” are located on the other Riemann sheets. They are the extraneous roots, which do not satisfy the original equation.

Example 2.41. Solve the pseudopolynomial equation with irrational degrees

$$s^{\sqrt{5}} + 25s^{\sqrt{3}} + 16s^{\sqrt{2}} - 3s^{0.4} + 7 = 0.$$

Solutions. Since the degrees here are irrational numbers, it is not possible to map it into an ordinary polynomial equation. Therefore, function `more_sols()` may be the only choice in finding the solutions to such an equation. The following commands can be used to solve the equation directly, where two complex solutions are found at $s = -0.0812 \pm 0.2880\mathrm{j}$.

```
>> f=@(s)s^sqrt(5)+25*s^sqrt(3)+16*s^sqrt(2)-3*s^0.4+7;
   more_sols(f,zeros(1,1,0),100+100i); x0=X(:)
   err=norm(f(x0(1)))
```

Substituting the solutions back to the original equation, the error found is 9.1551×10^{-16}, indicating that the solutions are accurate enough by numerical standards. It can be seen from this example that, even though the degree is changed into an irrational number, it does not add extra troubles in the solution process. The computational complexity is the same as that in the previous example.

2.5.3 A quasianalytical solver

In the solver `more_sols()`, the kernel tool is the function `fsolve()`. If it is replaced by the high-precision `vpasolve()` solver, a high-precision universal solver may be established:

```
function more_vpasols(f,X0,varargin)
[A,tlim]=default_vals({1000,60},varargin{:}); %default parameters
if length(A)==1, a=-0.5*A; b=0.5*A; else, a=A(1); b=A(2); end
ar=real(a); br=real(b); ai=imag(a); bi=imag(b);
X=X0; [i,n]=size(X0); tic
while (1), %infinite loop, press Ctrl+C to terminate
   x0=ar+(br-ar)*rand(1,n); %generate initial matrix
   if abs(imag(A))>1e-5, x0=x0+(ai+(bi-ai)*rand(1,n))*1i; end
   V=vpasolve(f,x0); %solve equation
   N=size(X,1); key=1; x=sol2vec(V);
   if length(x)>0 %if the solution set is not empty, judge it
   t=toc; if t>tlim, break; end %if no new solutions found
   for j=1:N, if norm(X(j,:)-x)<1e-5; key=0; break; end, end
   if key>0, i=i+1; X=[X; x]; %record new solutions found
      disp(['i=',int2str(i)]); assignin('base','X',X); tic
end, end, end
function v=sol2vec(A) %subfunction to convert solutions into vectors
v=[]; A=struct2cell(A);
for i=1:length(A), v=[v, Ai]; end %convert to row vector
```

The syntax of the function is `more_vpasols(`f`,`X_0`,`A`,`$t_{\lim}$`)`, where the low-level subfunction `sol2vec()` is written to convert a solution into a row vector. The input argument f is a symbolic row vector expression to express the simultaneous equations. The initial matrix X_0 can be assigned to `zeros(0,`n`)`, where n is the number of unknowns. The other input arguments are the same as those discussed for `more_sols()` function. The solution variable $X(i,:)$ stores the ith solution. It should be noted that the `more_vdpsols()` function is much slower than the `more_sols()` function, but the precision is much higher.

Example 2.42. Consider the simultaneous equations in Example 2.11. Find the quasi-analytical solutions within the interval $-2\pi \leqslant x, y \leqslant 2\pi$.

Solutions. The following statements can be used directly to solve the simultaneous equations:

```
>> syms x y; t=cputime;
   F=[x^2*exp(-x*y^2/2)+exp(-x/2)*sin(x*y);
      y^2*cos(x+y^2)+x^2*exp(x+y)];
   more_vpasols(F,zeros(0,2),4*pi); cputime-t %solution and time
```

To validate the precisions of the solutions, the roots in the expected regions should be extracted first, and we store them in $\boldsymbol{x}_0$ and $\boldsymbol{y}_0$. When the solutions are substituted back to the original equation, the norm of the error matrix is 7.79×10^{-32}. Compared with that in Example 2.36, the accuracy is much higher. The elapsed time is about half an hour, which is much longer than for the `more_sols()` solver. The number of roots found in the region of interest is only 105, similar to that in Figure 2.12, with some solutions missing.

```
>> x0=X(:,1); y0=X(:,2); ii=find(abs(x0)<2*pi & abs(y0)<2*pi);
   x0=x0(ii); y0=y0(ii); [x0 ii]=sort(x0); y0=y0(ii);
   double(norm(subs(F,{x,y},{x0,y0}))), size(x0)
   fimplicit(F,[-2*pi,2*pi]), hold on; plot(x0,y0,'o')
```

Example 2.43. Find the high-precision solutions to the equations in Example 2.39.

Solutions. The following statements can be tried, however, when the symbolic expression is used to describe the matrix equation, the expression f cannot be constructed. Therefore, the subsequent call to the solver `more_vpasols()` cannot be made. The high-precision solutions cannot be found. The numerical solver in Example 2.39 may be the only choice:

```
>> A=[2 1 9; 9 7 9; 6 5 3]; B=[0 3 6; 8 2 0; 8 2 8];
   C=[7 0 3; 5 6 4; 1 4 4]; D=[3 9 5; 1 2 9; 3 3 0];
   X=sym('x%d%d',3);
   f=expm(A*X)*funm(B*X,@sin)-C*X+D;
   more_vpasols(f,zeros(3,3,0),20);
```

2.6 Underdetermined algebraic equations

In the equations presented earlier, it is always assumed that the numbers of the unknowns and equations are the same. Such equations are normal. In this section, abnormal underdetermined equations are discussed.

Definition 2.11. If the numbers of unknowns and equations are the same, the equations are referred to as well-posed equations. If the number of equations is less, the equations are referred to as underdetermined equations; while if the number of unknowns is less, the equations are known as overdetermined.

The implicit function $f(x, y) = 0$ discussed earlier is, in fact, an underdetermined equation. If functions such as `ezplot()` or `fimplicit()` are used, all the points on the curves are the solutions of the underdetermined equations. In other words, underdetermined equations may have infinitely many solutions.

In some special cases, the implicit function plotting facilities may not be able to draw any curves, since the solutions may be isolated. In this case the solver `fsolve()` may be considered. Under the default setting, function `fsolve()` cannot be used to solve equations whose numbers of unknowns and equations are not equal. The solver should be set to `levenberg-marquardt`, i. e., the Levenberg–Marquardt algorithm, to solve the underdetermined equations. If function `more_sols()` is used, the same setting can be tried. In this section we shall explore underdetermined equations with isolated solutions.

Example 2.44. Solve the following underdetermined equation:

$$(4x_1^3 + 4x_1x_2 + 2x_2^2 - 42x_1 - 14)^2 + (4x_2^3 + 2x_1^2 + 4x_1x_2 - 26x_2 - 22)^2 = 0.$$

Solutions. If one wants to solve this equation manually, one can observe that the given equation can be converted into two independent equations. Therefore, the new equations have the same numbers in unknowns and equations, so that the solvers such as `more_sols()` can be used directly.

Manual conversion sometimes is not practical in real applications, since not all the underdetermined equations can be converted in this way. We are expecting to find a method to solve underdetermined equations.

If the following commands are used to draw the implicit functions, various warnings will be displayed. No curve can be drawn, indicating that the original equation may contain isolated solutions.

```
>> f=@(x1,x2)(4*x1.^3+4*x1.*x2+2*x2.^2-42*x1-14).^2+...
             (4*x2.^3+2*x1.^2+4*x1.*x2-26*x2-22).^2;
   fimplicit(f)
```

Now let us set the solver to Levenberg–Marquardt algorithm. Then we can try to solve the underdetermined equation. After some waiting, altogether 9 solutions can be found for the underdetermined equations.

```
>> ff=optimoptions('fsolve','Algorithm','levenberg-marquardt');
   F=@(x)f(x(1),x(2));        % anonymous function for the equation
   more_sols(F,zeros(2,1,0),10,eps,300,ff)
```

The nine solutions are $(-2.8051, 3.1313)$, $(3, 2)$, $(0.0867, 2.8843)$, $(3.3852, 0.0739)$, $(3.5844, -1.8481)$, $(-3.7793, -3.2832)$, $(-0.1280, -1.9537)$, $(-3.0730, -0.0813)$, and $(-0.2708, -0.9230)$, where the search of the forth solution may be quite time consuming.

2.7 Exercises

2.1 Prove that the manually simplified solution of x_1 in Example 2.26 is correct.

2.2 Solve the polynomial equation $x^4 + 14x^3 + 73.5x^2 + 171.5x + 150.0625 = 0$ and see whether the numerical method can be used to find its accurate solutions.

2.3 Check that for the solutions obtained in Example 2.38, those located far away do not satisfy the original equation.

2.4 Since complex matrices are involved, finding the solution of the equation in Example 2.34 with the quasianalytical method is rather time consuming. Solve the equation with the numerical method and see whether complex matrices bring extra trouble in the solution process.

2.5 Solve the following polynomial-type simultaneous equations and find the accuracy of the solutions:

$$(1)\quad \begin{cases} 24xy - x^2 - y^2 - x^2y^2 = 13, \\ 24xz - x^2 - z^2 - x^2z^2 = 13, \\ 24yz - y^2 - z^2 - y^2z^2 = 13, \end{cases}$$

$$(2)\quad \begin{cases} x^2y^2 - zxy - 4x^2yz^2 = xz^2, \\ xy^3 - 2yz^2 = 3x^3z^2 + 4xzy^2, \\ y^2x - 7xy^2 + 3xz^2 = x^4zy, \end{cases}$$

$$(3)\quad \begin{cases} x + 3y^3 + 2z^2 = 1/2, \\ x^2 + 3y + z^3 = 2, \\ x^3 + 2z + 2y^2 = 2/4. \end{cases}$$

2.6 Solve the following simultaneous equations:[23]

$$\begin{cases} x_1x_2 + x_1 - 3x_5 = 0, \\ 2x_1x_2 + x_1 + 3R_{10}x_2^2 + x_2x_3^2 + R_7x_2x_3 + R_9x_2x_4 + R_8x_2 - Rx_5 = 0, \\ 2x_2x_3^2 + R_7x_2x_3 + 2R_5x_3^2 + R_6x_3 - 8x_5 = 0, \\ R_9x_2x_4 + 2x_4^2 - 4Rx_5 = 0, \\ x_1x_2 + x_1 + R_{10}x_2^2 + x_2x_3^2 + R_7x_2x_3 + R_9x_2x_4 + R_8x_2 + R_5x_3^2 + R_6x_3 + x_4^2 = 1, \end{cases}$$

where $0.0001 \leqslant x_i \leqslant 100$, $i = 1, 2, 3, 4, 5$. The given constants are $R = 10$, $R_5 = 0.193$, $R_6 = 4.10622 \times 10^{-4}$, $R_7 = 5.45177 \times 10^{-4}$, $R_8 = 4.4975 \times 10^{-7}$, $R_9 = 3.40735 \times 10^{-5}$, and $R_{10} = 9.615 \times 10^{-7}$.

2.7 Solve the following equation:[23]

$$\frac{b}{T_0}Te^{c/T}-\frac{b(1+aT_0)}{aT_0}e^{c/T}+\frac{T}{T_0}-1=0,$$

where $100 \leqslant T \leqslant 1000$. The constants are $a = -1000/(3\Delta H)$, $b = 1.344\times10^9$, $c = -7548,1193$, $T_0 = 298$, and ΔH has three options, namely, $-50\,000$, $-35\,958$, and $-35\,510.3$.

2.8 Solve the following simultaneous equation:[23]

$$\begin{cases}0.5\sin x_1x_2-0.25x_2/\pi-0.5x_1=0,\\(1-0.25/\pi)\left[e^{2x_1}-e\right]+ex_2/\pi-2ex_1=0,\end{cases}$$

where $0.25 \leqslant x_1 \leqslant 1$, $1.5 \leqslant x_2 \leqslant 2\pi$.

(1) In [23], two solutions are provided. If the intervals are changed to $x_1 \in (-5,5)$, $x_2 \in (-10,10)$, find all the real solutions. Use graphical methods to validate the solutions and see if there are any unfound real solutions.

(2) Are there any complex solutions? If the imaginary part is restricted to $(-10,10)$, see how many complex solutions can be found.

2.9 Find the solutions of the following equation and validate the results:[45]

$$\begin{cases}t^{31}+t^{23}y+t^{17}x+t^{11}y^2+t^5xy+t^2x^2=0,\\t^{37}+t^{29}y+t^{19}x+t^{13}y^2+t^7xy+t^3x^2=0.\end{cases}$$

2.10 Solve the equations with extra parameters:

$$\begin{cases}x^2+ax^2+6b+3y^2=0,\\y=a+x+3.\end{cases}$$

2.11 Use graphical methods to solve the following equations, and validate the results:

(1) $f(x)=e^{-(x+1)^2+\pi/2}\sin(5x+2)$,

(2) $\begin{cases}(x^2+y^2+10xy)e^{-x^2-y^2-xy}=0,\\x^3+2y=4x+5.\end{cases}$

2.12 Use graphical and numerical methods to solve the following simultaneous equations in the interval $-2\pi \leqslant x,y \leqslant 2\pi$, and validate the results:[53]

$$\begin{cases}x^2e^{-xy^2/2}+e^{-x/2}\sin(xy)=0,\\y^2\cos(x+y^2)+x^2e^{x+y}=0.\end{cases}$$

2.13 Use numerical solvers to solve the equations in Exercise 2.11, and validate the solutions found.

2.14 For the following robot kinetic equations, see how many real solutions can be found:[23]

$$\begin{cases} 4.731\times 10^{-3}x_1x_3 - 0.3578x_2x_3 - 0.1238x_1 + x_7 - 1.637\times 10^{-3}x_2 - 0.9338x_4 = 0.3571, \\ 0.2238x_1x_3 + 0.7623x_2x_3 + 0.2638x_1 - x_7 - 0.07745x_2 - 0.6734x_4 - 0.6022 = 0, \\ x_6x_8 + 0.3578x_1 + 4.731\times 10^{-3}x_2 = 0, \\ -0.7623x_1 + 0.2238x_2 + 0.3461 = 0, \\ x_1^2 + x_2^2 - 1 = 0, \\ x_3^2 + x_4^2 - 1 = 0, \\ x_5^2 + x_6^2 - 1 = 0, \\ x_7^2 + x_8^2 - 1 = 0, \end{cases}$$

where $-1 \leqslant x_i \leqslant 1$, $i = 1, 2, \ldots, 8$. Validate the solutions. If the interval is enlarged, are there any new solutions? Are there any complex solutions?

2.15 Solve the following pseudopolynomial equation and validate the solutions:

$$x^{\sqrt{7}} + 2x^{\sqrt{3}} + 3x^{\sqrt{2}-1} + 4 = 0.$$

2.16 Find all solutions to the deformed Riccati equation, and validate the results:

$$\boldsymbol{AX} + \boldsymbol{XD} - \boldsymbol{XBX} + \boldsymbol{C} = \boldsymbol{0},$$

where

$$\boldsymbol{A} = \begin{bmatrix} 2 & 1 & 9 \\ 9 & 7 & 9 \\ 6 & 5 & 3 \end{bmatrix}, \quad \boldsymbol{B} = \begin{bmatrix} 0 & 3 & 6 \\ 8 & 2 & 0 \\ 8 & 2 & 8 \end{bmatrix}, \quad \boldsymbol{C} = \begin{bmatrix} 7 & 0 & 3 \\ 5 & 6 & 4 \\ 1 & 4 & 4 \end{bmatrix}, \quad \boldsymbol{D} = \begin{bmatrix} 3 & 9 & 5 \\ 1 & 2 & 9 \\ 3 & 3 & 0 \end{bmatrix}.$$

2.17 For the above given matrix, solve

$$\boldsymbol{AX} + \boldsymbol{XD} + \boldsymbol{CX}^2 - \boldsymbol{XBX} + \boldsymbol{X}^2\boldsymbol{C} + \boldsymbol{I} = \boldsymbol{0}.$$

2.18 Find the analytical solution of the following linear equations, and validate it:

$$\begin{bmatrix} 2 & -9 & 3 & -2 & -1 \\ 10 & -1 & 10 & 5 & 0 \\ 8 & -2 & -4 & -6 & 3 \\ -5 & -6 & -6 & -8 & -4 \end{bmatrix} \boldsymbol{X} = \begin{bmatrix} -1 & -4 & 0 \\ -3 & -8 & -4 \\ 0 & 3 & 3 \\ 9 & -5 & 3 \end{bmatrix}.$$

2.19 Solve the following linear simultaneous equations with the solver `solve()` and validate the solutions:

$$\begin{cases} x_1 + x_2 + x_3 + x_4 + x_5 = 1, \\ 3x_1 + 2x_2 + x_3 + x_4 - 3x_5 = 2, \\ x_2 + 2x_3 + 2x_4 + 6x_5 = 3, \\ 5x_1 + 4x_2 + 3x_3 + 3x_4 - x_5 = 4, \\ 4x_2 + 3x_3 - 5x_4 = 12. \end{cases}$$

2.20 Assume that the nonlinear matrix equation

$$\boldsymbol{A}\boldsymbol{X}^3 + \boldsymbol{X}^4\boldsymbol{D} - \boldsymbol{X}^2\boldsymbol{B}\boldsymbol{X} + \boldsymbol{C}\boldsymbol{X} - \boldsymbol{I} = \boldsymbol{0},$$

and matrices $\boldsymbol{A}$, $\boldsymbol{B}$, $\boldsymbol{C}$, and $\boldsymbol{D}$ are the same as those in Exercise 2.16. Find all the solutions of the equation. Assume that we have already found 77 real solutions, and altogether 3 351 complex solutions, and all of them are written in file data2ex1.mat. Try to continue solving the equation and see whether new solutions can be found.

3 Unconstrained optimization problems

Optimization technique is a very useful tool in modern scientific research. The so-called optimization means finding the values of independent variables such that the objective function is minimized or maximized. It is not exaggerating to say that when the ideas and solution methods of optimization are mastered, the level of the research can be boosted to a higher level. In the past, one could be very happy to find a solution of a certain problem. When the optimization concept is employed, it is natural to pursuit better solutions. Optimization problems can be classified into unconstrained and constrained optimization problems.

In this chapter, unconstrained optimization problems are introduced, and MATLAB based solutions are explored. In Section 3.1, the definitions and standard mathematical forms of unconstrained optimization problems are introduced. Then, analytical and graphical methods for unconstrained optimization problems are discussed, and the concepts and test methods of local and global optimum solutions are studied. A demonstrative example of univariate function is used to show the optimization solver and MATLAB implementations. In Section 3.2, we concentrate on the optimization solvers provided in the MATLAB Optimization Toolbox. Examples are used to demonstrate the solutions and skills needed for various unconstrained optimization problems. The use of gradient information in optimization solvers is also discussed. Further parallel computing in optimization problem solutions is discussed. In Section 3.3, the global optimum solutions of ordinary optimization problems are explored, and a new global optimum solver is implemented in MATLAB. A benchmark problem is used to test the efficiency of the global optimum solver. In Section 3.4, optimization problems with decision variable bounds are introduced, and attempts are made to find the global optimum solutions for univariate and multivariate optimization problems. In Section 3.5, we try to explore some application problems.

3.1 Introduction to unconstrained optimization problems

Unconstrained optimization problems are commonly encountered and are the simplest optimization problems. In this section, the standard mathematical model of unconstrained optimization problems is introduced. Graphical and analytical solutions to unconstrained optimization problems are discussed. The concepts of local and global optimum solutions are addressed. In this section, a simple univariate optimization problem is used to demonstrate the theory and practice.

3.1.1 The mathematical model of unconstrained optimization problems

Definition 3.1. The mathematical form of the unconstrained optimization problem is described by

$$\min_{\boldsymbol{x}} f(\boldsymbol{x}) \tag{3.1.1}$$

https://doi.org/10.1515/9783110667011-003

where $\boldsymbol{x} = [x_1, x_2, \dots, x_n]^\mathrm{T}$ are known as the decision variables. The scalar function $f(\cdot)$ is referred to as the objective function.

In the above mathematical description, the problem is how to select the components in the $\boldsymbol{x}$ vector, such that the value of the objective function $f(\boldsymbol{x})$ is minimized. Therefore, the problem is also known as a minimization problem. Since the components in the vector $\boldsymbol{x}$ can be selected arbitrarily, such optimization problem is also referred to as an unconstrained optimization problem.

In fact, the minimization description given above does not lose generality. If one wants to express a maximization problem, just multiplying the objective function by −1, the problem can be converted to a minimization problem. In the descriptions in this and subsequent chapters, the minimization problem is treated as standard.

3.1.2 Analytical solutions of unconstrained minimization problems

At the optimum point $\boldsymbol{x}^*$ of the minimization problem, the first-order derivatives of the objective function $f(\boldsymbol{x})$ with respect to each component in vector $\boldsymbol{x}$ are all zeros. Therefore, the following simultaneous equations:

$$\left.\frac{\partial f}{\partial x_1}\right|_{\boldsymbol{x}=\boldsymbol{x}^*} = 0, \quad \left.\frac{\partial f}{\partial x_2}\right|_{\boldsymbol{x}=\boldsymbol{x}^*} = 0, \quad \dots, \quad \left.\frac{\partial f}{\partial x_n}\right|_{\boldsymbol{x}=\boldsymbol{x}^*} = 0 \tag{3.1.2}$$

can be used to compute the extreme value points. In fact, the points obtained by solving the equations are not always minimum points, sometimes they are maximum points. The minimum points should have positive second-order derivatives. For univariate optimization problems, analytical solutions can be found in this way, while for multivariate problems, the optimization problem should be converted to multivariate nonlinear equations. The complexity of solving the optimization problem in this way may be higher than the direct solution approach. There is no need to use analytical methods for solving optimization problems.

3.1.3 Graphical solutions

If the simultaneous equations are known, they can be described by anonymous functions or symbolic expressions. Then, functions such as `fplot()` or `fimplicit()` can be used to draw the curves for the equations. The intersection information can be found so as to find the solutions of the equations.

Example 3.1. Consider the following objective function. Use analytical and graphical methods to solve the optimization problem for

$$f(t) = \mathrm{e}^{-3t}\sin(4t+2) + 4\mathrm{e}^{-0.5t}\cos(2t) - 0.5.$$

Solutions. The objective function should be expressed first, and its first-order derivative can be found. Function fplot() can be used to draw the function and its first-order derivative within the interval $t \in [0,4]$, as shown in Figure 3.1.

```
>> syms t;
   y=exp(-3*t)*sin(4*t+2)+4*exp(-0.5*t)*cos(2*t)-0.5; %objective fun
   y1=diff(y,t); fplot([y,y1],[0,4]) %first-order derivative
   line([0,4],[0,0]) %function and derivative curves
```

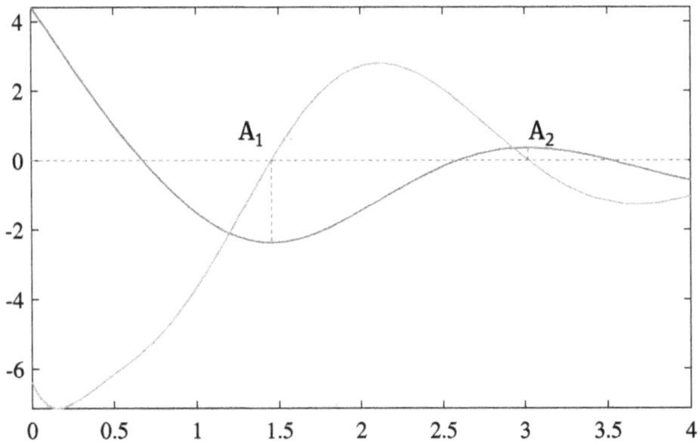

Figure 3.1: Graphical description of the objective function and its derivative.

The first-order derivative can be written as

$$f'(t) = -e^{-3t}(3\sin(4t+2) - 4\cos(4t+2)) - 2e^{-t/2}(\cos 2t + 4\sin 2t).$$

Solving the equation $f'(t) = 0$, two solutions can be found at $x_1 = 1.4528$ and $x_2 = 3.0190$. The second-order derivatives for the two points can be obtained as $z_1 = 7.8553$ and $z_2 = -3.646$.

```
>> x1=vpasolve(y1,2), x2=vpasolve(y1,3)
   y2=diff(y,2); z1=subs(y2,t,x1), z2=subs(y2,t,x2)
```

In fact, solving such equations is not simpler than solving the optimization problem itself. It can be seen from the graphical method that there are two points A_1 and A_2. To make the first-order derivatives equal to zero, it can be seen from the trends that A_2 corresponds to a negative second-order derivative, which corresponds to a maximum point, while A_1 point has a positive second-order derivative, which implies it is a minimum point.

Since the obtained functions are nonlinear, finding analytical solutions or using other equation solution methods may be more complicated than direct optimization computations. Therefore, unless for demonstration purposes, it is not recommended to use this method to solve the problem. Direct optimization approaches should be adopted instead.

3.1.4 Local and global optimum solutions

Ordinary methods used in numerical optimization solutions are usually searching methods. An initial search point should be selected, from it, using different searching methods, we find the next approximate point based on the information in the objective function. Then, from the new point, another approximate solution can be found. It is obvious that iterative methods should be used to minimize objective functions.

For univariate problems, since the objective function can be expressed graphically, one may consider the case where there is a ball placed on top of the curve and let it rolling down freely. The ball may eventually settle down at a certain point. The speed of the ball becomes zero. This point is the expected minimum point. Let us now consider an example where the concepts of local and global minimum points are addressed.

Example 3.2. Assume that the objective function is given below. Observe the initial values and possible optimum points. Discuss the concept of local and global optimum points for

$$y(t) = \mathrm{e}^{-2t}\cos 10t + \mathrm{e}^{-3t-6}\sin 2t, \quad 0 \leqslant t \leqslant 2.5.$$

Solutions. The following commands can be used to draw the curve of the objective function in the region of interest, as shown in Figure 3.2:

```
>> f=@(t)exp(-2*t).*cos(10*t)+exp(-3*(t+2)).*sin(2*t);
   fplot(f,[0,2.5]);     %evaluate and draw the objective function
```

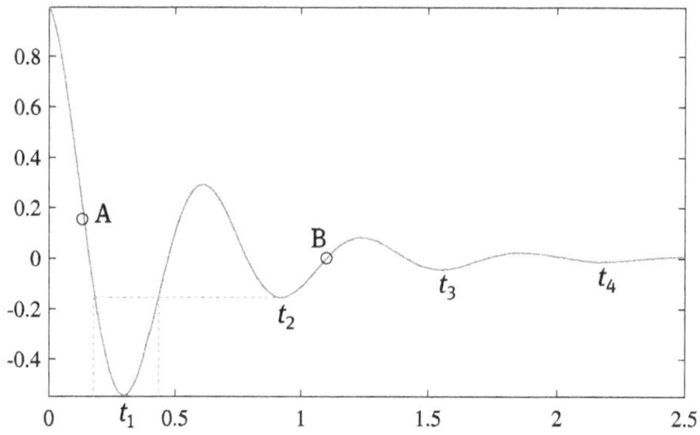

Figure 3.2: The function defined for $t \geqslant 0$.

Assume that point A is selected as the initial search point. The ball is rolling back and forth, and finally, it settles down at point t_1. If the initial point is selected as point B, the ball may eventually settle down at point t_2, where the first-order derivative is also zero. Therefore, this point is also a solution of the optimization problem. If the initial value is selected at a place where t is large, then the optimum solutions may be points t_3 and t_4.

It can be seen from the example that a given objective function may have many different optimum points, with possibly different values of the objective function. In this example, point t_1 is referred to as the global optimum point, for the given region of interest since the value of the objective function is the smallest in this region. If the ball is placed at point B, the settle down point is t_2. The other points such as t_2, t_3, and t_4 are referred to as local optimum points.

Definition 3.2. If in a neighborhood $\mathscr{R}$ there exists a point $\boldsymbol{x}^*$, such that $f(\boldsymbol{x}^*) \leqslant f(\boldsymbol{x})$ is satisfied for all points $\boldsymbol{x} \in \mathscr{R}$, then $\boldsymbol{x}^*$ is referred to as a local minimum solution.

Definition 3.3. If in a real number domain there exists a point $\boldsymbol{x}^*$ such that $f(\boldsymbol{x}^*) \leqslant f(\boldsymbol{x})$ holds for all points in this domain, then the decision variable $\boldsymbol{x}^*$ is referred to as a global optimum solution of the unconstrained problem.

In practical considerations, the global optimum point is commonly pursued, rather than the local optimum solutions. Acquiring the global optimum solution is the eventual goal. In fact, none of the methods known today ensure the global minimum solutions. More precisely, we should say that an algorithm is more likely to find the global optimum solutions. In real applications, local minimums are not of any use, since the objective function value at such points may be larger than numerous values at nonoptimum points. For instance, in Figure 3.2, it can be seen that any point in the interval $t \in (0.174, 0.434)$ yields better objective function values than the local minimum solution t_2.

Example 3.3. Consider again the problem in Example 3.2. If the region of interest is changed to $t \in (-0.5, 2.5)$, analyze again the global and minimum solutions.

Solutions. Now let us consider a larger domain of interest, for instance, $t \geqslant -0.5$. The following statements can be used to draw the curve of the objective function in the new domain of definition, as shown in Figure 3.3. It can be seen that the possible optimum solutions are t_0, t_1, t_2, t_3, and t_4. Now the point t_1 is no longer the global optimum solution, but a local minimum. The new global optimum solution becomes point t_0. It can be seen from the example that for the same objective function, since the domain of definition is selected differently, the global optimum solutions may also change.

```
>> f=@(t)exp(-2*t).*cos(10*t)+exp(-3*(t+2)).*sin(2*t);
   fplot(f,[-0.5,2.5]); ylim([-2,1.2])
```

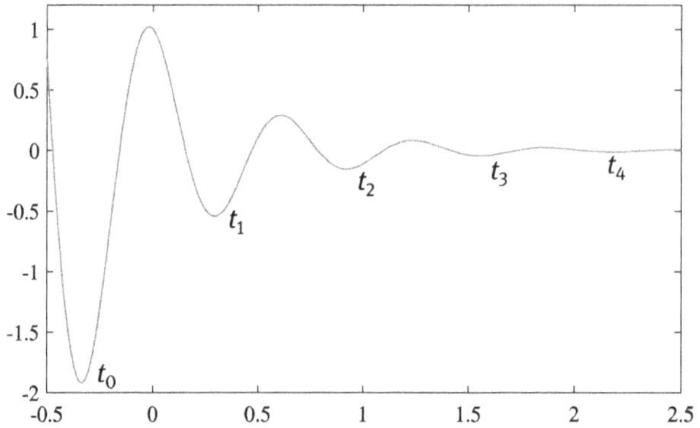

Figure 3.3: The domain of definition is $t \geqslant -0.5$.

If the domain of definition is changed to $t \in (-\infty, \infty)$, there is no genuine global optimum solution for the problem.

3.1.5 MATLAB implementation of optimization algorithms

There are various numerical methods which can be used to solve optimization problems. In this section a very simple one is introduced to demonstrate the solutions of optimization problems. A MATLAB implementation is also proposed.

Normally, for numerical optimization solvers, an initial point x_0 should be selected. Then, with iterative methods, we find the numerical solution of the problem. Assume that in the kth step in the iteration process, the decision variable is x_k. With the Taylor series expansion technique, we can use a quadratic function $g(x)$ to approximate the value at x as

$$g(x) = f(x_k) + f'(x_k)(x - x_k) + \frac{1}{2}f''(x_k)(x - x_k)^2. \tag{3.1.3}$$

Since we know that $f(x_k) = g(x_k)$, $f'(x_k) = g'(x_k)$, and $f''(x_k) = g''(x_k)$, we are not optimizing function $f(x)$, and consider merely the optimization of the quadratic function $g(x)$. The condition for the existence of an optimal point is $g'(x) = 0$, from which it is found that

$$0 = g'(x) = f'(x_k) + f''(x_k)(x - x_k). \tag{3.1.4}$$

Letting $x = x_{k+1}$, it can be found that

$$x_{k+1} = x_k - \frac{f'(x_k)}{f''(x_k)}. \tag{3.1.5}$$

If a backward difference method is used to evaluate the second-order derivative, it can be found that

$$x_{k+1} = x_k - \frac{x_k - x_{k-1}}{f'(x_k) - f'(x_{k-1})} f'(x_k). \tag{3.1.6}$$

In fact, when observing the formula closely, it is not difficult to find that this is the solution of the equation $f'(x) = 0$. The recursive formula is close to the Newton–Raphson iterative method. Since the second-order derivative is replaced by the backward difference method, the numerical solution of the equation can be found. We shall demonstrate the solution process of the optimization problems through the following example.

Example 3.4. Use the algorithm here to solve the problem in Example 3.1.

Solutions. If the optimization problem is to be solved, an anonymous function can be used to describe the derivative function. In Example 3.1, the analytical expression of its derivative is given. Therefore, the simple anonymous function representation can be used. Two initial points $t_0 = 1$ and $t_1 = 0.5$ can be selected, and a loop structure is used to implement the iterative process. The convergence test condition is set to $|t_{k+1} - t_k| < \epsilon$. Assuming that $\epsilon = 10^{-5}$, which means that the two consecutive search points are close enough, the loop can be terminated. With the following statements the optimum solution can be found:

```
>> df=@(t)-exp(-3*t).*(3*sin(4*t+2)-4*cos(4*t+2)) ...
         -2*exp(-t/2).*(cos(2*t)+4*sin(2*t));
   t0=1; t1=0.5; t=[t0,t1];
   while abs(t1-t0)>1e-5
      t2=t1-(t1-t0)/(df(t1)-df(t0))*df(t1);
      t0=t1; t1=t2; t=[t, t2];
   end
```

The intermediate solutions are obtained as follows. The optimum solution is $t^* = 1.4528$. It can be seen that for the simple problem listed here, only a few steps of iteration lead to the numerical optimum solutions:

$$\boldsymbol{t} = [1, 0.5, 1.7388, 1.4503, 1.4534, 1.4528, 1.4528].$$

With the intermediate results, the following statements can be used to describe the objective function and the derivative function curves, with the intermediate points labeled, as shown in Figure 3.4. From these points the intermediate of searching process is depicted

```
>> f=@(t)exp(-3*t).*sin(4*t+2)+4*exp(-0.5*t).*cos(2*t)-0.5;
   fplot(f,[0,4]), hold on
   fplot(df,[0,4]), plot(t,df(t),'o'), hold off
```

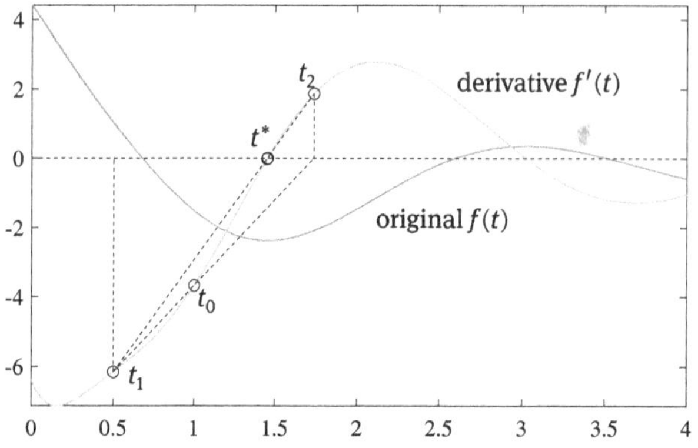

Figure 3.4: Intermediate results in the searching process.

Of course, this method may only work for univariate function optimization problems. For multivariate problems, the searching process may become much more complicated. Besides, since the first-order derivatives of the objective function (or gradient information) may not always be available, knowing merely the objective function, this method may be rather complicated to implement. In the later descriptions, these complicated low-level algorithms are not discussed. The leading edge solvers provided in MATLAB Optimization Toolbox will be exclusively used in the discussions.

3.2 Direct solutions of unconstrained optimization problem with MATLAB

Two effective unconstrained optimization problem solvers are provided in MATLAB. In this section, we shall discuss how to use these two solvers to solve unconstrained optimization problems directly. Discussion also includes the questions of how to increase the accuracy and efficiency of the solution process. Also parallel computation may be tried to solve optimization problems.

3.2.1 Direct solution methods

An optimization solver `fminsearch()` is provided in MATLAB, and yet another solver, `fminunc()`, is provided in the Optimization Toolbox. The syntaxes of the two functions are virtually the same:

$\boldsymbol{x}$=`fminsearch(Fun,`$\boldsymbol{x}_0$`)`,

`[`$\boldsymbol{x}$`,`f_0`,flag,out]=fminsearch(Fun,`$\boldsymbol{x}_0$`,opt)`,

`[`$\boldsymbol{x}$`,`f_0`,flag,out]=fminsearch(Fun,`$\boldsymbol{x}_0$`,opt,`p_1`,`p_2`,...)`, where Fun is the function handle for describing the objective function. It can be a MATLAB function or an anonymous function. The argument $\boldsymbol{x}_0$ is the initial search vector, opt is the control option object. Besides, additional parameters $p_1, p_2, \ldots$ are allowed, however, it is not recommended to use them. An alternative method will be introduced later. In the returned arguments, $\boldsymbol{x}$ is the optimum solution vector; f_0 is the value of objective function; and flag is the indicator of computation results. If flag is zero, then the solution process is unsuccessful, while if flag is 1, the solution process is successful. The returned argument out contains certain intermediate information, such as the number of iterations and so on.

In function fminsearch(), the improved simplex method presented in [39] is used, while in function fminunc(), quasi-Newton algorithm and trust region algorithms are used. Function fminsearch() and quasi-Newton algorithm do not require gradient information of the objective function, while in the trust region method used in fminunc(), the gradient information is expected.

It can be found that for many examples, the efficiency of the recent versions of fminunc() function improved a lot, and it is recommended to use such a solver wherever possible.

Now we shall use examples to demonstrate the solution of unconstrained optimization problems with numerical methods.

Example 3.5. For the given function $z = f(x,y) = (x^2 - 2x)e^{-x^2-y^2-xy}$, use MATLAB to find the minimum value, and explain its geometric meaning.

Solutions. Since the independent variables are x and y, in the optimization solvers, the independent variable vector $\boldsymbol{x}$ is expected. Therefore, variable substitution should be carried out first, to convert the problem into the standard form. For instance, assigning $x_1 = x$ and $x_2 = y$, the objective function can be manually modified as

$$z = f(\boldsymbol{x}) = (x_1^2 - 2x_1)e^{-x_1^2-x_2^2-x_1x_2}.$$

To solve the problem, the objective function is expressed first in MATLAB. Then, two methods can be used to describe the objective function. One is to write down a MATLAB function, the other is to define it using an anonymous function. Let us see the MATLAB function programming method. In this case, the objective function value y can be evaluated from the given vector $\boldsymbol{x}$:

```
function y=c3mopt(x)
y=(x(1)^2-2*x(1))*exp(-x(1)^2-x(2)^2-x(1)*x(2));
```

One may save the function to a file, such as c3mopt.m. After that the following commands can be used to solve the optimization problem, and the solution found is $\boldsymbol{x} = [0.6111, -0.3056]$. The returned flag is 1, indicating that the solution process is successful. It can be seen that the objective function is called 90 times. The number of iterations is 46.

```
>> [x,b,flag,c]=fminsearch(@c3mopt,x0)
```

The other way to describe the objective function is to use anonymous functions. The benefit of this method is that there is no need to create an actual MATLAB function file. A dynamically defined command is sufficient. Also the variables in MATLAB workspace can be used directly. With the following statements, the objective function can be defined with an anonymous function, and then the solution of the problem can be found. The results are exactly the same as those obtained earlier.

```
>> f=@(x)(x(1)^2-2*x(1))*exp(-x(1)^2-x(2)^2-x(1)*x(2));
   x0=[2; 1]; [x,b,flag,d]=fminsearch(f,x0)
```

The problem can also be solved with the function `fminunc()`, and the same results can be obtained. Meanwhile, it is known that the objective function is called 66 times, and the total number of iterations is 7.

```
>> [x,b,flag,d]=fminunc(f,[2; 1]) %another solver
```

Comparing the two methods, it is obvious that the efficiency of the function `fminunc()` is significantly higher than that of the function `fminsearch()`, since the numbers of function calls and iteration steps are significantly smaller than for the former. For unconstrained optimization problems, if the Optimization Toolbox is installed, it is recommended to use `fminunc()` function.

In fact, with the function `fsurf()`, the surface of the objective function can be drawn as shown in Figure 3.5. It can be seen that the minimum solution is the valley of the surface.

```
>> syms x y; f=(x^2-2*x)*exp(-x^2-y^2-x*y);
   fsurf(f,[-3 3, -2 2])
```

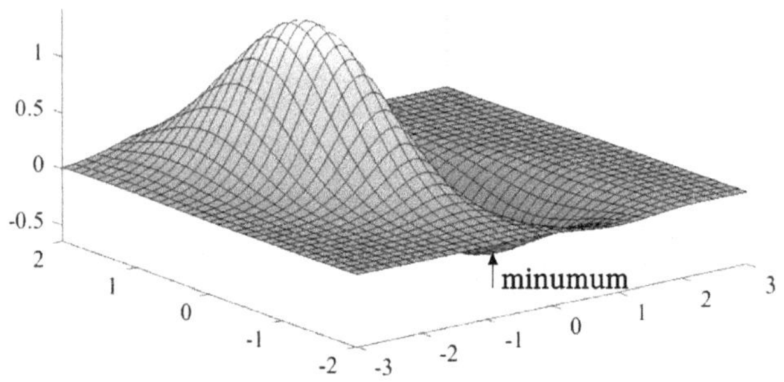

Figure 3.5: Surface of the objective function.

Example 3.6. Solve the De Jong benchmark problem

$$\min_{x} \sum_{i=1}^{20} x_i^2.$$

Solutions. It is obvious that the analytical solution is $x_i = 0$, $i = 1, 2, \ldots, 20$. Now an anonymous function is used to describe the objective function, and with the function `fminsearch()`, the optimization problem can be solved directly. In the description of the objective function, command `x(:)` is used to convert the vector $\boldsymbol{x}$ into a column vector, which means that no matter how the initial vector $\boldsymbol{x}_0$ is given, i. e., as a row or a column, the objective function can be evaluated in the same way.

```
>> f=@(x)x(:)'*x(:);
   x0=ones(20,1); [x0 f0 flag cc]=fminsearch(f,x0)
```

With the above solution commands, the value of the objective function is $f_0 = 0.0603$, so there is a relatively large discrepancy from the theoretical solution. A warning "Exiting, maximum allowed function calls are reached. Increase the `MaxFunEvals` option. Current function value is: 0.060339" is displayed. It can be seen that the searching process is abnormally terminated, since the number of objective function calls exceeded the specifications in `MaxFunEvals`. Meanwhile, note that the returned `flag` is 0, indicating that the solution process is unsuccessful.

3.2.2 Control options in optimization

The optimization solvers provided in MATLAB can be understood in this way: given an initial value vector, the values of $\|\boldsymbol{x}_{k+1} - \boldsymbol{x}_k\|$ and $|f(\boldsymbol{x}_{k+1}) - f(\boldsymbol{x}_k)|$ are computed. Generally speaking, when either condition below is satisfied, the searching process is terminated:

$$\|\boldsymbol{x}_{k+1} - \boldsymbol{x}_k\| < \varepsilon_1, \quad |f(\boldsymbol{x}_{k+1}) - f(\boldsymbol{x}_k)| < \varepsilon_2, \tag{3.2.1}$$

$$k_1 > k_{1\max}, \quad k_2 > k_{2\max} \tag{3.2.2}$$

where ε_1 and ε_2 are the user-specified error tolerances, which should be small positive values. The integer k_1 is the number of function calls, with $k_{1\max}$ being the maximum allowed function calls specified by the user; k_2 is the actual number of iteration steps, with $k_{2\max}$ being the maximum allowed number of iterations. The default values of the two options are $200n$, with n being the number of decision variables.

How can we modify the control options? Similar to the case discussed earlier when solving equation, function `optimset` should be called to load the control option template. Then, the relevant options can be modified. The major options in the `optimset` template are summarized in Table 3.1. They are provided in the structured variable, and its elements can be modified directly.

Table 3.1: Commonly used control options in optimization.

options	explanations
`Display`	display of intermediate results, which can be set to `'off'` for no display; `'iter'` for display at each iteration; `'notify'` means to prompt when not convergent (default); `'final'` for display final values
`FunValCheck`	check whether the objective function is valid, with options `'on'` and `'off'`. If the former is selected, error message appears if the objective function is complex or `NaN`; the default option is `'off'`
`GradObj`	when solving optimization problems, it is used to indicate whether the gradient information is available, with options `'off'` or `'on'`
`LargeScale`	indicates whether large-scale algorithm is used, with `'on'` and `'off'`
`MaxIter`	maximum numver of iterations, i. e., $k_{2\,\max}$
`MaxFunEvals`	maximum number of function calls, i. e., $k_{1\,\max}$
`OutputFcn`	used for handling intermediate results. A MATLAB function can be written to respond to actions
`TolFun`	error tolerance in the objective function, i. e., ε_2, the termination condition in (3.2.1)
`TolX`	error tolerance of the solution, i. e., ε_1, the termination condition in (3.2.1)

Normally, in the solution process, when both ε_1 and ε_2 conditions are satisfied, the searching process terminates normally. The returned `flag` is a positive number. If either $k_{1\,\max}$ or $k_{2\,\max}$ is reached, the searching process is terminated abnormally. The returned `flag` is 0. Therefore, we can consider the method of using the value `flag` to find meaningful optimal solutions.

In recent versions of Optimization Toolbox, a new set of control options is supported, with the function

```
opt=optimoptions('fminunc')
```

All new options can be listed, where the commonly used ones are given in Table 3.2. Some of the convergence conditions are combined, so as to make the control option setting more concise. Therefore, in the recent versions, the new control options are recommended. It should be noted that the new options cannot be used in the `fminsearch()` function. The option setting methods will be demonstrated later through examples.

Example 3.7. Solve the optimization problem in Example 3.5 with different algorithms and display the intermediate results.

Solutions. The solver `fminsearch()` can be called first, where the modified simplex method is used. Besides, in order to compare the algorithms, intermediate display is assigned, and the final objective function is −0.641423723509499.

```
>> f=@(x)(x(1)^2-2*x(1))*exp(-x(1)^2-x(2)^2-x(1)*x(2));
   ff=optimset; ff.Display='iter';
   x0=[2; 1]; [x,b,c,d]=fminsearch(f,x0,ff) % simplex method
```

Table 3.2: Control options in recent versions of the toolbox.

options	option explanations
`Algorithm`	algorithm selection, with `'quasi-newton'` for quasi-Newton algorithm (default one), and `'trust-region'` for trust region method. Note that the trust region algorithm requires the gradient information, otherwise the algorithm cannot be used
`Display`	the same as that in Table 3.1
`SpecifyObjectiveGradient`	indicates whether the gradient of the objective function is known, with options 0 and 1
`MaxIterations`	maximum allowed iteration steps with default being 400
`MaxFunctionEvaluations`	Maximum allowed function calls, with the default of 100 times the number of decision variables
`OutputFcn`	handling the intermediate results
`OptimalityTolerance`	unified error tolerance

The intermediate results are displayed below

```
 Iteration   Func-count     min f(x)         Procedure
     0            1              0
     1            3              0           initial simplex
     2            5     -0.000641131         expand
     3            7      -0.00181849         expand
     4            9       -0.0132889         expand
     5           11       -0.0654528         expand
     6           13       -0.0835065         reflect
    ... intermediate results omitted
    45           88        -0.641424         contract inside
    46           90        -0.641424         contract inside
```

If solver `fminunc()` is used, the default quasi-Newton algorithm is applied to solve the same problem. The value of the objective function is −0.641423726326, which is slightly better than that obtained earlier.

```
>> ff=optimoptions('fminunc','Display','iter');
   [x,b,c,d]=fminunc(f,x0,ff) %quasi-Newton algorithm
```

The intermediate results obtained are as shown below, and it can be seen that fewer iterations are made, which is more effective.

```
 Iteration  Func-count        f(x)        Step-size      optimality
     0           3              0                          0.00182
     1          24      -0.134531         872.458            0.324
     2          36      -0.134533           0.001            0.324
```

3	48	-0.623732	172	0.205
4	54	-0.641232	0.311866	0.0357
5	60	-0.641416	0.329315	0.00433
6	63	-0.641424	1	0.000218
7	66	-0.641424	1	1.49e-08

Example 3.8. Solve the problem in Example 3.6 to find accurate solutions.

Solutions. As discussed earlier, there were warning messages indicating that the maximum number of function calls is exceeded. A natural solution is to increase the value of `MaxFunEvals`, however, this may not be a good solution. An alternative way is to use the results as the initial value of $\boldsymbol{x}_0$ and start the searching process again. The value of `flag` should be tested and see whether its value is positive or not. If it is, then terminate the loop, otherwise continue the loop until the expected decision is found. In order to find accurate solutions, stricter error tolerance should be assigned.

```
>> f=@(x)x(:)'*x(:); x0=ones(20,1); flag=0; k=0;
   ff=optimset; ff.TolX=eps; ff.TolFun=eps;
   while flag==0, k=k+1
      [x0 f0 flag cc]=fminsearch(f,x0,ff);
   end, norm(x0)
```

It can be seen that after 19 loop executions ($k = 19$), it is found that $\|\boldsymbol{x}_0\| = 4.0405 \times 10^{-14}$, and the value of the objective function is as low as $f_0 = 1.6326 \times 10^{-27}$, which means that the problem is solved successfully.

Example 3.9. Solve again the optimization problem in Example 3.6 using the solver `fminunc()`.

Solutions. Assuming that the initial search point is selected randomly within the interval $(-1\,000, 1\,000)$, the following commands and `fminunc()` function can be used directly:

```
>> f=@(x)x(:)'*x(:); x0=-1000+2000*rand(20,1);
   ff=optimset; ff.TolX=eps; ff.TolFun=eps;
   [x0 f0 flag cc]=fminunc(f,x0,ff), norm(x0)
```

The error norm of $\boldsymbol{x}_0$ is $f_0 = 3.0924 \times 10^{-10}$. It can be seen that the solution obtained is close to the theoretical value. However, if a loop structure is used, no better solutions can be obtained.

It can be seen from this example that function `fminunc()` can be used to compute accurate solutions easily from the given initial values. Then, also function `fminsearch()` can be used to find even more accurate solutions. We need to assign $k_{1\max}$ and $k_{2\max}$ to very large values. For instance, when the following statements are called, the total number of function calls is 27 169, and the elapsed time is 0.44 seconds. The norm of the error is $\|\boldsymbol{x}_0\| = 5.7339 \times 10^{-13}$:

```
>> x1=-1000+2000*rand(20,1); x0=x1; k=0; tic
   [x0 f0 flag cc]=fminunc(f,x0,ff);
   ff.MaxFunEvals=100000; ff.MaxIter=100000; M=cc.funcCount;
   [x0 f0 flag cc]=fminsearch(f,x0,ff);
   toc, M=M+cc.funcCount, norm(x0)
```

If function `fminsearch()` is called, the number of objective function calls is 73 760, slightly higher than for the above method. The norm of the error is $\|\boldsymbol{x}_0\| = 9.8226 \times 10^{-16}$, total elapsed time is 0.56 seconds. The time needed and efficiency are lower than for the above method. But the accuracy is increased significantly.

```
>> x0=x1; tic, [x0 f0 flag cc]=fminsearch(f,x0,ff);
   toc, cc.funcCount, norm(x0)
```

3.2.3 Additional parameters

In some particular problems, the objective function may have direct relationship with $\boldsymbol{x}$. Besides, it may be related to other variables. These variables can be described by additional parameters. In this section, the use of additional parameters is demonstrated, and the method of avoiding additional parameters will be explored.

Example 3.10. Consider the extended Dixon problem

$$\min_{\boldsymbol{x}} \sum_{i=1}^{n/10}\left[(1-x_{10i-9})^2+(1-x_{10i})^2+\sum_{j=10i-9}^{10i-1}(x_j^2-x_{j+1})^2\right].$$

Using $n = 50$, solve the optimization problem.

Solutions. It is obvious that the analytical solution is $x_i = 1$, $i = 1, 2, \dots, n$. If n is a multiple of 10, the original decision variable vector $\boldsymbol{x}$ can be converted into a matrix $\boldsymbol{X}$ composed of 10 rows and $n/10$ columns. Therefore, the following vectorized programming pattern can be used in defining the objective function in MATLAB:

```
function y=c3mdixon(x,n)
m=n/10; X=reshape(x,10,m); y=0;
for i=1:m
   y=y+(1-X(1,i))^2+(1-X(10,i))^2+sum((X(1:8,i).^2-X(2:9,i)).^2);
end
```

Note that in the standard objective function syntax, variable n cannot be passed into the function. Therefore, n must be used as an additional parameter. Then, apart from the input argument $\boldsymbol{x}$, a second argument n is used.

With the objective function described, the following commands can be used to solve the optimization problem directly. Note that in the function call, the additional parameter n should be assigned. It must be the same as that in the MATLAB function describing the objective function. Otherwise the problem cannot be solved. After 38 loop executions, relatively high accuracy can be obtained. The norm of the error matrix is 1.2318×10^{-13}.

```
>> ff=optimset; ff.TolX=eps; ff.TolFun=eps;
   n=50; x0=10*rand(n,1); flag=0; k=0;
   while flag==0, k=k+1
      [x0,f0,flag,cc]=fminsearch(@c3mdixon,x0,ff,n); norm(x0-1)
   end, norm(x0-1)
```

If the solver `fminunc()` is used to substitute the solver `fminsearch()`, the precision may not be so high. Even with loop structures, accurate solutions cannot be found.

In recent versions of Optimization Toolbox, it seems that in many descriptions, the additional parameters are not fully supported. An alternative way is needed. That is, we need to write an interface to the MATLAB function with an anonymous function, so that the additional parameters can be passed into the MATLAB function. Since the MATLAB workspace variables can be used directly by the anonymous function, it is possible to solve the problem in this way. This will be demonstrated next in the example.

Example 3.11. Consider again the problem in Example 3.10. Solve the problem with the mechanism where additional parameters are not used.

Solutions. In Example 3.10, a MATLAB function `c3mdixon()` is written to describe the objective function, where an additional parameter n is used. In the structured variable description of optimization problems, which will be presented later, the additional parameters are not supported. Therefore, an anonymous interface should be written such that the additional parameters can be specified in the MATLAB workspace variables. In this way the following commands can be used in solving the unconstrained optimization problems directly:

```
>> n=50; f=@(x)c3mdixon(x,n); x0=rand(n,1);
   [x0,f0,flag,cc]=fminunc(f,x0)
```

3.2.4 Intermediate solution process

The control option `OutputFun` is mentioned in Table 3.1. The option can be used to link to the user-defined response function handle. The structure of the function is fixed, and will be demonstrated through the following example, to handle intermediate results.

Example 3.12. Find and display all the intermediate searching points while solving the problem in Example 3.5.

Solutions. In order to intercept the information in the optimization process, a switch structure is used to handle the output function. Note that the input and returned arguments in the function are fixed. The internal mechanism in MATLAB can be used to automatically generate the values of the input arguments.

```
function stop=c3myout(x,optimValues,state)
stop=false;
switch state  % switch structure to monitor the searching process
   case 'init', hold on                    % initialization response
   case 'iter', plot(x(1),x(2),'o') % iterative response
      text(x(1)+0.1,x(2),int2str(optimValues.iteration));
   case 'done', hold off                    % termination of the process
end
```

In each step of the iteration process, the intermediate results can be marked in the graphics window. To start the monitoring process, the `OutputFcn` option can be assigned to `@c3myout`. To demonstrate the solution process, the contour plots of the objective function should be drawn first. Then, selecting an initial search point at $\boldsymbol{x}_0 = [2, 1]^{\mathrm{T}}$, the following statements can be used to start the simulation process, and the intermediate points are superimposed on top of the contours, as shown in Figure 3.6. The intermediate search points are marked and numbered. If the distance between two intermediate points becomes small enough, it means that the point is close enough to the convergence point.

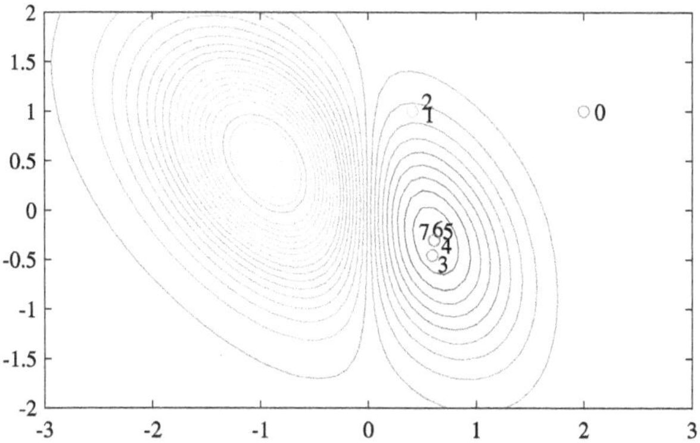

Figure 3.6: Intermediate solution process.

```
>> [x,y]=meshgrid(-3:0.1:3, -2:0.1:2); % mesh grid data
   z=(x.^2-2*x).*exp(-x.^2-y.^2-x.*y); contour(x,y,z,30);
   f=@(x)(x(1)^2-2*x(1))*exp(-x(1)^2-x(2)^2-x(1)*x(2));
   ff=optimset; ff.OutputFcn=@c3myout; x0=[2 1];
   x=fminunc(f,x0,ff)           % solve the optimization problem
```

3.2.5 Structured variable description of optimization problems

Structured variable format is also supported in MATLAB Optimization Toolbox to describe the whole optimization problems. This makes the description of optimization problems formal. A structured variable `problem` can be defined, whose members are described in Table 3.3. When the description is completed, the following commands can be used to solve the problem in the structured variable `problem`:

[$\boldsymbol{x}$,$f_{\rm m}$,flag,out]=fminunc(problem)

Note that the `fminsearch()` solver does not support the syntax. Ordinary syntax discussed earlier should be used when solving unconstrained optimization problems.

Table 3.3: Members in the unconstrained optimization structures.

member names	member explanations
`objective`	function handle
`x0`	initial decision variable vector
`options`	control options. The command `problem.options=optimset` can be used to assign default options; as before, the user may modify certain options and then assign them to the members in `options`
`solver`	should be set to `'fminunc'`

Example 3.13. Use a structured variable to solve again the problem in Example 3.5.

Solutions. The following commands can be used to set up the structured variable `P`. Then, function `fminunc()` can be called to solve directly the original problem. The result obtained is the same as that in Example 3.5.

```
>> P.solver='fminunc';
   P.options=optimset; % use structured variable to describe the problem
   P.objective=@(x)(x(1)^2-2*x(1))*exp(-x(1)^2-x(2)^2-x(1)*x(2));
   P.x0=[2; 1]; [x,b,c,d]=fminunc(P)    % direct solution
```

Example 3.14. Use a structured variable to describe and solve the optimization problem in Example 3.11.

Solutions. Since the additional parameters are not supported in the structured variables, the method in Example 3.11 should be adopted to convert the problem through an interface in anonymous function to function `c3mdixon()`, and n can be passed into the anonymous function from MATLAB workspace. In this way, structured variable can be used in describing unconstrained optimization problems. Now we can consider issuing the following commands to solve again the unconstrained optimization problem. The results obtained are exactly the same.

```
>> P.solver='fminunc'; n=50; P.options=optimset;
   P.objective=@(x)c3mdixon(x,n); %set up anonymous function handle
   P.x0=rand(n,1); [x,b,c,d]=fminunc(P) %solve directly
```

3.2.6 Gradient information

Sometimes the solution speed for optimization problems is low, and sometimes the method may not even find precise optimum solutions. Especially when there are too many variables, the gradient information of the objective function can be provided to speed up the solution process and improve the accuracy. While in real programming, the computation of the gradient may also need extra time, which may affect the overall speed. Therefore, in real applications one should consider whether it is worth using the gradient concepts.

When using MATLAB Optimization Toolbox to solve problems, the gradient functions should be described in the same file with the objective function. That is, we should let the objective function return two output arguments. The first is still the objective function, while the second is the gradient function. Meanwhile, the member `GradObj` in the control option should be set to `'on'`, or the `SpecifyObjectiveGradient` member in the new control template should be set to 1. The gradient information can then be used in the optimization problem solution process.

Example 3.15. Solve the unconstrained optimization problem for the Rosenbrock function $f(x_1, x_2) = 100(x_2 - x_1^2)^2 + (1 - x_1)^2$.

Solutions. The benchmark problem was proposed by the famous British scholar Howard Harry Resenbrock (1920–2010), and is used in testing and evaluating optimization solvers. It can be seen from the objective function that, since it is the sum of two squared terms, when $x_2 = x_1 = 1$, the entire objective function is 0. With the following statements, the three-dimensional contour of the objective function can be drawn, as shown in Figure 3.7:

```
>> [x,y]=meshgrid(0.5:0.01:1.5);            %mesh grid generation
   z=100*(y.^2-x).^2+(1-x).^2;              %objective function
   contour3(x,y,z,100), zlim([0,310])       %3D contours
```

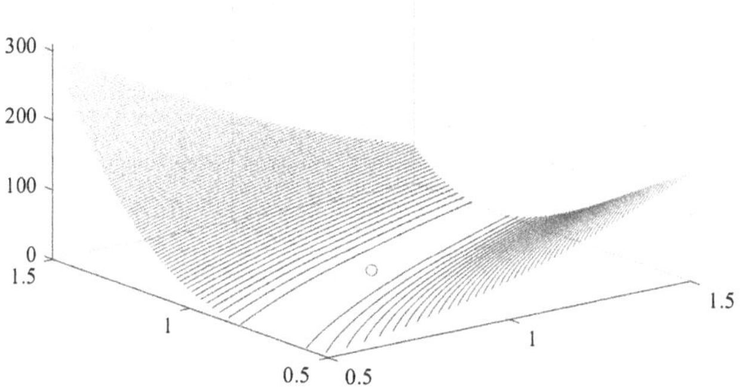

Figure 3.7: 3D contours of Rosenbrock function.

It can be seen from the obtained curves that the minimum point is located in the very narrow band. Therefore, Rosenbrock function is also known as a banana function. In this region, the changing of the objective function is very slow, which may bring troubles to the optimization process. The benchmark function is valuable in testing whether an optimization solver is good or not. Now let us use the following statements to solve the optimization problem. Function `fminunc()` can be tried to solve the problem, and its result can be used as the initial value for continued searching. Letting the loop execute 100 times, the final solution can be obtained.

```
>> f=@(x)100*(x(2)-x(1)^2)^2+(1-x(1))^2; %objective function
   ff=optimset; ff.TolX=eps; ff.TolFun=eps; x=[0;0];
   for k=1:100, k
      [x,f0,flag,cc]=fminunc(f,x,ff) %numerical solutions
   end
```

The final solution obtained is $\boldsymbol{x} = [0.999995588079578, 0.999991166415393]^{\mathrm{T}}$. It can be seen that after a period of waiting, the true value of $(1,1)$ cannot be precisely reached. The gradient information should be introduced. For the given Rosenbrock function, Symbolic Math Toolbox can be used to compute the gradient vector

```
>> syms x1 x2; f=100*(x2-x1^2)^2+(1-x1)^2;
   J=jacobian(f,[x1,x2]) %gradient computation
```

The gradient vector can be written as

$$\boldsymbol{J} = [-400(x_2 - x_1^2)x_1 - 2 + 2x_1, 200x_2 - 200x_1^2].$$

Now the objective function file can be used to express the gradient, and the new objective function can be written as

```
function [y,Gy]=c3fun3(x)
y=100*(x(2)-x(1)^2)^2+(1-x(1))^2; % two arguments
Gy=[-400*(x(2)-x(1)^2)*x(1)-2+2*x(1); 200*x(2)-200*x(1)^2];
```

The following statements can be used to solve the problem, and the solution obtained is $\boldsymbol{x} = [1.000000000000012, 1.000000000000023]^{\mathrm{T}}$.

```
>> ff.GradObj='on'; x=fminunc(@c3fun3,[0;0],ff) % with gradients
```

It can be seen that with gradient information, the solution speed is significantly increased, and the solution is very close to the true values. This solution cannot be reached with such a solver, if gradient information is not used. It can be seen that the gradient information is useful in some optimization problems. It should be pointed out that the computation and programming of gradient information may also need time.

If a structured variable is used to describe the optimization problem, the following commands can be written, and the same results can be obtained:

```
>> problem.solver='fminunc'; ff=optimset; problem.x0=[2; 1];
   problem.objective=@c3fun3; ff.GradObj='on';
   ff.TolX=eps; ff.TolFun=eps; problem.options=ff;
   [x,b,c,d]=fminunc(problem) % describe and solve the problem
```

If gradient information is not used, function `fminsearch()` can be used to solve the problem, and it can be seen that an accurate result can be found with such a solver, meaning that even though gradient information is not used, the problem can also be solved well.

```
>> ff.GradObj='off'; x=fminsearch(f,[0;0],ff) % direct solution
```

It should be pointed out that Rosenbrock function is an artificial one, designed to evaluate the optimization solvers. In order to accurately solve particular problems, the gradient of the objective function should be introduced. In real applications, many solvers do not need gradient information. With gradient information, the efficiency of some algorithms may not be increased at all.

Example 3.16. Consider again the optimization problem in Example 3.6. Since gradient information was not used, the searching process was slow. Use gradient information to solve again the same problem.

Solutions. From the given objective function, it can be seen that the gradient of the objective function is the vector $2[x_1, x_2, \dots, x_{20}] = 2\boldsymbol{x}$. Therefore, the gradient information can be expressed with the objective function such that two returned arguments are allowed. This cannot be implemented in the anonymous function description.

```
function [y,G]=c3mdej1(x)
y=x(:).'*x(:); G=2*x;
```

With the corresponding information, the following statements can be used to initiate the solution process directly. It can be seen that with gradient information, three iteration steps are needed to find the solution $\boldsymbol{x}$, whose norm is as low as 9.0317×10^{-31}, and the objective function is reduced to $f_0 = 8.1572 \times 10^{-61}$. The accuracy is by far higher than that of other numerical algorithms.

```
>> clear problem; problem.solver='fminunc';
   ff=optimset; problem.objective=@c3mdej1;
   ff.GradObj='on'; ff.TolX=eps; ff.TolFun=eps;
   problem.options=ff; problem.x0=-100+200*rand(20,1);
   [x,f0]=fminunc(problem), norm(x) %structured data
```

Example 3.17. Use trust region to solve again the problem in Example 3.5.

Solutions. If the trust region algorithm is used, then the gradients of the objective function should be computed. This can be obtained with the symbolic method.

```
>> syms x1 x2;
   f=exp(-x1^2-x1*x2-x2^2)*(x1^2-2*x1)
   G=simplify(jacobian(f,[x1,x2]))
```

The gradient of the objective function can be written as

$$\boldsymbol{G}(\boldsymbol{x}) = \mathrm{e}^{-x_1^2-x_1x_2-x_2^2}\begin{bmatrix} 2x_1 + 2x_1x_2 - x_1^2x_2 + 4x_1^2 - 2x_1^3 - 2 \\ (-x_1^2 + 2x_1)(x_1 + 2x_2) \end{bmatrix}.$$

With the objective function and its derivative function, the following MATLAB function can be written, with two arguments returned:

```
function [f,Gy]=c3mgrad(x)
u=exp(-x(1)^2-x(1)*x(2)-x(2)^2); f=u*(x(1)^2-2*x(1));
Gy=[2*x(1)+2*x(1)*x(2)-x(1)^2*x(2)+4*x(1)^2-2*x(1)^3-2;
    (-x(1)^2+2*x(1))*(x(1)+2*x(2))]*u;
```

With the corresponding information, the structured variable can be constructed, and the algorithm can be set to be the trust region algorithm. The problem can be tackled with the solver `fminunc()`:

```
>> clear problem; problem.solver='fminunc';
   ff=optimset; problem.objective=@c3mgrad;
   ff.Algorithm='trust-region'; ff.GradObj='on';
```

```
ff.Display='iter'; problem.options=ff;
problem.x0=[2;1]; [x,f0]=fminunc(problem)
```

The intermediate results are shown below

Iteration	f(x)	Norm of step	First-order optimality	CG-iterations
0	0		0.00182	
1	0	10	0.00182	1
2	0	2.5	0.00182	0
3	-0.0120683	0.625	0.0558	0
4	-0.46831	1.25	0.434	1
5	-0.46831	1.25226	0.434	1
6	-0.602878	0.313066	0.252	0
7	-0.640866	0.296542	0.0372	1
8	-0.641424	0.0310125	6.49e-05	1
9	-0.641424	6.10566e-05	6.96e-10	1

It can be seen from the results that the efficiency of the algorithm is lower than of that in Example 3.7, where quasi-Newton algorithm is used. It can be concluded that the gradient information is not necessary in solving optimization problems.

3.2.7 Optimization solutions from scattered data

In real applications, the mathematical form of the objective function may not be given. Only a set of samples in scattered form are available. Data interpolation method can be used to fit the objective function, so that the optimization problem can be solved. The univariate and multivariate functions with two independent variables are both considered. Even for multidimensional problems, the optimization problems will be considered.

(1) Univariate problems. If the samples are provided in the vectors $\boldsymbol{x}_0$ and $\boldsymbol{y}_0$, for an arbitrary vector $\boldsymbol{x}$, the objective function can be reconstructed from

f=@($\boldsymbol{x}$)interp1($\boldsymbol{x}_0$,$\boldsymbol{y}_0$,$\boldsymbol{x}$,'spline')

where the anonymous function can be used to compute the values of the objective function at $\boldsymbol{x}$, with the spline interpolation method.

(2) Two-dimensional problems. If the samples are provided in the vectors $\boldsymbol{x}_0$, $\boldsymbol{y}_0$, and $\boldsymbol{z}_0$, then let $p_1 = x$, $p_2 = y$. The interpolation of the objective function can be reconstructed with the following statements:

f=@($\boldsymbol{p}$)griddata($\boldsymbol{x}_0$,$\boldsymbol{y}_0$,$\boldsymbol{z}_0$,$\boldsymbol{p}$(1),$\boldsymbol{p}$(2),'v4')

(3) **Three-dimensional problems.** If the samples are provided in the vectors $\boldsymbol{x}_0$, $\boldsymbol{y}_0$, $\boldsymbol{z}_0$, and $\boldsymbol{v}_0$, we can let $p_1 = x$, $p_2 = y$, $p_3 = z$. The objective function handle can be defined as follows:

f=@($\boldsymbol{p}$)griddata($\boldsymbol{x}_0$,$\boldsymbol{y}_0$,$\boldsymbol{z}_0$,$\boldsymbol{v}_0$,$\boldsymbol{p}$(1),$\boldsymbol{p}$(2),$\boldsymbol{p}$(3),'v4')

(4) **Multivariate functions.** For multivariate problems, the interpolation can be carried out with `griddatan()` function:

f=@($\boldsymbol{p}$)griddatan([$\boldsymbol{x}_1$(:),$\boldsymbol{x}_2$(:),...,$\boldsymbol{x}_m$(:)],$\boldsymbol{z}$,$\boldsymbol{p}$)

With the objective function handle, the functions such as `fminunc()` can be used to solve the optimization problem. An illustrative example will be given next to show the solution of an optimization problem.

Example 3.18. Consider again the function in Example 3.5. Generate 200 uniformly distributed random samples in the intervals $x \in [-3, 3]$, $y \in [-2, 2]$. Then, based on the scattered samples, find the minimum value of the function and validate the results.

Solutions. A set of random samples can be generated first. Interpolation can be carried out to reconstruct the objective function with an anonymous function. Performing optimization, the optimum solution can be found as $x = 0.6069$ and $y = -0.3085$. It can be seen that the solution is quite close to the theoretical values in Example 3.5, indicating that the approach is practical.

```
>> x=-3+6*rand(200,1); y=-2+4*rand(200,1);
   z=(x.^2-2*x).*exp(-x.^2-y.^2-x.*y);     % generate scattered data
   f=@(p)griddata(x,y,z,p(1),p(2),'v4'); % objective function
   x=fminunc(f,[0,0])                      % optimization
```

3.2.8 Parallel computation in optimization problems

Normally, in the optimization process, a huge computational load may be encountered. The parallel computation facilities automatically launched by the Optimization Toolbox functions can be invoked. The following control options should be made:

```
options=optimoptions('fminunc','UseParallel',true)
```

If parallel computation option is selected, the parallel computing facilities are invoked, and they can be used to solve optimization problems. If one wants to shut down the parallel computing facilities, ordinary format can be used to solve optimization problems. We shall use next a large-scale optimization example to illustrate parallel computation methods.

Example 3.19. Consider again the problem in Example 3.11. If $n = 2000$, compare the efficiency of optimization with or without parallel computation facilities.

Solutions. If parallel computing facilities are to be used, the following statements should be employed. The elapsed time is about 107.49 seconds. The error obtained is 5.3493×10^{-6}. Four parallel workers can be assigned in MATLAB automatically to solve the problem.

```
>> problem.solver='fminunc'; n=2000;
   ff=optimoptions('fminunc'); ff.OptimalityTolerance=eps;
   ff.MaxFunctionEvaluations=1000000; ff.MaxIterations=1000000;
   ff.UseParallel=true; problem.options=ff; %with parallel
   problem.objective=@(x)c3mdixon(x,n);      %anonymous function
   x0=rand(n,1); problem.x0=x0;
   tic, [x,b,c,d]=fminunc(problem); toc, norm(x-1)
```

If parallel computation facilities are shut down, the following statements can be used directly. The same initial values are used for a fair comparison. The error obtained is the same as that obtained earlier. The elapsed time is around 147.22 seconds, slightly longer than in parallel computation.

```
>> ff.UseParallel=false; problem.options=ff;          %shut dow parallel
   tic, [x,b,c,d]=fminunc(problem); toc, norm(x-1) %compute again
```

3.3 Towards global optimum solutions

The global optimal solution problems have been introduced earlier. In this section, a benchmark problem with multiple valleys is used. The initial value dependency of the benchmark problem is demonstrated, then a MATLAB solver aiming at finding the global optimum solutions is presented, and the efficiency of the solver is illustrated.

Example 3.20. Consider a well known benchmark problem for unconstrained optimization problem – Rastrigin function problem[25] –

$$f(x_1,x_2) = 20 + x_1^2 + x_2^2 - 10(\cos 2\pi x_1 + \cos 2\pi x_2).$$

Draw the surface of the objective function, and solve the problem with traditional optimization solver. What may happen?

Solutions. The surface of the objective function can be drawn with the following statements, as shown in Figure 3.8. It can be seen that there are many peaks and valleys in the surface. The graphical solution to this problem is left as an exercise in Exercise 3.20.

```
>> f=@(x1,x2)20+x1.^2+x2.^2-10*(cos(2*pi*x1)+cos(2*pi*x2));
   fsurf(f) %surface of the objective function
```

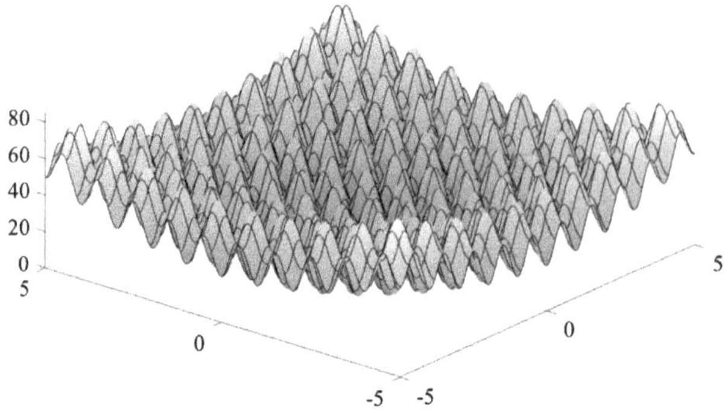

Figure 3.8: Rastrigin function surface.

The contours of the objective function can be obtained as shown in Figure 3.9. It can be seen from the plots that the point in the center is the global minimum point. There are many valleys, and they are all local minima. The minima around the global minimum point can be regarded as subminimum points.

```
>> fcontour(f,'MeshDensity',1000) %contour
```

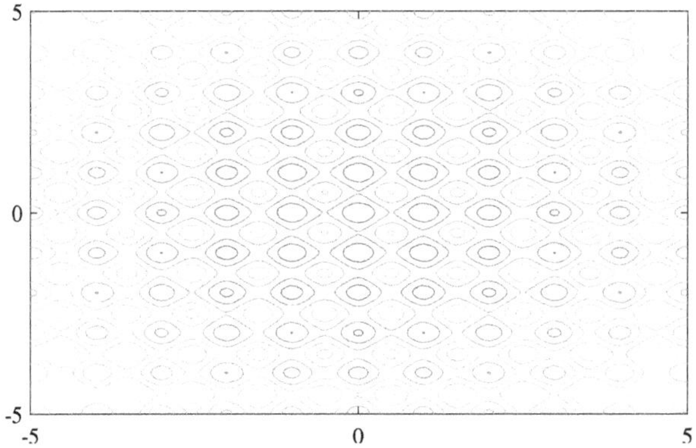

Figure 3.9: Contours of the objective function.

If several different initial points are tried, the following statements can be used to solve the optimization problem. When defining again the objective function, there is no need to rewrite it from low-level. An interface can be written based on the previously defined functions. With the anonymous function description of the objective function, the following initial values can be used in solving the optimization problem:

```
>> f1=@(x)f(x(1),x(2));   %redefine the objective function
   x1=fminunc(f1,[2,3]), f1(x1), x2=fminunc(f1,[-1,2]), f1(x2)
   x3=fminunc(f1,[8 2]), f1(x3), x4=fminunc(f1,[-4,6]), f1(x4)
```

The following results can be obtained:

$x_1 = [1.9602, 1.9602], f(x_1) = 7.8409,$

$x_2 = [-0.0000, 1.9602], f(x_2) = 3.9205,$

$x_3 = [7.8338, 1.9602], f(x_3) = 66.6213,$

$x_4 = [-3.9197, 5.8779], f(x_4) = 50.9570.$

It can be seen that the results obtained are all "optimal". There may exist significant differences. Some of the points are local minima. It can be seen that the traditional searching method depends heavily upon the selection of initial values. Most of the solutions are local minima. The global one at $x = [0, 0]$ was not found at all.

To avoid finding local minima, intelligent optimization methods are usually adopted, such as genetic algorithm or other optimization methods. These algorithms and solvers will be briefly described later in Chapter 9. Even the intelligent optimization solvers may not ensure global optimum solutions.

Similar to the ideas in equation solutions, the following new idea can be used in the global minimum search. At first, a random initial value can be generated in the interval of interest (a, b). Ordinary searching method can be used to find a solution x. Meanwhile, the value of the objective function is computed $f_1 = f(x)$. If the value is smaller than the recorded one, then the new one can be recorded. After N times in the solution process, it is likely that the global optimum solution can be found. Based on the idea, the following MATLAB function can be written to find the global optimum solutions:

```
function [x,f0]=fminunc_global(f,a,b,n,N,varargin)
k0=0; f0=Inf;
if strcmp(class(f),'struct'), k0=1; end %structured variable
for i=1:N, x0=a+(b-a).*rand(n,1); %loop to generate random point
   if k0==1, f.x0=x0; [x1 f1 key]=fminunc(f);       %solutions
   else, [x1 f1 key]=fminunc(f,x0,varargin{:}); end
   if key>0 & f1<f0, x=x1; f0=f1; end   %record better solutions
end
```

The syntax of the function is

`[x,f0]=fminunc_global(fun,a,b,n,N)`

where `fun` is the function handle of the objective function, which can be an anonymous or MATLAB function. Structured variables can be used to describe the whole problem. Arguments a and b can be used to describe the decision variable regions. The argument n is the number of decision variables, and N is the number of runs. If N is properly chosen, the returned variables x and f_0 may be the expected global optimum solution. If necessary, arguments a and b can be selected as vectors.

It is worth mentioning that, although arguments a and b can be selected, they are just used in automatically generating the bounds of the initial values. The final solution may exceed the range.

Example 3.21. Consider the unconstrained optimization problem in Example 3.20. Find the optimum solution and assess the new optimization solver.

Solutions. If the number of runs N is assigned to 50, it can be seen that each time, the global optimum solution $x_1 = x_2 = 0$ can be found.

```
>> f=@(x)20+x(1)^2+x(2)^2-10*(cos(2*pi*x(1))+cos(2*pi*x(2)));
   [x,f0]=fminunc_global(f,-2*pi,2*pi,2,50); % find the global solution
```

To better illustrate the solution process in the search for global optimal solutions, the solver can be executed 100 times, and it can be seen that the global minimum solution can be found every time.

```
>> F=[]; N=50; % a blank vector is created
   for i=1:100 % call the solver 100 times to evaluate the success rate
      [x,f0]=fminunc_global(f,-2*pi,2*pi,2,N); F=[F,f0];
   end
```

Of course, since uniformly distributed random numbers are used, the global solutions at $x_1 = x_2 = 0$ can be found easily. It is not quite fair to evaluate the behavior of the new solver. A more reasonable test function will be created to assess optimization solvers.

Example 3.22. Assume that the original Rastrigin function is modified as

$$f(x_1, x_2) = 20 + \left(\frac{x_1}{30} - 1\right)^2 + \left(\frac{x_2}{20} - 1\right)^2 - 10\left[\cos 2\pi\left(\frac{x_1}{30} - 1\right) + \cos 2\pi\left(\frac{x_2}{20} - 1\right)\right],$$

run the solver 100 times, and see what is the success rate of finding the global optimum solutions.

Solutions. It is obvious that the theoretical position of the new extremum is at $x_1 = 30$ and $x_2 = 20$. If the region of interest is extended to ± 100, the following test commands can be used:

```
>> f=@(x)20+(x(1)/30-1)^2+(x(2)/20-1)^2- ...
         10*(cos(2*pi*(x(1)/30-1))+cos(2*pi*(x(2)/20-1)));
   F=[]; tic, N=100;
   for i=1:100         % run the solver 100 times
      [x,f0]=fminunc_global(f,-100,100,2,N);
      F=[F,f0];  % record the solutions
   end, toc
```

It can be seen that out of 100 runs, 17 times we fail to get the global optimum solution at (30, 20), and during the other 83 trials, the global optimum solution is found. The success rate is 83 %, and total elapsed time is 57.07 seconds. It can be seen that the `fminunc_global()` function is reliable. Generally speaking, the global optimum solution can be found. If N is changed to 50, the total elapsed time is reduced to 26.23 seconds, and the success rate is reduced to 67 %. As will be demonstrated later in Chapter 9, the success rate here is several times higher than for the genetic algorithm solvers provided in MATLAB.

It can be seen that the solver discussed here is suitable for parallel computing. Unfortunately, when the parallel option is turned on, error messages will be displayed, indicating that the anonymous function handles cannot be used in resource assignment. Therefore, parallel facilities cannot be used for this problem.

3.4 Optimization with decision variable bounds

In the earlier presentation of this chapter, the theoretical unconstrained optimization problems are studied. In some particular applications, the "unconstrained optimization" problems with decision variable bounds are often encountered. For instance, the global optimum problem in different regions of interest studied in Example 3.2 is a problem with decision variable bounds. Such problems are not typical unconstrained optimization problems in mathematical terms. Therefore, in this section, the optimization problems with decision variable bounds are explored. Univariate optimization problems are addressed first, followed by the solutions of multivariate problems.

3.4.1 Univariate optimization problem

In this section univariate functions with decision variable bounds are studied. The definition is proposed first, then the direct solution with existing MATLAB functions and using global method is discussed.

Definition 3.4. The mathematical form of a univariate optimization problem with decision variable bounds is given by

$$\min_{x \text{ s.t. } x_\mathrm{m}\leqslant x\leqslant x_\mathrm{M}} f(x), \tag{3.4.1}$$

where x_m and x_M are respectively the lower and upper bounds of the decision variable. The notation s. t. stands for "subject to", followed by the relationships.

In MATLAB Optimization Toolbox, function `fminbnd()` is provided for solving a univariate function problem with decision variable bounds directly. The syntaxes of the function are

x=fminbnd(f,x_m,x_M)

[x,f_0,flag,out]=fminbnd(f,x_m,x_M,options)

[x,f_0,flag,out]=fminbnd(problem)

where f is the objective function handle, x_m and x_M are the lower and upper bounds of the decision variable x. Structured variable `problem` can be used in describing the optimization problem. The terms for x_m and x_M are named as `x1` and `x2`.

Example 3.23. Solve again the optimization problem in Example 3.3.

Solutions. The problem has decision variable bounds $t_m = -0.5$ and $t_M = 2.5$. Therefore, with the following commands, the problem can be solved, and the result is $t^* = 1.5511$. Unfortunately, it is not the global optimum in the region of interest. The global solution is labeled t_0 in Figure 3.3. The local optimum solution is t_3. In the function `fminbnd()`, initial values cannot be selected. Therefore, the solver cannot be used in finding global optimum solutions.

```
>> f=@(t)exp(-2*t).*cos(10*t)+exp(-3*(t+2)).*sin(2*t);
   tm=-0.5; tM=2.5; x=fminbnd(f,tm,tM)
```

Since there are no adjustable arguments in the function call, to find the global optimum solutions, it is necessary to divide the interval (x_m, x_M) into m subintervals. For each subinterval, a solution and subinterval the smallest objective function values can be found. The solution is likely to contain the global optimum solution of the problem. With such an algorithm, the following ideas are implemented in MATLAB. Function `fminsearchbnd()` is also called later for specific solutions.

```
function [x,f0]=fminbnd_global(f,xm,xM,n,m,varargin)
f0=Inf; M=ones(n,1);
if length(xm)==1, xm=xm*M; end, if length(xM)==1, xM=xM*M; end
for i=1:m
   if n==1, h=(xM-xm)/m;                 % univariate problem
      [x1,f1]=fminbnd(f,xm+(i-1)*h,xm+i*h,varargin{:});
   else, x0=xm+(xM-xm).*rand(n,1); % multivariate problem
      [x1 f1 key]=fminsearchbnd(f,x0,xm,xM,varargin{:});
   end      % generate random initial points with loops
   if f1<f0, x=x1; f0=f1; end % update the solutions
end
```

The syntax of the function is

[x, f_0]=fminbnd_global(f,x_m,x_M,1,m)

where m should be selected as a large value, such that it is likely to find the global optimum solution of the problem.

Example 3.24. Find the global optimum solution of the problem in Example 3.2.

Solutions. Selecting the number of subintervals as $m = 10$, the new solver can be called in finding the global optimum solutions. The solution obtained is $x = -0.3340$, and it can be seen that the global solution of t_0 in Figure 3.3 is indeed found.

```
>> f=@(t)exp(-2*t).*cos(10*t)+exp(-3*(t+2)).*sin(2*t);
   tm=-0.5; tM=2.5; x=fminbnd_global(f,tm,xM,1,10)
```

3.4.2 Multivariate optimization problems

If variable x in Definition 3.4 is changed to an $\boldsymbol{x}$ vector, the corresponding optimization problem is rewritten as

$$\min_{\boldsymbol{x} \text{ s.t. } \boldsymbol{x}_\mathrm{m}\leqslant\boldsymbol{x}\leqslant\boldsymbol{x}_\mathrm{M}} f(\boldsymbol{x}). \tag{3.4.2}$$

The problem described in (3.4.2) is to select $\boldsymbol{x}$ from the specified regions such that the objective function is minimized. The problem cannot be solved directly with the solvers such as `fminsearch()` and `fminbnd()`.

John D'Errico proposed a conversion method,[19] where new decision variables z_i are introduced such that

$$x_i = x_{\mathrm{m}_i} + \frac{1}{2}(x_{\mathrm{M}_i} - x_{\mathrm{m}_i})(\sin z_i + 1). \tag{3.4.3}$$

Therefore, the optimization problem with decision variable bounds on x_i can be mapped into the unconstrained problem of decision variables z_i. John D'Errico proposed a solver `fminsearchbnd()`, which extended the capabilities of the existing `fminsearch()` function, and can be used to solve the problem in (3.4.2) directly. The syntaxes of the function are

[$\boldsymbol{x}$,f_0,flag,out]=fminsearchbnd(f,$\boldsymbol{x}_0$,$\boldsymbol{x}_\mathrm{m}$,$\boldsymbol{x}_\mathrm{M}$)

[$\boldsymbol{x}$,f_0,flag,out]=fminsearchbnd(f,$\boldsymbol{x}_0$,$\boldsymbol{x}_\mathrm{m}$,$\boldsymbol{x}_\mathrm{M}$,opt)

[$\boldsymbol{x}$,f_0,flag,out]=fminsearchbnd(f,$\boldsymbol{x}_0$,$\boldsymbol{x}_\mathrm{m}$,$\boldsymbol{x}_\mathrm{M}$,opt,$p_1$,$p_2$,...)

If the upper or lower bounds are not specified, they are assigned automatically to empty vectors `[]`. Therefore, the conversion formula in (3.4.3) is adjusted accordingly, such that the problem can still be solved.

Example 3.25. Find the minimum value of the following function:

$$f(x,y) = (1.5 - x + xy)^2 + (2.25 - x + xy^2)^2 + (2.625 - x + xy^3)^2,$$

where $-4.5 \leqslant x, y \leqslant 4.5$.

Solutions. Letting $x_1 = x$ and $x_2 = y$, the objective function can be rewritten as

$$f(\boldsymbol{x}) = (1.5 - x_1 + x_1x_2)^2 + (2.25 - x_1 + x_1x_2^2)^2 + (2.625 - x_1 + x_1x_2^3)^2.$$

An anonymous function can be used to describe the objective function with dot operations. The surface of the objective function can be drawn as shown in Figure 3.10.

```
>> f=@(x,y)(1.5-x+x.*y).^2+(2.25-x+x.*y.^2).^2 ...
     +(2.625-x+x.*y.^3).^2;
   fsurf(f,[-4.5,4.5])
```

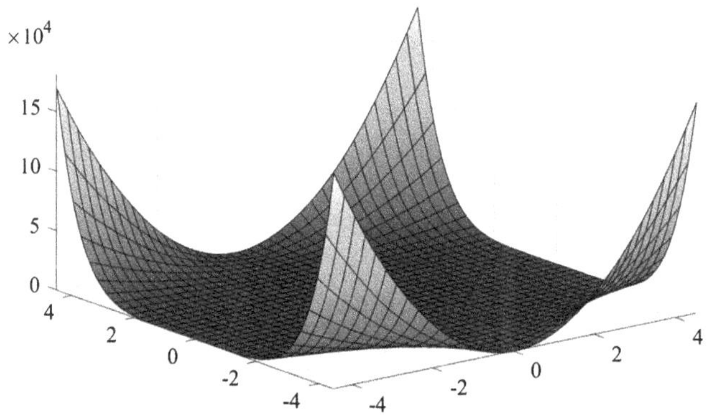

Figure 3.10: Objective function surface.

If the problem is to be solved, there is no need to write again the objective function. A modified version of the anonymous function can be written to describe $f(\boldsymbol{x})$. The function handle of f_0 can be created. The optimization problem can be solved, and the global optimum solution at $\boldsymbol{x}_1 = [3, 0.5]^{\mathrm{T}}$ can be found, with the objective function value of $f(\boldsymbol{x}_1) = 2.6871 \times 10^{-29}$.

```
>> f0=@(x)f(x(1),x(2));
   opt=optimset; opt.TolX=eps; ff.TolFun=eps;
   [x1,f1]=fminsearchbnd(f0,[0;0],-4.5*[1;1],4.5*[1;1],opt)
```

If the region of interest is changed to $-2 \leqslant x, y \leqslant 2$, the previously found solution is not located in this region. The following statements can be used to search again, and the optimum solution $\boldsymbol{x}_1 = [2, 0.1701]^{\mathrm{T}}$ can be found. The value of the objective function is $f_1 = 0.5233$.

```
>> [x1,f1]=fminsearchbnd(f0,[0;0],-2*[1;1],2*[1;1],opt)
   fsurf(f,[-2,2])
```

3.4.3 Global optimum solutions

For multiple-valley or large range searching problems, the `fminsearchbnd()` function discussed earlier may fail to find the global optimum solutions. Global optimum solvers are expected. The solver `fminbnd_global()` discussed earlier can be tried to find the global optimum solutions.

Example 3.26. If the interval is enlarged to $-500 \leqslant x, y \leqslant 500$, solve again the problem in Example 3.25.

Solutions. If function `fminsearchbnd()` is called, in the following statements, it is likely find the solution of $\boldsymbol{x}_1 = [-50, 1.0195]^\mathrm{T}$ and $f_0 = 0.4823$. Meanwhile, a warning message "Maximum function calls exceeded, increase `MaxFunEvals` option" is displayed, meaning that the solution process is unsuccessful.

```
>> f=@(x,y)(1.5-x+x.*y).^2+(2.25-x+x.*y.^2).^2 ...
      +(2.625-x+x.*y.^3).^2;
   f0=@(x)f(x(1),x(2));
   opt=optimset; opt.TolX=eps; ff.TolFun=eps;
   [x1,f1]=fminsearchbnd(f0,500*rand(2,1),-500*[1;1],500*[1;1],opt)
```

With the repeated use the last statement above, almost half the time the global solution of $[3, 0.5]$ cannot be obtained.

The solver `fminbnd_global()` can be called to tackle the original problem, with repeated calls of the following function, and it always finds the global optimum solution:

```
>> [x1,f1]=fminbnd_global(f0,-500,500,2,5,opt)
```

3.5 Application examples of optimization problems

Several application examples are presented in this section. Linear regression problem is presented first. If experimental data are acquired, we build up the mathematical model from the data. This is related to the least squares solution of overdetermined linear equations. If the function model is complicated, least-squares curve fitting techniques are also demonstrated. Besides, optimization-based boundary value problems involving differential equations are studied.

In fact, the eventual target of the examples is to demonstrate how to convert a seeming unrelated problem into an optimization problem. Through the modeling and solution processes of an optimization problem, we try to find the result of the original problem.

3.5.1 Solutions of linear regression problems

Assume that a certain linear combination of functions is defined as

$$g(x) = c_1 f_1(x) + c_2 f_2(x) + c_3 f_3(x) + \cdots + c_n f_n(x), \tag{3.5.1}$$

where $f_1(x), f_2(x), \ldots, f_n(x)$ are known functions, and $c_1, c_2, \ldots, c_n$ are undetermined coefficients. It is assumed that the measured samples are (x_1, y_1), $(x_2, y_2), \ldots$, (x_m, y_m), so the following linear equation can be set up:

$$\boldsymbol{A}\boldsymbol{c} = \boldsymbol{y}, \tag{3.5.2}$$

where

$$\boldsymbol{A} = \begin{bmatrix} f_1(x_1) & f_2(x_1) & \cdots & f_n(x_1) \\ f_1(x_2) & f_2(x_2) & \cdots & f_n(x_2) \\ \vdots & \vdots & \ddots & \vdots \\ f_1(x_m) & f_2(x_m) & \cdots & f_n(x_m) \end{bmatrix}, \quad \boldsymbol{y} = \begin{bmatrix} y_1 \\ y_2 \\ \vdots \\ y_m \end{bmatrix}, \tag{3.5.3}$$

and $\boldsymbol{c} = [c_1, c_2, \ldots, c_n]^\mathrm{T}$. Therefore, the least squares solution can be found from $\boldsymbol{c} = \boldsymbol{A}\backslash\boldsymbol{y}$.

Example 3.27. Assume that a set of measured data (x_i, y_i) is given in Table 3.4. The prototype function is given below. Use the data and prototype function to find the undetermined coefficients c_i if

$$y(x) = c_1 + c_2 \mathrm{e}^{-3x} + c_3 \cos(-2x)\mathrm{e}^{-4x} + c_4 x^2.$$

Table 3.4: Measured data in Example 3.27.

x_i	0	0.2	0.4	0.7	0.9	0.92	0.99	1.2	1.4	1.48	1.5
y_i	2.88	2.2576	1.9683	1.9258	2.0862	2.109	2.1979	2.5409	2.9627	3.155	3.2052

Solutions. The data in the table can be used to fit the c_i coefficients in the prototype function. The subfunctions can be computed through dot operations such that matrix $\boldsymbol{A}$ can be created. Least squares command can be used to find the values of the undetermined coefficients:

```
>> x=[0,0.2,0.4,0.7,0.9,0.92,0.99,1.2,1.4,1.48,1.5]';
   y=[2.88;2.2576;1.9683;1.9258;2.0862;2.109;2.198;
      2.541;2.9627;3.155;3.2052];   % samples input
   A=[ones(size(x)) exp(-3*x),cos(-2*x).*exp(-4*x) x.^2];
   c=A\y; c1=c' % least squares solution
```

The fitting parameters obtained are $\boldsymbol{c}^\mathrm{T} = [1.22, 2.3397, -0.6797, 0.87]$. A densely distributed $\boldsymbol{x}$ vector can be substituted into the prototype function, and the fitted curve is shown in Figure 3.11. The fitting is satisfactory.

```
>> x0=[0:0.01:1.5]';
   B=[ones(size(x0)) exp(-3*x0) cos(-2*x0).*exp(-4*x0) x0.^2];
   y1=B*c; plot(x0,y1,x,y,'x') %draw fitting curve
```

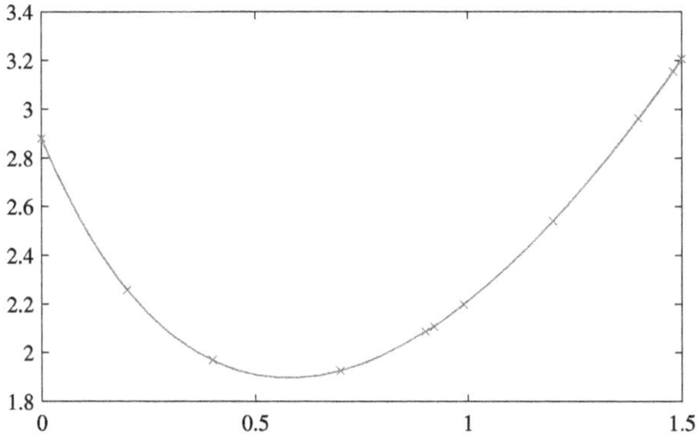

Figure 3.11: Original data and fitted curve.

3.5.2 Least-squares curve fitting

In the linear regression problem studied earlier, functions $f_i(x)$ were assumed to be given functions, therefore, the problem could be converted into a linear equation problem. In practice, functions $f_i(x)$ may also contain undetermined coefficients. Therefore, the problem cannot be solved from linear equations. It should be converted into an optimization problem.

Definition 3.5. For a given set of data x_i, y_i, $i = 1, 2, \dots, m$, assume that the data come from a prototype function $\hat{y}(x) = f(\boldsymbol{a}, x)$, where $\boldsymbol{a}$ is the vector of undetermined coefficients. The target of least-squares curve fitting is to find the coefficients, and to convert the problem into the following optimization problem:

$$J = \min_{\boldsymbol{a}} \sum_{i=1}^{m} [y_i - \hat{y}(x_i)]^2 = \min_{\boldsymbol{a}} \sum_{i=1}^{m} [y_i - f(\boldsymbol{a}, x_i)]^2. \tag{3.5.4}$$

Low-level solution of the optimization problem is possible. Alternatively, MATLAB function `lsqcurvefit()` can be used directly to solve the least-squares fitting problem. The syntax of the function is

$[\boldsymbol{a}, J_{\rm m}, \texttt{cc}, \texttt{flag}, \texttt{out}] = \texttt{lsqcurvefit}(\texttt{Fun}, \boldsymbol{a}_0, \boldsymbol{x}, \boldsymbol{y}, \texttt{options})$

where Fun is the prototype function handle, which can be a MATLAB or anonymous function. The argument $\boldsymbol{a}_0$ contains the initial values of the undetermined coefficients. Vectors $\boldsymbol{x}$ and $\boldsymbol{y}$ are the original input and output vectors, respectively. For multivariate function fitting, the argument $\boldsymbol{x}$ can be a matrix. This will be demonstrated later through examples. The argument options is the control template in the Optimization Toolbox. With such a function, the optimal undetermined coefficient vector $\boldsymbol{a}$ and the value of the objective function J_m can be returned. If flag is positive, the solution process is successful.

Example 3.28. Assume that a set of samples are generated in vectors $\boldsymbol{x}$ and $\boldsymbol{y}$.

```
>> x=0:0.4:10; %generate samples
   y=0.12*exp(-0.213*x)+0.54*exp(-0.17*x).*sin(1.23*x);
```

and assume that the data come from the prototype function

$$y(x) = a_1 \mathrm{e}^{-a_2 x} + a_3 \mathrm{e}^{-a_4 x} \sin(a_5 x),$$

where a_i are undetermined coefficients. Use least-squares curve fitting method to find the undetermined coefficients such that the value of the objective function is minimized.

Solutions. It is obvious that the linear regression method cannot be used in solving this problem. Least-squares solution method should be used instead.

Based on the given prototype function, the following anonymous function can be established. From that, the following commands can be used to find the undetermined coefficients:

```
>> f=@(a,x)a(1)*exp(-a(2)*x)+a(3)*exp(-a(4)*x).*sin(a(5)*x);
   a0=[1,1,1,1,1]; [xx,res]=lsqcurvefit(f,a0,x,y); %fitting
   x1=0:0.01:10; y1=f(xx,x1); plot(x1,y1,x,y,'o')  %comparison
```

The undetermined coefficients obtained are $\boldsymbol{c} = [0.12, 0.213, 0.54, 0.17, 1.23]$, and the fitting error is 1.7928×10^{-16}. It can be seen that the precision of the coefficients is rather high. The samples and the fitting curve can also be drawn, as shown in Figure 3.12. The fitting is satisfactory.

In fact, it is not hard to solve the optimization problem with low-level commands. A new anonymous function can be written to describe the objective function. The solution obtained is exactly the same as that obtained above.

```
>> F=@(a)norm(f(a,x)-y); x1=fminunc(F,a0)
```

Example 3.29. Assume that a set of measured data is given in Table 3.5. It is known that the prototype function is $y(x) = ax + bx^2\mathrm{e}^{-cx} + d$. Use the least squares method to find the values of a, b, c, and d.

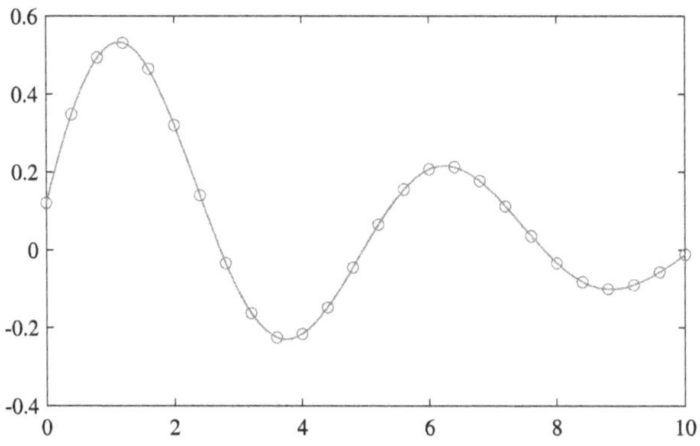

Figure 3.12: Fitting quality demonstration.

Table 3.5: Measured data in Example 3.29.

x_i	0.1	0.2	0.3	0.4	0.5	0.6	0.7	0.8	0.9	1
y_i	2.3201	2.6470	2.9707	3.2885	3.6008	3.9090	4.2147	4.5191	4.8232	5.1275

Solutions. The data in Table 3.5 can be loaded directly into MATLAB workspace with the following statement:

```
>> x=0.1:0.1:1; % input samples
   y=[2.3201,2.6470,2.9707,3.2885,3.6008,3.9090,...
      4.2147,4.5191,4.8232,5.1275];
```

Letting $a_1 = a$, $a_2 = b$, $a_3 = c$, $a_4 = d$, the prototype function can be rewritten as

$$y(x) = a_1x + a_2x^2\mathrm{e}^{-a_3x} + a_4.$$

It can be seen that anonymous function can be used to describe the prototype function. The following commands can be used to compute the undetermined coefficients, and the result is $\boldsymbol{a} = [3.1001, 1.5027, 4.0046, 2]^{\mathrm{T}}$. Note that if a loop structure is not used, the real values cannot be reached.

```
>> f=@(a,x)a(1)*x+a(2)*x.^2.*exp(-a(3)*x)+a(4);
   a=[1;2;2;3]; % prototype function and initial values
   while (1)
      [a,f0,cc,flag]=lsqcurvefit(f,a,x,y); if flag>0, break;
   end, end % least squares solutions
```

The following statements can be used to compute the fitted function values at different x, and draw the two curves in the same coordinates, as shown in Figure 3.13,

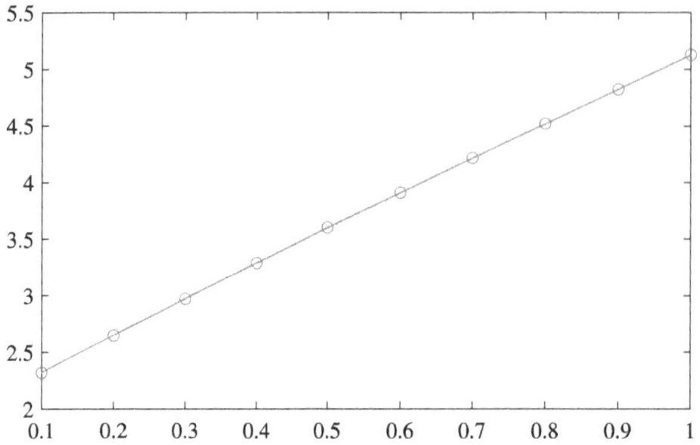

Figure 3.13: Fitting quality comparisons.

indicating that the fitting is satisfactory.

```
>> y1=f(a,x); plot(x,y,x,y1,'o') % fitting quality comparison
```

If a function has several independent variables, and the prototype function is given by $\boldsymbol{z} = f(\boldsymbol{a}, x_1, x_2, \ldots, x_m)$, function `lsqcurvefit()` can still be used to fit the coefficients $\boldsymbol{a}$, where $\boldsymbol{a} = [a_1, a_2, \ldots, a_n]$. An anonymous or MATLAB function can be written to describe the prototype function. Function `lsqcurvefit()` can then be called directly to compute vector $\boldsymbol{a}$. Multivariate least-squares curve fitting problem is demonstrated through examples.

Example 3.30. Assume that the prototype function with three independent variables is given by

$$v = a_1 x^{a_2 x} + a_3 y^{a_4(x+y)} + a_5 z^{a_6(x+y+z)},$$

and a set of input–output data is provided in file c3data1.dat. The first three columns are the data for independent variables x, y, and z, and the fourth column is the returned vector. Use the least squares method to compute the undetermined coefficients a_i.

Solutions. To solve this type problem, the first step is to introduce a vectorized decision variable vector $\boldsymbol{x}$. For instance, letting $x_1 = x$, $x_2 = y$, $x_3 = z$, the prototype function can be rewritten as

$$v = a_1 x_1^{a_2 x_1} + a_3 x_2^{a_4(x_1+x_2)} + a_5 x_3^{a_6(x_1+x_2+x_3)}.$$

Since the data are provided in the text file, function `load()` can be used to input the data into MATLAB workspace. With submatrix extraction commands, the data can be assigned to matrix $\boldsymbol{X}$ and column vector $\boldsymbol{v}$. The following commands can be use to find the values of the undetermined coefficients. The results obtained are $\boldsymbol{a} = [0.1, 0.2, 0.3, 0.4, 0.5, 0.6]$, such that the fitting error is minimized. The error is 1.0904×10^{-7}. Note that in the anonymous function description, the ith independent variable should be represented as $\boldsymbol{X}(:,i)$:

```
>> f=@(a,X)a(1)*X(:,1).^(a(2)*X(:,1))+...
          a(3)*X(:,2).^(a(4)*(X(:,1)+X(:,2)))+...
          a(5)*X(:,3).^(a(6)*(X(:,1)+X(:,2)+X(:,3))); %prototype
   XX=load('c3data1.dat'); X=XX(:,1:3); v=XX(:,4);    %load samples
   a0=[2 3 2 1 2 3]; [a,f,err,flag]=lsqcurvefit(f,a0,X,v)
```

In fact, the data in the file were generated by assuming $\boldsymbol{a} = [0.1, 0.2, 0.3, 0.4, 0.5, 0.6]$. Therefore, the undetermined coefficients can be accurately obtained with the fitting method.

3.5.3 Shooting method in boundary value differential equations

Analytical and numerical solutions of differential equations will be thoroughly studied in Volume V of the series. Here only a simple example is given to show the use of optimization through shooting method, and how to convert a differential equation problem into an optimization problem, to find the solutions.

For ordinary differential equations, the initial value problems are well studied, that is, if the mathematical model of the differential equation is known. The formal mathematical description of an initial value differential equation problem is given by

$$\boldsymbol{x}'(t) = \boldsymbol{f}(t, \boldsymbol{x}), \quad \text{for given } \boldsymbol{x}(t_0). \tag{3.5.5}$$

Therefore, function `ode45()` can be used to find the numerical solutions of the differential equation directly.

In real applications, if some components in $\boldsymbol{x}(t_0)$ are known, and also some other components in vector $\boldsymbol{x}(t_n)$ are given, how can we solve the original differential equation? The shooting method is the commonly adopted method.

The idea of shooting method is stated as follows. The unknown initial values in $\boldsymbol{x}(t_0)$ should be assigned, such that the original problem can be converted into an initial value problem. Function `ode45()` can be used to solve the differential equation numerically, and the final values $\hat{\boldsymbol{x}}(t_n)$ of the equation can be found. The error between $\hat{\boldsymbol{x}}(t_n)$ and the given values $\boldsymbol{x}(t_n)$ can be measured. Based on the error, the initial value $\boldsymbol{x}(t_0)$ can be adjusted, and the differential equation is solved again. The process can be implemented in a loop structure. The original problem can be converted into an optimization problem. The error norm of some of the components in $\hat{\boldsymbol{x}}(t_n)$ and $\boldsymbol{x}(t_n)$ can be used as an objective function. The compatible initial values can be used as the decision variables. Finally, the boundary value problem can be solved numerically.

Example 3.31. Consider the following ordinary differential equation:

$$\begin{cases} x_1'(t) = x_2(t), \\ x_2'(t) = -x_1(t) - 3x_2(t) + e^{-5t}, \\ x_3'(t) = x_4(t), \\ x_4'(t) = 2x_1(t) - 4x_2(t) - 3x_3(t) - 4x_4(t) - \sin t, \end{cases}$$

with given boundary conditions $x_1(0) = 1$, $x_2(0) = 2$, $x_3(10) = -0.021677$, and $x_4(10) = 0.15797$. Solve the differential equation within the interval $t \in (0, 10)$.

Solutions. Variables $x_3(0)$ and $x_4(0)$ can be used as decision variables. The original problem can be converted into an optimization problem:

$$\min_{x_3(0),x_4(0)} \|x_3(10) - \hat{x}_3(10)\| + \|x_4(10) - \hat{x}_4(10)\|.$$

Letting $y_1 = x_3(0)$ and $y_2 = x_4(0)$, the mathematical model of the optimization problem can be formulated as

$$\min_{\boldsymbol{y}} \|y_1 - \hat{x}_3(10)\| + \|y_2 - \hat{x}_4(10)\|,$$

and $\hat{x}_3(10)$, $\hat{x}_4(10)$ are the last two terms in the final values of the states, when the initial values are selected as $[x_1(0), x_2(0), y_1, y_2]^{\mathrm{T}}$. The objective function can be described by the following MATLAB function:

```
function z=c3mode(y,f,x0,xn)
x0=[x0(1:2); y]; [t,x]=ode45(f,[0,10],x0);
z=norm([x(end,3:4)]'-xn);
```

Note that since there are intermediate variables, an anonymous function cannot be used in the model description. Besides, f, $\boldsymbol{x}_0$, and $\boldsymbol{x}_{\mathrm{n}}$ are additional parameters, and conversion later is used to transform the problem without additional parameters, through the newly created anonymous function g.

Another anonymous function, f, is used to describe the differential equation. The following statements can be issued to solve the optimization problem:

```
>> f=@(t,x)[x(2); -x(1)-3*x(2)+exp(-5*t);
            x(4); 2*x(1)-3*x(2)-3*x(3)-4*x(4)-sin(t)];
   x0=[1; 2]; xn=[-0.021677; 0.15797];
   g=@(x)c3mode(x,f,x0,xn); % convert the problem with extra parameters
   x2=rand(2,1); x3=fminunc(g,x2) % solve optimization problem
   [t,x]=ode45(f,[0,10],[x0; x3]); plot(t,x), x(end,:)
```

It is found that the equivalent initial conditions, $x_3(0)$ and $x_4(0)$, can be obtained as $x_3(0) = 0.101584621163685$, $x_4(0) = -2.318606769109767$. Under the equivalent initial conditions, the differential equation can be solved numerically and the solutions are shown in Figure 3.14. It can be seen that the final values are as expected, so the solution process is successful.

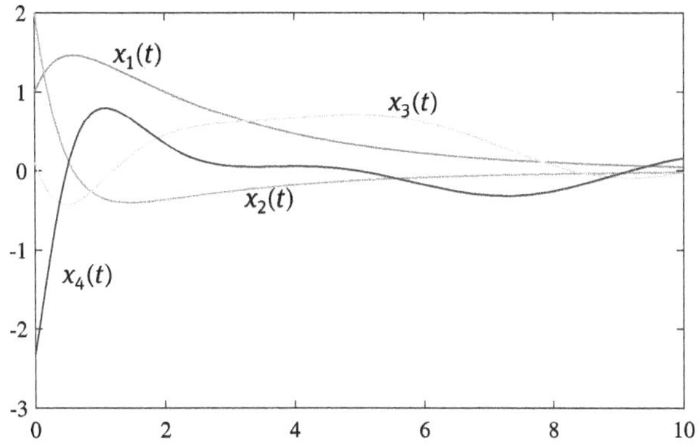

Figure 3.14: Solution of a differential equation.

In fact, the backward formulation method may not have unique solutions. If the initial decision variables are selected as $x_3(0) = -1$, $x_4(0) = 1$, a different equation solution can be found, as shown in Figure 3.15. It can be seen that the final values are also satisfied. Every time when $\boldsymbol{x}_2$ is set to a different initial value, a set of consistent initial conditions is found. All these solutions satisfy the target final value problem.

```
>> [t,x]=ode45(f,[0,10],[x0; -1; 1]; x(end,:)
```

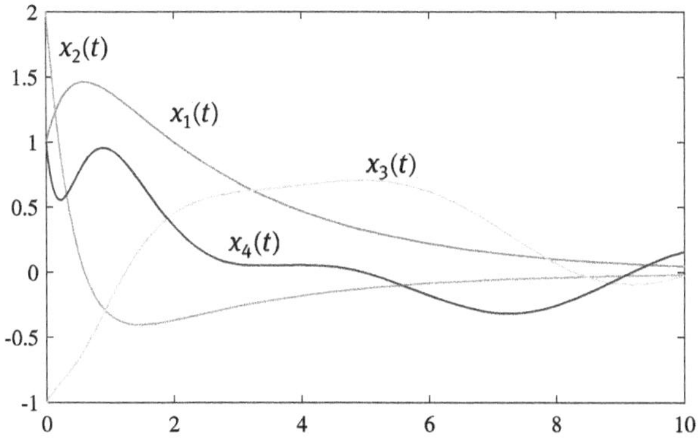

Figure 3.15: Numerical solution for different initial values.

3.5.4 Converting algebraic equations into optimization problems

Numerical solutions of nonlinear algebraic equations were proposed in Chapter 2. In fact, an equation solution problem can also be converted into an unconstrained optimization problem. Assuming that an equation $\boldsymbol{f}(\boldsymbol{x}) = \mathbf{0}$ is given, what condition should $\boldsymbol{x}$ satisfy? Of course, the conditions should be $f_i(\boldsymbol{x}) = 0$, so it is natural to convert the problem into an optimization problem, where the objective function is the sum of squared values of the equations. The target is to find the decision vector $\boldsymbol{x}$, such that the objective function is minimized:

$$\min_{\boldsymbol{x}} f_1^2(\boldsymbol{x}) + f_2^2(\boldsymbol{x}) + \cdots + f_n^2(\boldsymbol{x}) = \min_{\boldsymbol{x}} \sum_{i=1}^{n} f_i^2(\boldsymbol{x}). \tag{3.5.6}$$

Example 3.32. Solve the following simultaneous equations with two independent variables:[23]

$$\begin{cases} 4x_1^3 + 4x_1x_2 + 2x_2^2 - 42x_1 = 14, \\ 4x_2^3 + 2x_1^2 + 4x_1x_2 - 26x_2 = 22. \end{cases}$$

Solutions. It can be seen that the equations are the converted form of the underdetermined equation studied in Example 2.44. It is natural to convert the problem into the following unconstrained optimization problem:

$$\min_{\boldsymbol{x}} \left(4x_1^3 + 4x_1x_2 + 2x_2^2 - 42x_1 - 14\right)^2 + \left(4x_2^3 + 2x_1^2 + 4x_1x_2 - 26x_2 - 22\right)^2.$$

Anonymous functions can be written to define the two equations, and the implicit function plotting facilities can be used to draw the two equations, as shown in Figure 3.16. It can be seen from the curves that there are nine real solutions. Of course, with `more_sols()` function, all the nine solutions are found in Example 2.44.

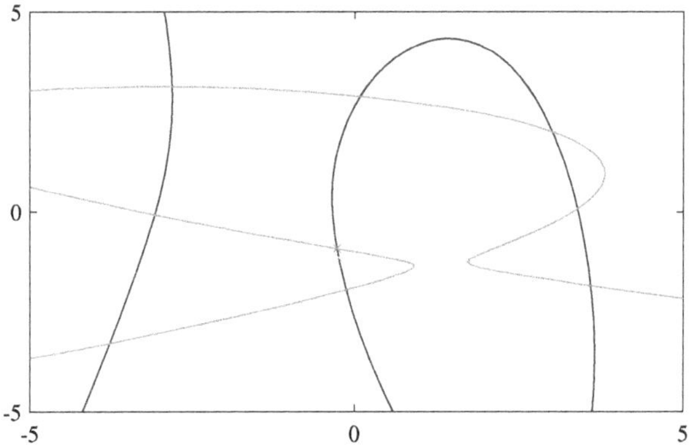

Figure 3.16: Illustration of simultaneous equation solutions.

```
>> f1=@(x1,x2)4*x1.^3+4*x1.*x2+2*x2.^2-42*x1-14;
   f2=@(x1,x2)4*x2.^3+2*x1.^2+4*x1.*x2-26*x2-22;
   fimplicit(f1), hold on, fimplicit(f2), hold off
```

Now let us consider solving the problem with an optimization technique. Based on the two anonymous functions created above, a new anonymous function can be constructed to describe the objective function. Selecting an initial value, one solution of the optimization problem can be found. If the initial values are modified, a different solution may be found.

```
>> f=@(x)(f1(x(1),x(2)))^2+(f2(x(1),x(2)))^2;
   [x,f0]=fminunc(f,rand(2,1));
```

From the viewpoints of efficiency and accuracy, the method is clearly worse than the dedicated equation solver `fsolve()`. The example is used here to demonstrate how to convert an equation solution problem into an optimization problem.

It can be seen through the examples that when the concept of optimization is introduced, some problems can be deliberately converted into optimization problems, by choosing a meaningful objective function. The problem can be solved by the solvers presented in this chapter.

3.6 Exercises

3.1 Find the value of c such that $\int_0^1 (e^x - cx)^2 dx$ is minimized.

3.2 Solve the following unconstrained optimization problem:

$$\min_{x} \begin{array}{l} 100(x_2 - x_1^2)^2 + (1-x_1)^2 + 90(x_4 - x_3^2) + (1-x_3^2)^2 \\ \quad +10.1[(x_2-1)^2 + (x_4-1)^2] + 19.8(x_2-1)(x_4-1). \end{array}$$

3.3 Find the global valley of the following function:

$$f(x_1,x_2) = -\frac{\sin(0.1+\sqrt{(x_1-4)^2+(x_2-9)^2})}{1+(x_1-4)^2+(x_2-9)^2}.$$

3.4 Solve the Griewangk benchmark problem for $n = 20$, namely

$$\min_{x}\left(1+\sum_{i=1}^{n}\frac{x_i^2}{4\,000}-\prod_{i=1}^{n}\cos\frac{x_i}{\sqrt{i}}\right),\ \text{where } x_i \in [-600,600].$$

3.5 Solve the Ackley benchmark problem:[1]

$$\min_{x}\left[20+10^{-20}\exp\left(-0.2\sqrt{\frac{1}{p}\sum_{i=1}^{p}x_i^2}\right)-\exp\left(\frac{1}{p}\sum_{i=1}^{p}\cos 2\pi x_i\right)\right].$$

3.6 Solve the Kursawe benchmark problem:

$$J = \min_{\boldsymbol{x}} \sum_{i=1}^{p} |x_i|^{0.8} + 5\sin^3 x_i + 3.5828, \quad \text{where } p = 2 \text{ or } p = 20,$$

3.7 Find the global minimum solution of the extended Freudenstein–Roth function, with $n = 20$:

$$f(x) = \sum_{i=1}^{n/2} (-13 + x_{2i-1} + ((5 - x_{2i})x_{2i} - 2)x_{2i})^2 + (-29 + x_{2i-1} + ((x_{2i} + 1)x_{2i} - 14)x_{2i})^2.$$

The initial value can be selected as $\boldsymbol{x}_0 = [0.5, -2, \ldots, 0.5, -2]^{\mathrm{T}}$, or other possible values. The analytical solution is $\boldsymbol{x}^* = [5, 4, \ldots, 5, 4]^{\mathrm{T}}$, $f_{\text{opt}} = 0$. When the searching region is enlarged, what is the global minimum solution?

3.8 Solve the minimum problem for the extended trigonometric function problem for $n = 20$:

$$f(x) = \sum_{i=1}^{n} \left[\left(n - \sum_{j=1}^{n} \cos x_j \right) + i(1 - \cos x_i) - \sin x_i \right]^2.$$

The initial point can be selected as $\boldsymbol{x}_0 = [1/n, 1/n, \ldots, 1/n]^{\mathrm{T}}$, and the analytical solutions are $x_i = 0$, with $f_{\text{opt}} = 0$.

3.9 Solve the extended Rosenbrock function problem for $n = 20$:

$$f(x) = \sum_{i=1}^{n/2} 100(x_{2i} - x_{2i-1}^2)^2 + (1 - x_{2i-1})^2.$$

An initial point is selected as $\boldsymbol{x}_0 = [-1.2, 1, \ldots, -1.2, 1]$, and the analytical solution is $x_i = 1$, $f_{\text{opt}} = 0$. If $x_{2i} - x_{2i-1}^2$ is substituted to $x_{2i} - x_{2i-1}^3$, the problem is converted into an extended White–Holst problem. Solve the problem again.

3.10 Solve the extended Beale test function problem for $n = 20$:

$$f(x) = \sum_{i=1}^{n/2} (1.5 - x_{2i-1}(1 - x_{2i}))^2 + (2.25 - x_{2i-1}(1 - x_{2i}^2)^2)^2 + (2.625 - x_{2i-1}(1 - x_{2i}^3)^2)^2.$$

An initial value is selected as $\boldsymbol{x}_0 = [1, 0.8, \ldots, 1, 0.8]^{\mathrm{T}}$, and the analytical solution is unknown.

3.11 Solve the two Raydan problems for $n = 20$:

$$f_1(x) = \sum_{i=1}^{n} \frac{i}{10}(\mathrm{e}^{x_i} - x_i), \quad f_2(x) = \sum_{i=1}^{n} (\mathrm{e}^{x_i} - x_i).$$

An initial value is $\boldsymbol{x}_0 = [1, 1, \ldots, 1]^{\mathrm{T}}$; the analytical solution is $x_i = 1$, $f_{1\text{opt}} = \sum_{i=1}^{n} i/10$, $f_{2\text{opt}} = n$.

3.12 Solve the extended Miele–Cantrell function problem for $n = 20$:

$$f(x) = \sum_{i=1}^{n/4} (e^{4i-3} - x_{4i-2})^2 + 100(x_{4i-2} - x_{4i-1})^6 + \tan^4(x_{4i-1} - x_{4i}) + x_{4i-3}^8.$$

An initial value is $\boldsymbol{x}_0 = [1, 2, 2, 2, \ldots, 1, 2, 2, 2]^\mathrm{T}$; the analytical solution is $\boldsymbol{x}^* = [0, 1, 1, 1, \ldots, 0, 1, 1, 1]^\mathrm{T}$, $f_\mathrm{opt} = 0$.

3.13 Many of the above exercises assume that $n = 20$. If $n = 200$, solve again the above exercises.

3.14 Find the global minimum value of Hartmann function,

$$f(\boldsymbol{x}) = -\sum_{i=1}^{4} \alpha_i \exp\left[-\sum_{j=1}^{3} a_{ij}(x_j - p_{ij})^2\right],$$

where $\boldsymbol{\alpha} = [1, 1, 2, 3, 3, 2]^\mathrm{T}$ and

$$\boldsymbol{A} = \begin{bmatrix} 3 & 10 & 30 \\ 0.1 & 10 & 35 \\ 3 & 10 & 30 \\ 0.1 & 10 & 35 \end{bmatrix}, \quad \boldsymbol{P} = 10^{-4} \times \begin{bmatrix} 3\,689 & 1\,170 & 2\,673 \\ 4\,699 & 4\,387 & 7\,470 \\ 1\,091 & 8\,732 & 5\,547 \\ 381.5 & 5\,743 & 8\,828 \end{bmatrix}.$$

3.15 Find the global minimum of the Schwefel function, with $n = 20$:

$$f(\boldsymbol{x}) = 418.9829n - \sum_{i=1}^{n} x_i \sin\sqrt{|x_i|}.$$

The searching interval is $-500 \leqslant x_i \leqslant 500$, with the analytical solution of $x_i = 1$, and optimum objective function being $f_\mathrm{opt} = 0$. Validate the result graphically for $n = 2$.

3.16 Find the global minimum value of the Eggholder function, and validate graphically the result if

$$f(x, y) = -(y + 47)\sin\sqrt{\left|\frac{x}{2} + (y + 47)\right|} - x\sin\sqrt{|x - (y + 47)|}.$$

3.17 Find the minimum value of the following function in the interval $-2 \leqslant x \leqslant 11$, and validate the results graphically:

$$f(x) = x^6 - \frac{52}{25}x^5 + \frac{39}{80}x^4 + \frac{71}{10}x^3 - \frac{79}{20}x^2 - x + \frac{1}{10}.$$

3.18 Find the maximum value of the function $f(x) = x\sin 10\pi x + 2$ in the intervals $(-1, 2)$ and $(-10, 20)$, respectively, and validate the results using graphical methods.

3.19 For $-10 \leqslant x, y \leqslant 10$, find the global minimum function value and show graphically the solution if

$$f(x,y) = \sin^2 3\pi x + (x-1)^2(1+\sin^2 3\pi y) + (y-1)^2(1+\sin^2 2\pi y).$$

3.20 The function in Example 3.20 has many peaks and valleys. The function values of the valleys are different. Use the graphical method to find the global optimum valley. (Hint: use front and side views to find the values of x_1 and x_2).

3.21 If $0 \leqslant x, y \leqslant 1$, find the global maximum value of the following function, and validate the result using graphical methods:

$$f(x,y) = \sin 19\pi x + \frac{x}{1.7} + \sin 19\pi y + \frac{y}{1.7} + 2.$$

3.22 Find the global minimum function values within $-100 \leqslant x, y \leqslant 100$ for

$$f(x,y) = 0.5 + \frac{\cos^2(\sin|x^2-y^2|) - 0.5}{(1+0.001(x^2+y^2))^2}.$$

3.23 Assume that a set of data is given in Table 3.6. It is known that the data comes from the following prototype function:

$$y(x) = \frac{1}{\sqrt{2\pi}\sigma} \mathrm{e}^{-(x-\mu)^2/2\sigma^2}.$$

Use the least squares method to determine the values of μ and σ. Draw the function obtained and observe the fitting quality.

Table 3.6: Data for Exercise 3.23.

x_i	−2	−1.7	−1.4	−1.1	−0.8	−0.5	−0.2	0.1	0.4	0.7	1	1.3
y_i	0.1029	0.1174	0.1316	0.1448	0.1566	0.1662	0.1733	0.1775	0.1785	0.1764	0.1711	0.1630
x_i	1.6	1.9	2.2	2.5	2.8	3.1	3.4	3.7	4	4.3	4.6	4.9
y_i	0.1526	0.1402	0.1266	0.1122	0.0977	0.0835	0.0702	0.0577	0.0469	0.0373	0.0291	0.0224

3.24 Assume that the measured data are provided in Table 3.7. Find the minimum solution of (x, y) in the region $(0.1, 0.1) \sim (1.1, 1.1)$ via the interpolation method.

3.25 Assume that in a chemical process, the relationship between the pressure P and temperature T satisfies the condition $P = \alpha \mathrm{e}^{\beta T}$. The parameters α and β are unknown. From the experimental data in Table 3.8, compute the undetermined coefficients α and β.[6]

3.26 Assume that a set of data is supplied in the data file data3ex5.dat. The first column contains the information of scattered x, the second column is for y, and the third for the function values. Compute the undetermined coefficients $a \sim e$ from the following prototype function:

$$f(x,y) = (ax^2 - bx)\mathrm{e}^{-cx^2-dy^2-exy}.$$

Table 3.7: Data for Exercise 3.24.

y_i	x_1 0.1	x_2 0.2	x_3 0.3	x_4 0.4	x_5 0.5	x_6 0.6	x_7 0.7	x_8 0.8	x_9 0.9	x_{10} 1	x_{11} 1.1
0.1	0.8304	0.8273	0.8241	0.8210	0.8182	0.8161	0.8148	0.8146	0.8158	0.8185	0.8230
0.2	0.8317	0.8325	0.8358	0.8420	0.8513	0.8638	0.8798	0.8994	0.9226	0.9496	0.9801
0.3	0.8359	0.8435	0.8563	0.8747	0.8987	0.9284	0.9638	1.0045	1.0502	1.1000	1.1529
0.4	0.8429	0.8601	0.8854	0.9187	0.9599	1.0086	1.0642	1.1253	1.1904	1.2570	1.3222
0.5	0.8527	0.8825	0.9229	0.9735	1.0336	1.1019	1.1764	1.2540	1.3308	1.4017	1.4605
0.6	0.8653	0.9105	0.9685	1.0383	1.118	1.2046	1.2937	1.3793	1.4539	1.5086	1.5335
0.7	0.8808	0.9440	1.0217	1.1118	1.2102	1.3110	1.4063	1.4859	1.5377	1.5484	1.5052
0.8	0.8990	0.9828	1.0820	1.1922	1.3061	1.4138	1.5021	1.5555	1.5573	1.4915	1.346
0.9	0.9201	1.0266	1.1482	1.2768	1.4005	1.5034	1.5661	1.5678	1.4889	1.3156	1.0454
1.0	0.9438	1.0752	1.2191	1.3624	1.4866	1.5684	1.5821	1.5032	1.3150	1.0155	0.6248
1.1	0.9702	1.1279	1.2929	1.4448	1.5564	1.5964	1.5341	1.3473	1.0321	0.6127	0.1476

Table 3.8: Experimental data in Exercise 3.25.

temperature T (°C)	20	25	30	35	40	50	60	70
pressure P(mmHg)	15.45	19.23	26.54	34.52	48.32	68.11	98.34	120.45

3.27 For the differential equations $x_1'(t) = x_2(t)$, $x_2'(t) = 2x_1(t)x_2(t)$, and initial boundary values $x_1(0) = -1$, $x_1(\pi/2) = 1$, find the numerical solution to the differential equations.

3.28 For a given ordinary differential equation given below, find parameters α and β, and solve the differential equation:

$$x_1' = 4x_1 - \alpha x_1 x_2, \quad x_2' = -2x_2 + \beta x_1 x_2$$

with known conditions $x_1(0) = 2$, $x_2(0) = 1$, $x_1(3) = 4$, and $x_2(3) = 2$.

3.29 In Example 3.32, an algebraic equation solution problem was converted into an optimization problem. Run the function 100 times, also run the `fsolve()` solver 100 times. Compare the efficiency and accuracy of the two methods.

3.30 Convert the equation solution problems in Exercises 2.12 and 2.14 into optimization problems, and compare the accuracy and speed of equation solvers.

4 Linear and quadratic programming

In contrast to the unconstrained optimization problems, from this chapter on, constrained optimization problems are studied. Mathematical and physical explanation of constrained optimization problems are introduced first, then we concentrate on the linear and quadratic programming problems. In the subsequent chapters, nonlinear and other programming problems are introduced.

Definition 4.1. The general mathematical description of a constrained optimization problem is

$$\min_{\boldsymbol{x} \text{ s.t. } \boldsymbol{g}(\boldsymbol{x}) \leqslant \boldsymbol{0}} f(\boldsymbol{x}), \tag{4.0.1}$$

where $\boldsymbol{x} = [x_1, x_2, \dots, x_n]^{\mathrm{T}}$. The inequalities $\boldsymbol{g}(\boldsymbol{x}) \leqslant 0$ are referred to as constraints.

Definition 4.2. All the solutions $\boldsymbol{x}$ satisfying the constraints $\boldsymbol{g}(\boldsymbol{x}) \leqslant \boldsymbol{0}$ are referred to as the feasible solutions of the optimization problem.

The physical interpretation of the mathematical model of an constrained optimization problem is as follows: we strive to find the decision vector $\boldsymbol{x}$, satisfying all the constraints $\boldsymbol{g}(\boldsymbol{x}) \leqslant 0$, such that the objective function $f(\boldsymbol{x})$ is minimized. In real applications, the constraints may be complicated. They can be equalities or inequalities, linear or nonlinear. Sometimes they may not be described with mathematical formulas.

It can be seen from the constraints that there seem to be limitations in $\boldsymbol{g}(\boldsymbol{x}) \leqslant \boldsymbol{0}$, since sometimes the constraints may include $\geqslant$ inequalities. In fact, if there are $\geqslant$ inequalities, both sides of the inequalities can be multiplied by -1, such that they can be converted to the standard $\leqslant$ inequalities.

Constrained optimization problems are usually referred to as mathematical programming problems. Mathematical programming problems may further be classified as linear, nonlinear, integer programming, and so on.

In Section 4.1, linear programming problems are introduced, and their mathematical descriptions are formulated. Through simple examples, the most important simplex method and graphical solutions are addressed. In Section 4.2, a dedicated solver provided in MATLAB is illustrated for various linear programming problems. A demonstration is made to show linear programming application in transportation problems. In Section 4.3, a problem-based description of linear programming problems supported in MATLAB is introduced. Direct description and solution of various linear programming problems are illustrated. In Section 4.4, the mathematical modeling and solution of quadratic programming problems are proposed. In Section 4.5, the linear matrix inequality method is presented, and linear programming problem solutions with linear matrix inequalities are proposed.

https://doi.org/10.1515/9783110667011-004

4.1 An introduction to linear programming

A linear programming (LP) problem is the most widely encountered constrained programming problem. In the linear programming problems, the objective function and constraints can all be expressed as linear expressions of the decision variables in $\boldsymbol{x}$.

A comprehensive historical review of linear programming can be found in [40]. The simplex method in linear programming was proposed by American scholar George Bernard Dantzig (1914–2005) in 1947. In 1953, a revised simplex method was proposed. The simplex method was selected as one of the "Top 10 Algorithms of the Century". It provided a major mathematical tool for the work of several Nobel prize winners, for instance, for the 1973 Nobel Economics Prize winner Wassily Wassilyevich Leontief's (1905–1999) input–output model. The Soviet Union scholar Leonid Vitaliyevich Kantorovich (1912–1986) and American scholar Tjalling Charles Koopmans (1910–1985) were the Nobel Economics Prize winners in 1975, for their contribution to the resource allocation theory based on linear programming techniques. Linear programming algorithms are often of high efficiency. Reference [7] published in 1992 reported linear programming solutions with 12 753 313 decision variables.

In this section, a mathematical model of linear programming problems is proposed. A graphical interpretation of linear programming problems with two decision variables is given. A simple example is given to show the simplex method in solving standard linear programming problems. Also a conversion demonstration is given to transform linear programming problems into standard linear programming problems.

4.1.1 Mathematical model of linear programming problems

In Definition 4.1, if the objective function and constraints are linear expressions of the decision variable $\boldsymbol{x}$, the optimization problem becomes a linear programming problem. In this section, the definition of linear programming is mathematically introduced, then demonstrations are given to show solutions of linear programming problems.

Definition 4.3. The standard form of a linear programming problem is

$$\min_{\boldsymbol{x} \text{ s.t. } \begin{cases} \boldsymbol{A}\boldsymbol{x}\leqslant\boldsymbol{B} \\ \boldsymbol{A}_{\rm eq}\boldsymbol{x}=\boldsymbol{B}_{\rm eq} \\ \boldsymbol{x}_{\rm m}\leqslant\boldsymbol{x}\leqslant\boldsymbol{x}_{\rm M} \end{cases}} \boldsymbol{f}^{\rm T}\boldsymbol{x}. \tag{4.1.1}$$

Here the constraints are classified into linear equality constraints $\boldsymbol{A}_{\rm eq}\boldsymbol{x} = \boldsymbol{B}_{\rm eq}$, linear inequality constraints $\boldsymbol{A}\boldsymbol{x} \leqslant \boldsymbol{B}$, as well as decision variable upper and lower bounds, $\boldsymbol{x}_{\rm M}$ and $\boldsymbol{x}_{\rm m}$, such that $\boldsymbol{x}_{\rm m} \leqslant \boldsymbol{x} \leqslant \boldsymbol{x}_{\rm M}$.

For inequality constraints, the standard form in MATLAB is the $\leqslant$ inequalities. If a certain inequality is a $\geqslant$ one, both sides should multiplied by −1 to convert it into a $\leqslant$ one. In this book, the inequalities are always converted to $\leqslant$ ones.

4.1.2 Graphical solutions of linear programming problems

A linear programming problem with two decision variables can be used in a graphical demonstration. In this section, the feasible region of linear programming problem is demonstrated first, then the concept of a convex set is presented, and it is shown that linear programming is a convex problem. The global optimum solution of a linear programming problem is located among the vertices of the convex feasible region.

Example 4.1. Draw the feasible region for the following problem with graphical methods:

$$\min_{x,y \text{ s.t. } \begin{cases} -x+2y\leqslant 4 \\ 3x+2y\leqslant 12 \\ x\geqslant 0,\ y\geqslant 0 \end{cases}} -x-2y.$$

Solutions. Two straight lines for the constraints $-x+2y = 4$ and $3x+2y = 12$ are drawn first in the x–y plane. The two lines and axes $x = 0$ and $y = 0$ enclose the feasible region, as shown in Figure 4.1.

```
>> syms x; fplot([2+x/2, 6-3*x/2],[-1,5]), axis([-1 5 -1 5])
   line([-1,5],[0,0]), line([0,0],[-1,5])
   hold on; fill([0,0,2,4,0],[0,2,3,0,0],'g'), hold off
```

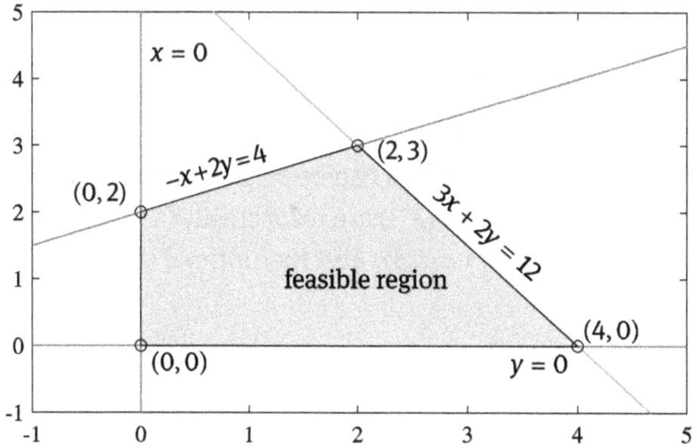

Figure 4.1: Feasible region.

Definition 4.4. A convex set is a geometric concept. If every point on the segment joining any two points a and b from the set lies in it, the set is referred to as a convex set. Otherwise it is a nonconvex set.

The graphical interpretation of convex and nonconvex sets is given in Figure 4.2. In Figure 4.2(a), all the points on the segment between any two points lie in the same set, so the set is convex; while in Figure 4.2(b), there is a hole in the set. Two points can be found such that some points on the joining line segment are not in the set. Therefore, it is not a convex set. In Figure 4.2(c), since there exists a concave part, two points can be found such that some points on the joining line are not in the set. Therefore, it is a nonconvex set.

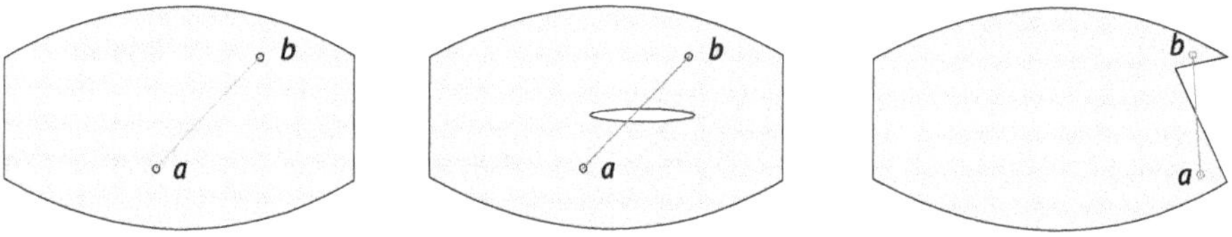

(a) convex set **(b)** nonconvex set (with a hole) **(c)** nonconvex set

Figure 4.2: Demonstration of convex and nonconvex sets.

Theorem 4.1. *If there exist feasible solutions in a linear programming problem, the feasible region is a convex set.*

Theorem 4.2. *If the feasible region is bounded, the global optimum solution of a linear programming problem lies on a vertex of the feasible region.*[51]

With the theoretic explanation, if the vertices of the feasible region can be found, and the number of vertices is bounded, the objective function values at the vertices can be compared, such that the linear programming problem can be solved.

Example 4.2. Solve the linear programming problem in Example 4.1.

Solutions. For the convex feasible set in Figure 4.1, if the objective function values at the vertices are found, the global optimum solution can be obtained. The objective function values at vertices $(0,0)$, $(4,0)$, $(0,2)$, and $(2,3)$ are respectively 0, -4, -4, and -8. It can be seen that the optimum solution is $(2,3)$, and the optimal value of the objective function is -8.

4.1.3 Introduction to the simplex method

It has been shown that the global optimum solution of a linear programming problem is the smallest value of objective function at the vertices of the feasible region. For linear programming problems with two decision variables, the graphical method can be used to find all the vertices and compare the objective function values at these points to find the solutions. For multivariate linear programming problems, it is difficult to find all the vertices. A suitable algorithm is expected for solving the problems.

The original form of a linear programming problem is the standard linear programming problem. In this section, the standard form is presented, then the conversion method is formulated. Finally, the commonly used simplex method is demonstrated with an example.

Definition 4.5. The mathematical form of a standard linear programming problem is

$$\min \quad f_1x_1 + f_2x_2 + \cdots + f_nx_n. \quad \boldsymbol{x}\ \text{s.t.}\ \begin{cases} a_{11}x_1+a_{12}x_2+\cdots+a_{1n}x_n=b_1 \\ a_{21}x_1+a_{22}x_2+\cdots+a_{2n}x_n=b_2 \\ \vdots \\ a_{m1}x_1+a_{m2}x_2+\cdots+a_{mn}x_n=b_m \\ x_j\geqslant 0,\ j=1,2,\ldots,n \end{cases} \tag{4.1.2}$$

From the given standard linear programming problem, based on the linear equalities, several decision variables can be eliminated. The optimum solution of the linear programming problem can be found.

Example 4.3. Solve the standard linear programming problem[40]

$$\min \quad x_1 + 2x_2 - x_4 + x_5. \quad \boldsymbol{x}\ \text{s.t.}\ \begin{cases} 2x_1+x_2+3x_3+2x_4=5 \\ x_1-x_2+2x_3-x_4+3x_5=1 \\ x_1-2x_3-2x_5=-1 \\ x_i\geqslant 0,\ i=1,2,\ldots,5 \end{cases}$$

Solutions. If the constraints are written in matrix form, it is found that $\boldsymbol{Ax} = \boldsymbol{B}$, where

$$\boldsymbol{A} = \begin{bmatrix} 2 & 1 & 3 & 2 & 0 \\ 1 & -1 & 2 & -1 & 3 \\ 1 & 0 & -2 & 0 & -2 \end{bmatrix}, \quad \boldsymbol{B} = \begin{bmatrix} 5 \\ 1 \\ -1 \end{bmatrix}.$$

Now the matrices $\boldsymbol{A}$ and $\boldsymbol{B}$ can be input into the computer. Matrix $\boldsymbol{C}$ can be constructed from $\boldsymbol{C} = [\boldsymbol{A}, \boldsymbol{B}]$. With MATLAB, the reduced row echelon form of $\boldsymbol{C}$ can be found

```
>> A=[2,1,3,2,0; 1,-1,2,-1,3; 1,0,-2,0,-2];
   B=[5; 1; -1]; C=[A,B]; D=rref(sym(C))
```

With function `rref()`, the standard form of $\boldsymbol{C}$ can be found as

$$\boldsymbol{D} = \begin{bmatrix} 1 & 0 & 0 & 2/11 & -4/11 & 7/11 \\ 0 & 1 & 0 & 15/11 & -19/11 & 14/11 \\ 0 & 0 & 1 & 1/11 & 9/11 & 9/11 \end{bmatrix}.$$

Letting $x_4 = x_5 = 0$, a feasible solution of the problem can be found $\boldsymbol{x}_1 = [7/11, 14/11, 9/11, 0, 0]$, with the objective function value of $f_1 = 35/11$.

With $\boldsymbol{D}$ matrix, it can be found that $x_1 = -2x_4/11 + 4x_5/11 + 7/11$, $x_2 = -15x_4/11 + 19x_5/11 + 14/11$, $x_3 = -x_4/11 - 9x_5/11 + 9/11$. The objective function can be written as

$$f = -\frac{2}{11}x_4 + \frac{4}{11}x_5 + \frac{7}{11} - \frac{30}{11}x_4 + \frac{38}{11}x_5 + \frac{28}{11} - x_4 + x_5 = \frac{35}{11} - \frac{43}{11}x_4 + \frac{53}{11}x_5.$$

The first simplex tableau can be established, as shown in Table 4.1. In fact, it can be seen from the f expression that, in order to reduce the objective function value, x_5 should be made as small as possible, and x_4 as large as possible. Therefore, letting $x_5 = 0$, the formulas for x_1, x_2 and x_3 can be found, by solving the inequalities:

$$\begin{cases} x_1 = -2x_4/11 + 7/11 \geqslant 0, \\ x_2 = -15x_4/11 + 14/11 \geqslant 0, \\ x_3 = -x_4/11 + 9/11 \geqslant 0, \end{cases} \Rightarrow \begin{cases} x_4 \leqslant 7/2, \\ x_4 \leqslant 14/15, \\ x_4 \leqslant 9. \end{cases}$$

Table 4.1: Simplex tableau (I).

x_1	x_2	x_3	x_4	x_5	f	right side
1	0	0	2/11	-4/11		7/11
0	1	0	15/11	-19/11		14/11
0	0	1	1/11	9/11		9/11
0	0	0	-43/11	53/11	-1	-35/11

It is found that $x_4 = \min(7/2, 14/15, 9) = 14/15$. An improved feasible solution can be found as $\boldsymbol{x}_2 = [7/15, 0, 11/15, 14/15, 0]^{\mathrm{T}}$. The value of the objective function is $f_2 = -7/15$. Of course, this one is better than f_1 obtained above.

If x_1, x_4, and x_3 are selected as pivots, the columns in $\boldsymbol{C}$ can be swapped, then, with function `rref()`, the reduced row echelon form can be found, and they can be swapped back.

```
>> C1=C(:,[1,4,3,2,5,6]); D1=rref(C1); D1=D1(:,[1,4,3,2,5,6])
```

The obtained reduced row echelon form becomes

$$\boldsymbol{D}_1 = \begin{bmatrix} 1 & -2/15 & 0 & 0 & -2/15 & 7/15 \\ 0 & 11/15 & 0 & 1 & -19/15 & 14/15 \\ 0 & -1/15 & 1 & 0 & 14/15 & 11/15 \end{bmatrix}.$$

Considering the value of the objective function, the new simplex tableau is shown in Table 4.2. Letting $x_2 = 0$, it is found that

Table 4.2: Simplex tableau (II).

x_1	x_2	x_3	x_4	x_5	f	right side
1	−2/15	0	0	−2/15		7/15
0	11/15	0	1	−19/15		14/15
0	−1/15	1	0	14/15		11/15
0	43/15	0	0	2/15	−1	7/15

$$\begin{cases} x_1 = 7/15 + 2x_5/15 \geqslant 0, \\ x_4 = 14/15 + 19x_5/15 \geqslant 0, \\ x_3 = 11/15 - 14x_5/15 \geqslant 0, \end{cases} \Rightarrow \begin{cases} x_5 \geqslant -7/2, \\ x_5 \geqslant -19/14, \\ x_5 \leqslant 11/14. \end{cases}$$

We have $x_5 = 11/14$. Substituting it back to the above equations, a better feasible solution can be found at $\boldsymbol{x}_3 = [4/7, 0, 0, 27/14, 11/14]^{\mathrm{T}}$, with the objective function of $f_3 = -4/7$, which is smaller than f_2.

Selecting x_1, x_4, and x_5 as pivots, with the following statements:

```
>> C2=C(:,[1,4,5,2,3,6]); D2=rref(C2); D2=D2(:,[1,4,5,2,3,6])
```

the matrix obtained is

$$\boldsymbol{D}_2 = \begin{bmatrix} 1 & -1/7 & 1/7 & 0 & 0 & 4/7 \\ 0 & 9/14 & 19/14 & 1 & 0 & 27/14 \\ 0 & -1/14 & 15/14 & 0 & 1 & 11/14 \end{bmatrix}.$$

With the objective function value, a new simplex tableau shown in Table 4.3 can be obtained. It can be seen that the optimum solution found is $\boldsymbol{x} = [4/7, 0, 0, 27/14, 11/14]^{\mathrm{T}}$, with the value of the objective function $f = -4/7$.

Table 4.3: Simplex tableau (III).

x_1	x_2	x_3	x_4	x_5	f	right side
1	−1/7	1/7	0	0		4/7
0	9/14	19/14	1	0		27/14
0	−1/14	15/14	0	1		11/14
0	20/7	1/7	0	0	−1	4/7

With the solution pattern of the standard linear programming problem, if linear inequality constraints are involved, some relaxation variables can be introduced, such that the problem can be converted into the standard linear programming problem. The simplex method can be used to solve the new problem. An example is given below to show the use of the relaxation variables, and the solutions of linear programming problems with linear inequality constraints.

Example 4.4. Consider the problem in Example 4.3. If the first two constraints are changed into the ⩽ inequalities, then

$$\min \quad x_1 + 2x_2 - x_4 + x_5.$$
$$\boldsymbol{x} \text{ s.t. } \begin{cases} 2x_1+x_2+3x_3+2x_4 \leqslant 5 \\ x_1-x_2+2x_3-x_4+3x_5 \leqslant 1 \\ x_1-2x_3-2x_5=-1 \\ x_i \geqslant 0,\ i=1,2,\dots,5 \end{cases}$$

Convert the problem into a standard linear programming problem.

Solutions. Since there are two inequality constraints, two relaxation variables $x_6 \geqslant 0$ and $x_7 \geqslant 0$ should be introduced. Then, $2x_1 + x_2 + 3x_3 + 2x_4 \leqslant 5$ can be converted into $2x_1 + x_2 + 3x_3 + 2x_4 + x_6 = 5$. Similar action should be made to the next inequality constraint. The original problem can then be converted into a standard linear programming problem:

$$\min \quad x_1 + 2x_2 - x_4 + x_5.$$
$$\boldsymbol{x} \text{ s.t. } \begin{cases} 2x_1+x_2+3x_3+2x_4+x_6=5 \\ x_1-x_2+2x_3-x_4+3x_5+x_7=1 \\ x_1-2x_3-2x_5=-1 \\ x_i \geqslant 0,\ i=1,2,\dots,7 \end{cases}$$

4.2 Direct solutions of linear programming problems

Although there exist strict mathematical formulas in the simplex method, it is still rather complicated to solve a multivariate linear programming problem using low-level commands. Computer solvers should be used to replace manual work, to find the solutions of linear programming problems directly. In this section, a solver provided in the Optimization Toolbox in MATLAB is demonstrated for solving linear programming problems directly.

4.2.1 A linear programming problem solver

Linear programming is the simplest, while most commonly used, mathematical programming problem. There are many ways to solve linear programming problems. The simplex method and its variations are the most effective. In Optimization Toolbox of MATLAB, a dedicated solver for linear programming problems, `linprog()`, is provided with the following syntaxes:

`[`$\boldsymbol{x}$`,`$f_{\rm opt}$`,flag,out]=linprog(`$\boldsymbol{f}$`,`$\boldsymbol{A}$`,`$\boldsymbol{B}$`)`

`[`$\boldsymbol{x}$`,`$f_{\rm opt}$`,flag,out]=linprog(`$\boldsymbol{f}$`,`$\boldsymbol{A}$`,`$\boldsymbol{B}$`,`$\boldsymbol{A}_{\rm eq}$`,`$\boldsymbol{B}_{\rm eq}$`)`

`[`$\boldsymbol{x}$`,`$f_{\rm opt}$`,flag,out]=linprog(`$\boldsymbol{f}$`,`$\boldsymbol{A}$`,`$\boldsymbol{B}$`,`$\boldsymbol{A}_{\rm eq}$`,`$\boldsymbol{B}_{\rm eq}$`,`$\boldsymbol{x}_{\rm m}$`,`$\boldsymbol{x}_{\rm M}$`)`

[$\boldsymbol{x}$,$f_{\rm opt}$,flag,out]=linprog($\boldsymbol{f}$,$\boldsymbol{A}$,$\boldsymbol{B}$,$\boldsymbol{A}_{\rm eq}$,$\boldsymbol{B}_{\rm eq}$,$\boldsymbol{x}_{\rm m}$,$\boldsymbol{x}_{\rm M}$,options)

[$\boldsymbol{x}$,$f_{\rm opt}$,flag,out]=linprog(problem)

where $\boldsymbol{f}$, $\boldsymbol{A}$, $\boldsymbol{B}$, $\boldsymbol{A}_{\rm eq}$, $\boldsymbol{B}_{\rm eq}$, $\boldsymbol{x}_{\rm m}$, and $\boldsymbol{x}_{\rm M}$ are the same as those in Definition 4.1. If any of the constraints does not exist, empty matrices `[]` should be used to hold the places. The argument `options` is the control option. After the optimization process, the results are returned in vector $\boldsymbol{x}$. The objective function value is returned in argument $f_{\rm opt}$. Since linear programming problems are convex, if a solution can be found, it is the global optimum solution. There is no need to assign initial search points.

It can be seen from the mathematical expression that $\boldsymbol{f}$ should be a column vector. While in practice, fault tolerance manipulation is carried out in MATLAB, such that if $\boldsymbol{f}$ is by mistake given as a row vector, the internal facilities in the solver will change it automatically into a column vector. There is no impact on the solution process and results.

In the control option parameter in `Algorithm`, the actual algorithm can be chosen to be the default `'dual-simplex'`, or taken as `'interior-point'` or `'interior-point-legacy'`. Suitable algorithms should be selected.

Function `linprog()` allows using structured variable `problem` to describe the linear programming problem. All the members in the structured variable are listed in Table 4.4.

Table 4.4: Members in linear programming structured variable.

member names	explanation of members
`f`	coefficient vector of objective function $\boldsymbol{f}$
`Aineq, bineq`	linear inequality constraints in matrix form, $\boldsymbol{A}$ and $\boldsymbol{b}$
`Aeq, beq`	linear equality constraints in matrix form, $\boldsymbol{A}_{\rm eq}$ and $\boldsymbol{b}_{\rm eq}$. If some of the constraints are missing, they can be set to empty matrices
`ub,lb`	upper and lower bounds of decision variables, $\boldsymbol{x}_{\rm M}$ and $\boldsymbol{x}_{\rm m}$
`options`	control options. As in earlier examples, the user may assign the control options, then assign them to the `options` member
`solver`	must be assigned as `'linprog'`

Examples are given here to demonstrate the solution methods of linear programming problems.

Example 4.5. Consider the standard linear programming problem in Example 4.3, and the linear programming problem in Example 4.4. Solve the problems directly with `linprog()` function.

Solutions. It can be seen from Example 4.3 that the objective function coefficient vector is $\boldsymbol{f} = [1, 2, 0, -1, 1]^{\rm T}$, and the matrices in the equality constraints are

$$A_{eq} = \begin{bmatrix} 2 & 1 & 3 & 2 & 0 \\ 1 & -1 & 2 & -1 & 3 \\ 1 & 0 & -2 & 0 & -2 \end{bmatrix}, \quad B_{eq} = \begin{bmatrix} 5 \\ 1 \\ -1 \end{bmatrix}.$$

Since there are no inequality constraints, $\boldsymbol{A}$ and $\boldsymbol{B}$ can be set to empty matrices. Besides the lower bounds can be described by a zero vector. The following commands can be used to solve the linear programming problem directly:

```
>> A=[]; B=[]; f=[1,2,0,-1,1];
   Aeq=[2 1 3 2 0; 1 -1 2 -1 3; 1 0 -2 0 -2]; Beq=[5;1;-1];
   xm=zeros(5,1); [x f0 flag]=linprog(f,A,B,Aeq,Beq,xm)
   err=norm(x-[4/7,0,0,27/14,11/14]'), f0=f0+4/7
```

If the symbolic form is used to express the solutions, then $\boldsymbol{x} = [4/7, 0, 0, 27/14, 11/14]^{\mathrm{T}}$, and the objective function is $f_0 = -4/7$. It can be seen that the result obtained is the same as that derived in Example 4.3. The solution method here is simple and straightforward. The norm of the error is 3.3307×10^{-16}. It can be seen that under the default setting, the accuracy of the solver is very high, so that it can be used to find the solutions reliably.

If the other two algorithms are selected, the errors obtained are respectively $e_1 = 7.0655 \times 10^{-9}$ and $e_2 = 1.2472 \times 10^{-8}$, with the precision lower than for the default algorithm. Therefore, if there is no special need, it is not recommended to select other algorithms. The default one may lead to ideal results.

```
>> ff=optimset; ff.Algorithm='interior-point';
   [x f0 flag]=linprog(f,A,B,Aeq,Beq,xm,[],ff)
   e1=norm(x-[4/7,0,0,27/14,11/14]'), f1=f0+4/7
   ff.Algorithm='interior-point-legacy';
   [x f0 flag]=linprog(f,A,B,Aeq,Beq,xm,[],ff)
   e2=norm(x-[4/7,0,0,27/14,11/14]'), f2=f0+4/7
```

Now let us consider the inequality problem in Example 4.4. In the original example, relaxation variables were introduced, so that it can be converted into a standard linear programming problem. Then, the simplex method can be used to solve it. This method is rather complicated. The solver `linprog()` can be used to solve the problem directly. From the new linear programming problem, it is not hard to write down the matrix form of the linear equality and inequality conditions:

$$A = \begin{bmatrix} 2 & 1 & 3 & 2 & 0 \\ 1 & -1 & 2 & -1 & 3 \end{bmatrix}, \quad B = \begin{bmatrix} 5 \\ 1 \end{bmatrix},$$

$$A_{eq} = \begin{bmatrix} 1 & 0 & -2 & 0 & -2 \end{bmatrix}, \quad B_{eq} = [-1].$$

Now the following commands can be used to solve the original problem. The result obtained is $\boldsymbol{x} = [0, 0, 0, 5/2, 1/2]$, and the objective function is $f_0 = -2$. It can be seen that if the existing solver in MATLAB is used, there is no need to consider trivial things. The optimal solution of the original problem can be found directly.

```
>> A=Aeq(1:2,:); B=Beq(1:2); Aeq=Aeq(3,:); Beq=Beq(3);
   [x f0 flag]=linprog(f,A,B,Aeq,Beq,xm), x=sym(x), f0=sym(f0)
```

Example 4.6. Solve the following linear programming problem:

$$\min \quad -2x_1 - x_2 - 4x_3 - 3x_4 - x_5.$$
$$\boldsymbol{x} \text{ s.t. } \begin{cases} 2x_2+x_3+4x_4+2x_5\leqslant 54 \\ 3x_1+4x_2+5x_3-x_4-x_5\leqslant 62 \\ x_1,x_2\geqslant 0, x_3\geqslant 3.32, x_4\geqslant 0.678, x_5\geqslant 2.57 \end{cases}$$

Solutions. From the given mathematical formula, we see that the problem is not given standard form of a linear programming problem, and that the objective function can be expressed as a coefficient vector $\boldsymbol{f} = [-2,-1,-4,-3,-1]^{\mathrm{T}}$. There are two inequality constraints, so we can set up two matrices:

$$\boldsymbol{A} = \begin{bmatrix} 0 & 2 & 1 & 4 & 2 \\ 3 & 4 & 5 & -1 & -1 \end{bmatrix}, \quad \boldsymbol{B} = \begin{bmatrix} 54 \\ 62 \end{bmatrix}.$$

Besides, since there are no equality constraints, $\boldsymbol{A}_{\mathrm{eq}}$ and $\boldsymbol{B}_{\mathrm{eq}}$ can be set as empty matrices. It is also seen from the mathematical form that the lower bounds of $\boldsymbol{x}$ can be expressed as $\boldsymbol{x}_{\mathrm{m}} = [0,0,3.32,0.678,2.57]^{\mathrm{T}}$. There are no upper bounds so they can also be set to an empty matrix. With the previous analysis, the following MATLAB commands can be used to solve the linear programming problem. The solution found is $\boldsymbol{x} = [19.785,0,3.32,11.385,2.57]^{\mathrm{T}}$, $f_{\mathrm{opt}} = -89.5750$:

```
>> f=-[2 1 4 3 1]'; A=[0 2 1 4 2; 3 4 5 -1 -1]; B=[54; 62]; Ae=[];
   Be=[]; xm=[0,0,3.32,0.678,2.57];  %input the matrices and vectors
   ff=optimset; ff.TolX=eps;          %control parameters
   [x,f_opt,key,c]=linprog(f,A,B,Ae,Be,xm,[],[],ff) %solve problem
```

It can be seen from the result that, since `key` is 1, the solution process is successful. Only with two iterations, the linear programming problem is solved. It can be seen that the solver is rather powerful, and the linear programming problem can be solved easily.

Example 4.7. Describe the linear programming problem in Example 4.6 with a structured variable, and solve the problem again.

Solutions. The following structured variable `P` can be used to describe the linear programming problem. Some of the members need not to be assigned, and it is not necessary to set them as empty matrices. For instance, `Aeq` member may not be declared. There is no need also to assign upper bounds.

```
>> P.f=-[2 1 4 3 1]';
   P.Aineq=[0 2 1 4 2; 3 4 5 -1 -1]; P.Bineq=[54; 62];
   P.lb=[0,0,3.32,0.678,2.57]; P.solver='linprog';
   ff=optimset; ff.TolX=eps; P.options=ff;      % control options
   [x,f_opt,key,c]=linprog(P) % structured variable and problem solutions
```

Example 4.8. Solve the linear programming problem with four decision variables:

$$\max_{\boldsymbol{x} \text{ s.t. } \begin{cases} x_1/4-60x_2-x_3/50+9x_4\leqslant 0 \\ -x_1/2+90x_2+x_3/50-3x_4\geqslant 0 \\ x_3\leqslant 1, x_1\geqslant -5, x_2\geqslant -5, x_3\geqslant -5, x_4\geqslant -5 \end{cases}} \quad 3x_1/4 - 150x_2 + x_3/50 - 6x_4.$$

Solutions. The original problem is a maximization problem. We need to convert it into a minimization problem, that is, multiply the objective function by −1, which is rewritten as $-3x_1/4 + 150x_2 - x_3/50 + 6x_4$. Besides, the constraints can be converted into the standard form:

$$\min_{\boldsymbol{x} \text{ s.t. } \begin{cases} x_1/4-60x_2-x_3/50+9x_4\leqslant 0 \\ x_1/2-90x_2-x_3/50+3x_4\leqslant 0 \\ x_3\leqslant 1, x_1\geqslant -5, x_2\geqslant -5, x_3\geqslant -5, x_4\geqslant -5 \end{cases}} \quad -3x_1/4 + 150x_2 - x_3/50 + 6x_4.$$

The objective function can be written as a vector $\boldsymbol{f}^{\mathrm{T}} = [-3/4, 150, -1/50, 6]$.

Now let us considering the bounds. It can be seen that the lower and upper bounds can be expressed in vector forms as

$$\boldsymbol{x}_{\mathrm{m}} = [-5, -5, -5, -5]^{\mathrm{T}}, \quad \boldsymbol{x}_{\mathrm{M}} = [+\infty, +\infty, 1, +\infty]^{\mathrm{T}},$$

where `Inf` can be used to express $+\infty$. In the constraints, the second one is expressed as $\geqslant$, so multiplying both sides by −1, the inequality can be converted to a $\leqslant$ inequality. Therefore, the constraints can be expressed as

$$\boldsymbol{A} = \begin{bmatrix} 1/4 & -60 & -1/50 & 9 \\ 1/2 & -90 & -1/50 & 3 \end{bmatrix}, \quad \boldsymbol{B} = \begin{bmatrix} 0 \\ 0 \end{bmatrix}.$$

Since there are no equality constraints, we set $\boldsymbol{A}_{\mathrm{eq}}$=`[]`, $\boldsymbol{B}_{\mathrm{eq}}$=`[]` to express empty matrices. Finally, the following commands can be used to solve the linear programming problem:

```
>> f=[-3/4,150,-1/50,6]; Aeq=[]; Beq=[];                 % objective function
   A=[1/4,-60,-1/50,9; 1/2,-90,-1/50,3]; B=[0;0]; % constraints
   xm=[-5;-5;-5;-5]; xM=[Inf;Inf;1;Inf];                  % bounds
   F=optimset; F.TolX=eps;                                % precision setting
   [x,f0,key,c]=linprog(f,A,B,Aeq,Beq,xm,xM,[0;0;0;0],F) % solve
```

It can be seen that with 10 iterations, the optimal solution can be found as $x = [-5, -0.1947, 1, -5]^{\mathrm{T}}$. The optimum solution is of high accuracy. The last statement can be replaced by the following statements, and the same results are obtained. Note that, since the `P` variable may exist in MATLAB workspace, `clear` command should be used to remove `P`, otherwise the members in `P` defined earlier may impact the solution process.

```
>> clear P; P.f=f; P.Aineq=A; P.Bineq=B; P.solver='linprog';
   P.lb=xm; P.ub=xM; P.options=F; linprog(P) %structured variable
```

Example 4.9. Solve the following linear programming problem:

$$\max \quad -22x_1 + 5x_2 - 7x_3 - 10x_4 + 8x_5 + 8x_6 - 9x_7.$$
$$x \text{ s.t. } \begin{cases} 3x_1-2x_3-2x_4+3x_7\leqslant 4\\ 2x_1+3x_2+x_3+3x_6+x_7\leqslant 1\\ 2x_1+4x_2-4x_3+2x_4-3x_5+2x_6+2x_7\leqslant 2\\ 2x_2-2x_3-3x_5+2x_6+2x_7\geqslant -4\\ x_2-2x_3-x_4+5x_6+x_7=2\\ 5x_1+x_2-x_3+x_4-5x_6-x_7\geqslant -3\\ 5x_1-3x_2+x_3+2x_4+3x_5+2x_6+x_7=-2\\ x_1-2x_2+2x_3-3x_4+x_5+6x_6+4x_7\leqslant 3\\ 3x_2-5x_3-x_4+3x_5+3x_6+3x_7\leqslant 2\\ x_i\geqslant -5,\ i=1,2,\dots,7 \end{cases}$$

Solutions. Since maximization is to be performed, the objective function should be multiplied by −1, to convert the problem into a minimization one. The coefficient vector is $\boldsymbol{f} = [22, -5, 7, 10, -8, -8, 9]$.

The fifth and seventh constraints are linear equalities, and the matrix and vector involved are

$$\boldsymbol{A}_{\text{eq}} = \begin{bmatrix} 0 & 1 & -2 & -1 & 0 & 5 & 1\\ 5 & -3 & 1 & 2 & 3 & 2 & 1 \end{bmatrix}, \quad \boldsymbol{b}_{\text{eq}} = \begin{bmatrix} 2\\ -2 \end{bmatrix}.$$

In the remaining 7 constraints, the fourth and sixth are $\geqslant$ inequalities, so we should multiply both sides by −1, so that they can be converted into $\leqslant$ inequalities. The matrix form of the constraints is

$$\boldsymbol{A} = \begin{bmatrix} 3 & 0 & -2 & -2 & 0 & 0 & 3\\ 2 & 3 & 1 & 0 & 0 & 3 & 1\\ 2 & 4 & -4 & 2 & -3 & 2 & 2\\ 0 & -2 & 2 & 0 & 3 & -2 & -2\\ -5 & -1 & 1 & -1 & 0 & 5 & 1\\ 1 & -2 & 2 & -3 & 1 & 6 & 4\\ 0 & 3 & -5 & -1 & 3 & 3 & 3 \end{bmatrix}, \quad \boldsymbol{b} = \begin{bmatrix} 4\\ 1\\ 2\\ 4\\ 3\\ 3\\ 2 \end{bmatrix}.$$

With these matrices and vectors, the following commands can be used to solve the linear programming problem directly:

```
>> f=[22,-5,7,10,-8,-8,9];
   Aeq=[0,1,-2,-1,0,5,1; 5,-3,1,2,3,2,1]; beq=[2; -2];
   A=[3,0,-2,-2,0,0,3; 2,3,1,0,0,3,1;
      2,4,-4,2,-3,2,2; 0,-2,2,0,3,-2,-2;
      -5,-1,1,-1,0,5,1; 1,-2,2,-3,1,6,4; 0,3,-5,-1,3,3,3];
   b=[4; 1; 2; 4; 3; 3; 2]; xm=-5*ones(7,1);
   [x,f0,flag]=linprog(f,A,b,Aeq,beq,xm)
```

Since `flag` is 1, the solution process is successful. The global optimal solution of the linear programming problem is

$$\boldsymbol{x} = [-1.9505, -3.1237, -5.0000, 3.0742, 0.1432, 0.4648, -4.1263]^{\mathrm{T}},$$

and the value of objective function is $f_0 = -73.5521$.

4.2.2 Linear programming problems with multiple decision vectors

In the previously discussed linear programming problem, the decision variable is a vector $\boldsymbol{x}$. For problems with multiple decision vectors, a single decision vector can be redefined. Suitable matrices and vectors can be constructed to describe the objective function and constraints, such that the problem can be converted into the standard form. Function `linprog()` can be called directly. Here examples are used to show how to solve such problems.

Example 4.10. Solve the following linear programming problem:[46]

$$\max \quad 30x_1 + 40x_2 + 20x_3 + 10x_4 - (15s_1 + 20s_2 + 10s_3 + 8s_4).$$
$$\boldsymbol{x},\boldsymbol{s}\ \text{s.t.}\ \begin{cases} 0.3x_1+0.3x_2+0.25x_3+0.15x_4\leqslant 1\,000 \\ 0.25x_1+0.35x_2+0.3x_3+0.1x_4\leqslant 1\,000 \\ 0.45x_1+0.5x_2+0.4x_3+0.22x_4\leqslant 1\,000 \\ 0.15x_1+0.15x_2+0.1x_3+0.05x_4\leqslant 1\,000 \\ x_1+s_1=800,\ x_2+s_2=750 \\ x_3+s_3=600,\ x_4+s_4=500 \\ x_j\geqslant 0,\ s_j\geqslant 0,\ j=1,2,3,4 \end{cases}$$

Solutions. In this problem, the two vectors $\boldsymbol{x}$ and $\boldsymbol{s}$ are the decision variables, not merely $\boldsymbol{x}$. The problem cannot be solved directly with the existing `linprog()` function. A unified decision vector should be constructed, by letting $x_5 = s_1$, $x_6 = s_2$, $x_7 = s_3$, and $x_8 = s_4$. The original problem can be converted manually into the following form:

$$\min_{\boldsymbol{x}\text{ s.t.}\begin{cases}0.3x_1+0.3x_2+0.25x_3+0.15x_4\leqslant 1\,000\\0.25x_1+0.35x_2+0.3x_3+0.1x_4\leqslant 1\,000\\0.45x_1+0.5x_2+0.4x_3+0.22x_4\leqslant 1\,000\\0.15x_1+0.15x_2+0.1x_3+0.05x_4\leqslant 1\,000\\x_1+x_5=800\\x_2+x_6=750\\x_3+x_7=600\\x_4+x_8=500\\x_j\geqslant 0,\ j=1,2,\ldots,8\end{cases}} -30x_1-40x_2-20x_3-10x_4+15x_5+20x_6+10x_7+8x_8.$$

From this model, it is not difficult to extract the coefficient vector

$$\boldsymbol{f}=[-30,-40,-20,-10,15,20,10,8].$$

Also the linear inequality and equality constraints can be expressed in matrix form:

$$\boldsymbol{A}=\begin{bmatrix}0.3&0.3&0.35&0.15&0&0&0&0\\0.25&0.35&0.3&0.1&0&0&0&0\\0.45&0.5&0.4&0.22&0&0&0&0\\0.15&0.15&0.1&0.05&0&0&0&0\end{bmatrix},\quad \boldsymbol{b}=\begin{bmatrix}1000\\1000\\1000\\1000\end{bmatrix},$$

$$\boldsymbol{A}_{\mathrm{eq}}=\begin{bmatrix}1&0&0&0&1&0&0&0\\0&1&0&0&0&1&0&0\\0&0&1&0&0&0&1&0\\0&0&0&1&0&0&0&1\end{bmatrix},\quad \boldsymbol{b}_{\mathrm{eq}}=\begin{bmatrix}800\\750\\600\\500\end{bmatrix}.$$

With the matrices and vectors, the following commands can be used to solve the original problem directly:

```
>> f=[-30,-40,-20,-10,15,20,10,8];
   A=[0.3,0.3,0.35,0.15,0,0,0,0; 0.25,0.35,0.3,0.1,0,0,0,0;
      0.45,0.5,0.4,0.22,0,0,0,0; 0.15,0.15,0.1,0.05,0,0,0,0];
   Aeq=[1,0,0,0,1,0,0,0; 0,1,0,0,0,1,0,0;
        0,0,1,0,0,0,1,0; 0,0,0,1,0,0,0,1];
   b=[1000; 1000; 1000; 1000]; beq=[800; 750; 600; 500];
   xm=zeros(8,1); [x,f0,flag]=linprog(f,A,b,Aeq,beq,xm)
```

The solution to the linear programming problem is $\boldsymbol{x}=[800,750,387.5,500,0,0,212.5,0]$, that is, $x_1=800$, $x_2=750$, $x_3=387.5$, $x_4=500$, $s_1=s_2=s_4=0$, and $s_3=212.5$. The result is the same as that in [46].

4.2.3 Linear programming with double subscripts

In some particular application fields, the decision variables are not described by vectors, they are described by matrices. A double subscript decision variable problem is encountered. A new vectorized decision variable is to be defined, such that the problem can be converted into a standard form. The solver `linprog()` can be used to solve the original problem. After solution, the decision variables should be transformed back to restore the decision variable matrix.

Example 4.11. Solve the linear programming problem with two subscripts:

$$\min \quad 2\,800(x_{11}+x_{21}+x_{31}+x_{41})+4\,500(x_{12}+x_{22}+x_{32})+6\,000(x_{13}+x_{23})+7\,300x_{14}$$
$$\boldsymbol{x} \text{ s.t. } \begin{cases} x_{11}+x_{12}+x_{13}+x_{14}\geqslant 15\\ x_{12}+x_{13}+x_{14}+x_{21}+x_{22}+x_{23}\geqslant 10\\ x_{13}+x_{14}+x_{22}+x_{23}+x_{31}+x_{32}\geqslant 20\\ x_{14}+x_{23}+x_{32}+x_{41}\geqslant 12\\ x_{ij}\geqslant 0\ (i=1,2,3,4,j=1,2,3,4)\end{cases}$$

Solutions. It is obvious that the problem cannot be solved with the methods discussed earlier. The double subscript decision variable matrix should be converted to the one with a single subscript. To do this, new variables should be selected. For instance, let $x_1 = x_{11}$, $x_2 = x_{12}$, $x_3 = x_{13}$, $x_4 = x_{14}$, $x_5 = x_{21}$, $x_6 = x_{22}$, $x_7 = x_{23}$, $x_8 = x_{31}$, $x_9 = x_{32}$, and $x_{10} = x_{41}$. The problem can be manually rewritten as

$$\min \quad 2\,800(x_1+x_5+x_8+x_{10})+4\,500(x_2+x_6+x_9)+6\,000(x_3+x_7)+7\,300x_4.$$
$$\boldsymbol{x} \text{ s.t. } \begin{cases} -(x_1+x_2+x_3+x_4)\leqslant -15\\ -(x_2+x_3+x_4+x_5+x_6+x_7)\leqslant -10\\ -(x_3+x_4+x_6+x_7+x_8+x_9)\leqslant -20\\ -(x_4+x_7+x_9+x_{10})\leqslant -12\\ x_i\geqslant 0,\ i=1,2,\dots,10\end{cases}$$

The following commands can be used to solve the problem:

```
>> f([1,5,8,10])=2800; f([2,6,9])=4500; f([3,7])=6000; f(4)=7300;
   A=-[1 1 1 1 0 0 0 0 0 0; 0 1 1 1 1 1 1 0 0 0; %constraints
       0 0 1 1 0 1 1 1 1 0; 0 0 0 1 0 0 1 0 1 1];
   B=-[15; 10; 20; 12]; xm=[0 0 0 0 0 0 0 0 0 0]; Aeq=[]; Beq=[];
   [x,f0,flag]=linprog(f,A,B,Aeq,Beq,xm) %direct solution
```

The result obtained is $\boldsymbol{x} = [5, 0, 0, 10, 0, 0, 0, 8, 2, 0]$. Transforming the decision variables back to the double subscript form, it is found that $x_{11} = 5$, $x_{14} = 10$, $x_{31} = 8$, $x_{32} = 2$, the remain independent variables are all zero. The value of the objective function is 118 400.

4.2.4 Transportation problem

Transportation problem is a typical problem in operations research. Before presenting the material of the section, an example is given. Then definition and solution methods of transportation problem are presented.

Example 4.12. Assume that a manufacture has factories in three different cities, and there are warehouses in four different cities. The transportation cost is shown in Table 4.5.[27] In each warehouse, there are allocations. From each factory there is an amount of product to be transported. Design a transportation plan such that the total cost is minimized.

Table 4.5: Cost per vehicle.

city	transportation costs and variables								output
	A	variable	B	variable	C	variable	D	variable	s_i
1	464	x_{11}	513	x_{12}	654	x_{13}	867	x_{14}	75
2	352	x_{21}	416	x_{22}	690	x_{23}	791	x_{24}	125
3	995	x_{31}	682	x_{32}	388	x_{33}	685	x_{34}	100
allocation s_i	80		65		70		85		

Solutions. In the table, the cost as well as the assigned variables are provided. It is obvious that the total cost can be computed as follows:

$$\begin{aligned} f(\boldsymbol{X}) = {} & 464x_{11} + 513x_{12} + 654x_{13} + 867x_{14} + 352x_{21} + 416x_{22} + 690x_{23} + 791x_{24} \\ & + 995x_{31} + 682x_{32} + 388x_{33} + 685x_{34}. \end{aligned}$$

Apart from the objective function, the following constraints must be satisfied:

$$\begin{gathered} x_{11} + x_{12} + x_{13} + x_{14} = 75, \quad x_{21} + x_{22} + x_{23} + x_{24} = 125, \quad x_{31} + x_{32} + x_{33} + x_{34} = 100, \\ x_{11} + x_{21} + x_{31} = 80, \quad x_{12} + x_{22} + x_{32} = 65, \quad x_{13} + x_{23} + x_{33} = 70, \quad x_{14} + x_{24} + x_{34} = 85. \end{gathered}$$

How to select x_{ij} such that the total cost $f(\boldsymbol{X})$ is minimized, when the constraints are satisfied?

Definition 4.6. The standard transportation problem is described by the problem in Table 4.6.[27] There are m suppliers, and n goods to transport. Assume that j is a goods

Table 4.6: Typical form of the transportation problem.

goods	transportation cost				demands
	1	2	⋯	n	s_i
1	c_{11}	c_{12}	⋯	c_{1n}	s_1
2	c_{21}	c_{22}	⋯	c_{2n}	s_2
⋮	⋮	⋮	⋮	⋮	⋮
m	c_{m1}	c_{m2}	⋯	c_{mn}	s_m
supplies	d_1	d_2	⋯	d_n	

number, and i is a supplier number. The cost of the ith goods transported to the jth supplier is c_{ij}. The total demand of the jth goods is d_j, and the total supply of the ith goods is s_i. The transportation problem is to decide how many goods to transport such that the transportation costs are minimized.

Definition 4.7. If a mathematical model is used to describe the problem, the decision variables x_{ij} are assigned according to the costs c_{ij}. Therefore, the transportation problem can be described by the linear programming problem with double subscripts:

$$\min \quad \sum_{i=1}^{m}\sum_{j=1}^{n} c_{ij}x_{ij}. \tag{4.2.1}$$

$$\boldsymbol{X} \text{ s.t. } \begin{cases} \sum_{j=1}^{n} x_{ij}=s_i,\ i=1,2,\dots,m \\ \sum_{i=1}^{m} x_{ij}=d_j,\ j=1,2,\dots,n \\ x_{ij}\geqslant 0,\ i=1,2,\dots,m,\ j=1,2,\dots,n \end{cases}$$

The direct solution of double subscript linear programming problem is rather complicated and error-prone. With the above introduction, a MATLAB function is written to automatically convert the problem into a single subscript one. The user needs to enter matrix $\boldsymbol{C}$ and vectors $\boldsymbol{s}$ and $\boldsymbol{d}$.

The syntax of the function is $\boldsymbol{X}$=`transport_linprog(`$\boldsymbol{C}$,$\boldsymbol{s}$,$\boldsymbol{d}$`)`. With such a function, the optimal solution can be obtained in $\boldsymbol{X}$ matrix, where the integer linear programming problem will be discussed later.

```
function [x,f0,flag]=transport_linprog(C,s,d,intkey)
[m,n]=size(C); A=[]; B=[];
for i=1:n, Aeq(i,(i-1)*m+1:i*m)=1; end
for i=1:m, Aeq(n+i,i:m:n*m)=1; end
xm=zeros(1,n*m); f=C(:); Beq=[s(:); d(:)];
if nargin==3 %linear programming problem
   [x,f0,flag]=linprog(f,A,B,Aeq,Beq,xm);
else            %integer linear programming
   [x,f0,flag]=intlinprog(f,1:n*m,A,B,Aeq,Beq,xm); x=round(x);
end
x=reshape(x,m,n); %convert the vector back to the matrix
```

Example 4.13. Solve the transportation problem in Example 4.12, and interpret the physical meaning of the results.

Solutions. The cost matrix $\boldsymbol{C}$ can be established from the table, which is the 3×4 matrix in the center of Table 4.5. With the last column in the table, the demand column vector $\boldsymbol{d}$ can be set up, while framing the last row, the supply row vector $\boldsymbol{c}$ can be constructed. With the matrix and vectors, function `transport_linprog()` can be called directly to build the transportation plan:

```
>> C=[464,513,654,867; 352,416,690,791; 995,682,388,685];
   s=[80,65,70,85]; d=[75; 125; 100];
   [x0,f0]=transport_linprog(C,s,d)
```

The following results can be obtained:

$$x_0 = \begin{bmatrix} 0 & 20 & 0 & 55 \\ 80 & 45 & 0 & 0 \\ 0 & 0 & 70 & 30 \end{bmatrix}, \quad f_0 = 152535.$$

The remaining task for the user is to interpret the returned results. It can be seen from the returned x_0 matrix that, from factory 1, 20 trucks should be transported to warehouse B and 55 trucks to D; from factory 2, 80 trucks should be sent to A and 45 trucks to B; while from factory 3, 70 trucks should be sent to C, and 30 trucks should be sent to D. The minimum total costs can be found as 152 535.

Example 4.14. Suppose a department store wants to import clothes from cities I, II and III. There are four different styles of clothes, labeled A, B, C, and D. The demands are respectively 1 500, 2 000, 3 000, and 3 500. The maximum supplies of the clothes in the three cities are 2 500, 2 500, and 5 000. The profit of each type of clothes is given in Table 4.7. Design an import plan such that the profit is maximized.

Table 4.7: Profit of each type of clothes.

cities	different types of clothes				total supply
	A	B	C	D	s_i
I	10	5	6	7	2 500
II	8	2	7	6	2 500
III	9	3	4	8	5 000
demands d_i	1 500	2 000	3 000	3 500	

Solutions. The original problem is a profit maximization one. The objective function should be multiplied by −1 to convert it to a minimization problem. Besides, the lower bounds of the decision variables are zeros. The problem can be solved with the following statements:

```
>> C=[10 5 6 7; 8 2 7 6; 9 3 4 8]; %matrices and vectors
   s=[2500 2500 5000]; d=[1500 2000 3000 3500];
   X=transport_linprog(-C,s,d) %solve the problem
   f=sum(C(:).*X(:))            %maximum profit
```

With the above function call, the result obtained is

$$X = \begin{bmatrix} 0 & 2\,000 & 500 & 0 \\ 0 & 0 & 2\,500 & 0 \\ 1\,500 & 0 & 0 & 3\,500 \end{bmatrix}, \quad f = 72\,000.$$

The meaning is that the store should import 2 000 of clothes B and 500 of clothes C from city I; import 2 500 of clothes C from city II; as well as import 1 500 of clothes A and 3 500 of clothes D from city III; the maximum profit is then $f = 72\,000$.

If the total demands are changed to $\boldsymbol{d} = [1\,500, 2\,500, 3\,000, 3\,500]$, the decision variables obtained may be fractional. Such solutions are, of course, not practical. Integer programming should be introduced, and the problems will revisited later.

4.3 Problem-based description and solution of linear programming problems

The linear programming problem solver `linprog()` is fully demonstrated in the above discussion. It can be seen that the problem can be treated easily using the solver. From a practical point of view, there are a many considerations to be made in some problems, since before tackling problems on a computer, standard linear programming form should be achieved manually. This is usually not a simple topic. Is there a simpler way to handle the original problem without bothering with the standard form conversions?

In some other applications, there are text files created by other optimization problems. How can we use MATLAB to solve the problems directly?

In this section, the method of reading the general-purpose MPS file describing linear programming problems into MATLAB environment is presented. Then, the problem-based description methods supported by recent versions of MATLAB are addressed. With them, linear programming problems can be described in a simpler and more straightforward manner.

4.3.1 MPS file for linear programming problems

MPS (mathematical programming system) format was the file format used earlier by IBM to express optimization problems. It can be handled now by many optimization software packages. Linear programming problem can be described directly by MPS files. There are many benchmark problems for linear programming expressed in MPS files. In this section, an MPS file for a simple example is written and explained. Then, the MATLAB solutions of linear programming problems described in MPS files are demonstrated.

Example 4.15. Understand the MPS file description to the linear programming problem in Example 4.1:

$$\min_{x,y \text{ s. t. } \begin{cases} -x+2y\leqslant 4 \\ 3x+2y\leqslant 12 \\ x\geqslant 0,\ y\geqslant 0 \end{cases}} \quad -x-2y.$$

The following MPS file can be written:

```
NAME mytest.mps // problem name
ROWS            // each row describes objective function and constraints
   N   obj      // first row, N stands for no relationship expression
   L   c1       // first constraint, L for ⩽, G for ⩾, E for equality
   L   c2       // second constraint
COLUMNS         // about the decision variable, maximum 5 in each row
   x   obj   -1    c1  -1 // x appears, position and coefficient
   x   c2    3            // x also appears in the second constraint
   y   obj   -2    c1  2  // y variable description
   y   c2    2
RHS                       // data to describe the right-hand side
   rhs c1    4     c2 12  // data for the RHS of two constraints
BOUNDS                    // bounds of decision variables
   LO  x    0   y   0     // LO for lower bounds, UP for upper
ENDATA
```

Solutions. We do not attempt to explain in detail how to write MPS files. Only an example is used here to show linear programming problem format. It may help the reader understand MPS descriptions. In fact, comparing the mathematical expression and model, it is not difficult to understand the MPS file format.

A MATLAB function *p*=`mpsread('filename')` is provided in the Optimization Toolbox. It may load the linear programming problem described in an MPS file into MATLAB workspace as a structured variable *p*. With such a variable, the `linprog()` solver can be used to solve the problem directly.

Examples are provided below to show the use of the command, and also the solution of linear programming problems.

Example 4.16. Load the MPS file in Example 4.15 and solve the linear programming problem.

Solutions. To solve the linear programming problem described by an MPS file, the first thing to do is load it into MATLAB workspace, then call the `linprog()` solver directly to find the solution. For this particular problem, the following commands can be issued. The solution obtained is $\boldsymbol{x} = [2,3]$, which is exactly the same as that obtained in Example 4.1.

```
>> p=mpsread('mytest.mps'); x=linprog(p)
```

Example 4.17. An MPS file `eil33.2.mps` is provided in MATLAB, containing 4 516 variables, with 32 equality constraints. It is an integer linear programming problem. The file is with 25 660 lines of source code. Load the problem into MATLAB and solve it. It can be modified into an ordinary linear programming problem, then it can be solved by a MATLAB solver. Find the solution of the problem, measure the elapsed time, and state the objective function value.

Solutions. Integer programming problems will be presented in later chapters. In the structured variable, the `solver` member is set to `'intlinprog'`. If an ordinary linear programming problem is expected, it should be modified to `'linprog'`. Then, the solver `linprog()` can be called directly to solve the problem. It can be seen that a linear programming problem with 4 516 decision variables can be solved within 0.043 seconds, the optimal objective function value is 811.2790.

```
>> p=mpsread('eil33.2.mps');     % load linear programming problem model
   p.solver='linprog'; p.options=optimset;
   tic, [x,f0,flag]=linprog(p); toc, size(x), f0
```

4.3.2 Problem-based description of linear programming problems

Apart from the matrix–vector method discussed earlier to present objective functions and constraints for linear programming problems, a problem-based format is provided in MATLAB starting with version R2017b. Many problems needing manual conversions can be entered into MATLAB directly, without bothering with manual work. Currently this description method only applies to linear and quadratic programming problems.

The procedures of the problem-based description are:

(1) Problem creation. Command `optimproblem()` can be used to initiate a blank optimization problem, with the syntax given below

```
prob=optimproblem('ObjectiveSense','max')
```

If `'ObjectiveSense'` option is not given, the default minimization problem is created.

(2) Decision variable definition. Function `optimvar()` can be used with the following syntax:

$\boldsymbol{x}$=`optimvar('x',`n`,`m`,`k`,'LowerBound',`$\boldsymbol{x}_\mathrm{m}$`)`

where n, m, k are used to define three-dimensional arrays. If k is not given, a decision matrix $\boldsymbol{x}$ of size $n \times m$ is defined. If m is 1, a decision vector of $n \times 1$ is defined. If $\boldsymbol{x}_\mathrm{m}$ is a scalar, all the lower bounds of the decision variables are set to the same value. The `LowerBound` option may be simplified as `Lower`. By a similar method, the `UpperBound` option, simplified as `Upper`, can also be defined.

With the above definitions, the `prob` problem can be described with the objective function and constraints assigned. The specific format is demonstrated through the following examples.

(3) **Problem solution.** With `prob` variable, function `sols=solve(prob)` can be called to solve the optimization problem directly. The result is returned in `sols`, whose $\boldsymbol{x}$ and other members are the actual solution of the problem. Command `options=optimset()` can be used to assign control options, so that `sols=solve(prob,'options',options)` command can be used to solve the linear programming problem.

Example 4.18. Consider again the linear programming problem in Example 4.8. For convenience of presentation, it is rewritten here:

$$\max_{\boldsymbol{x} \text{ s.t.} \begin{cases} x_1/4-60x_2-x_3/50+9x_4\leqslant 0 \\ -x_1/2+90x_2+x_3/50-3x_4\geqslant 0 \\ x_3\leqslant 1, x_1\geqslant -5, x_2\geqslant -5, x_3\geqslant -5, x_4\geqslant -5 \end{cases}} 3x_1/4-150x_2+x_3/50-6x_4.$$

Use problem-based approach to describe and solve the problem.

Solutions. There are many differences between the model and the standard form. With the traditional method, these places must be modified manually. For instance, the maximization problem, the $\geqslant$ inequality problem, the matrix extraction problem, and so on. A simple and straightforward way of problem-based description is a better solution. The result obtained is the same as that in Example 4.8.

```
>> P=optimproblem('ObjectiveSense','max'); % maximization
   x=optimvar('x',4,1,'LowerBound',-5);    % decision variables
   P.Objective=3*x(1)/4-150*x(2)+x(3)/50-6*x(4);
   P.Constraints.cons1=x(1)/4-60*x(2)-x(3)/50+9*x(4) <= 0;
   P.Constraints.cons2=-x(1)/2+90*x(2)+x(3)/50-3*x(4) >= 0;
   P.Constraints.cons3=x(3) <= 1;          % upper bounds
   sols=solve(P); x0=sols.x
```

Example 4.19. Solve again the problem in Example 4.9, where the problem is rewritten here:

$$\max_{\boldsymbol{x} \text{ s.t.} \begin{cases} 3x_1-2x_3-2x_4+3x_7\leqslant 4 \\ 2x_1+3x_2+x_3+3x_6+x_7\leqslant 1 \\ 2x_1+4x_2-4x_3+2x_4-3x_5+2x_6+2x_7\leqslant 2 \\ 2x_2-2x_3-3x_5+2x_6+2x_7\geqslant -4 \\ x_2-2x_3-x_4+5x_6+x_7=2 \\ 5x_1+x_2-x_3+x_4-5x_6-x_7\geqslant -3 \\ 5x_1-3x_2+x_3+2x_4+3x_5+2x_6+x_7=-2 \\ x_1-2x_2+2x_3-3x_4+x_5+6x_6+4x_7\leqslant 3 \\ 3x_2-5x_3-x_4+3x_5+3x_6+3x_7\leqslant 2 \\ x_i\geqslant -5,\ i=1,2,\ldots,7 \end{cases}} -22x_1+5x_2-7x_3-10x_4+8x_5+8x_6-9x_7.$$

Solutions. Using the problem-based description format, the original linear programming problem can easily be expressed with MATLAB commands. Unlike in Example 4.9, where manual manipulation is needed, the commands used here are simpler and more straightforward. With `solve()` command, the result obtained is the same as in Example 4.9. Note that the constraint names can be arbitrarily assigned. No regulations must be followed.

```
>> P=optimproblem('ObjectiveSense','max'); %maximization
   x=optimvar('x',7,1,'LowerBound',-5);    %decision variables
   P.Objective=-22*x(1)+5*x(2)-7*x(3)-10*x(4)+8*x(5)+8*x(6)-9*x(7);
   P.Constraints.c1=3*x(1)-2*x(3)-2*x(4)+3*x(7)<=4;
   P.Constraints.c2=2*x(1)+3*x(2)+x(3)+3*x(6)+x(7)<=1;
   P.Constraints.c3=2*x(1)+4*x(2)-4*x(3)+2*x(4)- ...
                    3*x(5)+2*x(6)+2*x(7)<=2;
   P.Constraints.c4=2*x(2)-2*x(3)-3*x(5)+2*x(6)+2*x(7)>=-4;
   P.Constraints.c5=x(2)-2*x(3)-x(4)+5*x(6)+x(7)==2;
                                                %double equal signs
   P.Constraints.c6=5*x(1)+x(2)-x(3)+x(4)-5*x(6)-x(7)>=-3;
   P.Constraints.a=5*x(1)-3*x(2)+x(3)+2*x(4)+3*x(5)+2*x(6)+x(7)==-2;
   P.Constraints.b=x(1)-2*x(2)+2*x(3)-3*x(4)+x(5)+6*x(6)-4*x(7)<=3;
   P.Constraints.c9=3*x(2) -5*x(3)-x(4)+3*x(5)+3*x(6)+3*x(7)<=2;
   sols=solve(P); x0=sols.x
```

Example 4.20. Use problem-based format to describe and solve the linear programming problem in Example 4.10, expressed again as follows:

$$\max_{\boldsymbol{x},\boldsymbol{s}\ \text{s. t.}\left\{\begin{array}{l}0.3x_1+0.3x_2+0.25x_3+0.15x_4\leqslant 1\,000\\ 0.25x_1+0.35x_2+0.3x_3+0.1x_4\leqslant 1\,000\\ 0.45x_1+0.5x_2+0.4x_3+0.22x_4\leqslant 1\,000\\ 0.15x_1+0.15x_2+0.1x_3+0.05x_4\leqslant 1\,000\\ x_1+s_1=800,\ x_2+s_2=750\\ x_3+s_3=600,\ x_4+s_4=500\\ x_j\geqslant 0,\ s_j\geqslant 0,\ j=1,2,3,4\end{array}\right.} 30x_1+40x_2+20x_3+10x_4-(15s_1+20s_2+10s_3+8s_4).$$

Solutions. In Example 4.10, manual conversion was needed to rewrite the problem in standard form. Then, the solver `linprog()` could be used to solve it. If the problem-based solution pattern is used, there is no need to modify it manually. Two decision vectors can be declared so that the problem can be described with the following commands. Then, the solution can be found, which is the same as in Example 4.10.

```
>> P=optimproblem('ObjectiveSense','max'); %maximization
   x=optimvar('x',4,1,'LowerBound',0);     %decision vector
   s=optimvar('s',4,1,'LowerBound',0);     %another one
```

```
P.Constraints.c1=0.3*x(1)+0.3*x(2)+0.25*x(3)+0.15*x(4)<=1000;
P.Constraints.c2=0.25*x(1)+0.35*x(2)+0.3*x(3)+0.1*x(4)<=1000;
P.Constraints.c3=0.45*x(1)+0.5*x(2)+0.4*x(3)+0.22*x(4)<=1000;
P.Constraints.c4=0.15*x(1)+0.15*x(2)+0.1*x(3)+0.05*x(4)<=1000;
P.Constraints.c5=x(1)+s(1)==800;
P.Constraints.c6=x(2)+s(2)==750;
P.Constraints.c7=x(3)+s(3)==600;
P.Constraints.c8=x(4)+s(4)==500;
P.Objective=30*x(1)+40*x(2)+20*x(3)+10*x(4) ...
            -(15*s(1)+20*s(2)+10*s(3)+8*s(4));
sols=solve(P); x0=sols.x, s0=sols.s
```

With problem-based description, matrix operations can be used in describing the objective function and constraints. For instance, constraints c5~c8 can be simplified with the following command:

```
>> P.Constraints.cnew=x+s==[800; 750; 600; 500];
```

Example 4.21. Solve again the linear programming problem with double subscripts in Example 4.11. The original problem is presented again here:

$$\min \quad 2800(x_{11}+x_{21}+x_{31}+x_{41})+4500(x_{12}+x_{22}+x_{32})+6\,000(x_{13}+x_{23})+7\,300x_{14}.$$

$$\boldsymbol{x} \text{ s.t. } \begin{cases} x_{11}+x_{12}+x_{13}+x_{14}\geqslant 15 \\ x_{12}+x_{13}+x_{14}+x_{21}+x_{22}+x_{23}\geqslant 10 \\ x_{13}+x_{14}+x_{22}+x_{23}+x_{31}+x_{32}\geqslant 20 \\ x_{14}+x_{23}+x_{32}+x_{41}\geqslant 12 \\ x_{ij}\geqslant 0 \ (i=1,2,3,4, j=1,2,3,4) \end{cases}$$

Solutions. With a problem-based description, the decision matrix can be declared and used directly. Therefore, the decision variables are given in the 4×4 matrix. The following commands can be used to describe the problem and solve it directly:

```
>> P=optimproblem;
   x=optimvar('x',4,4,'LowerBound',0);
   P.Objective=2800*(x(1,1)+x(2,1)+x(3,1)+x(4,1))+...
      4500*(x(1,2)+x(2,2)+x(3,2))+6000*(x(1,3)+x(2,3))+7300*x(1,4);
   P.Constraints.cons1=x(1,1)+x(1,2)+x(1,3)+x(1,4) >= 15;
   P.Constraints.c2=x(1,2)+x(1,3)+x(1,4)+x(2,1)+x(2,2)+x(2,3)>=10;
   P.Constraints.c3=x(1,3)+x(1,4)+x(2,2)+x(2,3)+x(3,1)+x(3,2)>=20;
   P.Constraints.cons4=x(1,4)+x(2,3)+x(3,2)+x(4,1) >= 12;
   sols=solve(P); x0=sols.x
```

The optimal solution and objective function are:

$$\boldsymbol{x}=\begin{bmatrix}3&0&0&12\\0&0&0&0\\8&0&0&0\\0&0&0&0\end{bmatrix},\quad f_{\mathrm{opt}}=118\,400.$$

It can be seen that although the results obtained are different from those in Example 4.11, they yield the same objective function value. In other words, the problem does not have a unique solution. Both results, perhaps more, are solutions of the original problem.

Example 4.22. Solve the transportation problem in Example 4.12.

Solutions. The cost matrix $\boldsymbol{C}$ can be created, then, with the problem-based method, the original problem can be entered and solved. The result is exactly the same as in Example 4.12. Note that in the constraints description, `sum(x,1)` stands for the sum of all the elements in each column in $\boldsymbol{x}$. The result is a row vector. Command `sum(x,2)` stands for the sum of row elements, and the result is a column vector.

```
>> C=[464,513,654,867; 352,416,690,791; 995,682,388,685];
   s=[80,65,70,85]; d=[75; 125; 100];
   P=optimproblem; x=optimvar('x',3,4,'Lower',0);
   P.Objective=sum(sum(C.*x)); P.Constraints.c1=sum(x,1)==s;
   P.Constraints.c2=sum(x,2)==d; sol=solve(P); x0=sol.x
```

Example 4.23. Consider the Klee–Minty test problem.[49] Find the optimum solution when $n=20$ and $a=2$:

$$\max \sum_{j=1}^{n} a^{n-j}x_j.$$
$$\boldsymbol{x}\ \text{s.t.}\begin{cases}2\sum\limits_{j=1}^{i-1}a^{i-j}x_j+x_i\leqslant(a^2)^{i-1},\ i=1,2,\dots,n\\ x_j\geqslant 0,\ j=1,2,\dots,n\end{cases}$$

Solutions. In the original problem, $a=10$ and $n=25$. Under the double precision scheme, $a=10$ may be too large, and normal manipulation may not be possible. Therefore, $a=2$ and $n=25$ are considered. Problem-based method can be used to describe the linear programming problem. Since there are too many constraints, a loop structure is used to model the constraints. Therefore, the following commands can be used to solve the problem directly:

```
>> P=optimproblem('ObjectiveSense','max'); n=25;
   x=optimvar('x',n,1,'LowerBound',0); a=2;
   P.Objective=a.^[n-1:-1:0]*x; P.Constraints.c1= x(1)<=1;
   for i=2:n
      cons(i-1,1)=2*a.^[i-1:-1:0]*x(1:i)+x(i)<=(a^2)^(i-1);
```

```
end
P.Constraints.c2=cons; sols=solve(P); x0=sols.x
```

4.3.3 Conversions in linear programming problems

Up to now, three methods were presented to describe the modeling of linear programming problems. The ordinary matrix and vector method, structured variables method, and problem-based method. The former two are essentially the same. It is natural to convert from one format to another. The conversions are simple and straightforward.

The problem-based method is completely different from the other two methods. The decision variables must be declared first then described with expressions. The benefit of this method is that no manual conversion is needed. The expressions can be used directly to describe the original optimization problem. Three functions are provided in Optimization Toolbox to handle manipulations to this type of model:

(1) **Problem display.** Function `showproblem(`P`)` can be used to display the optimization problem. Command `showconstr(`P`)` can be used to display individual constraints.
(2) **Problem conversion.** One can convert a problem-based model P through command p=`prob2struct(`P`)` into a structured variable p.
(3) **Storing the problem.** Command `writeproblem()` saves the problem into a text file.

The following examples are used to demonstrate these MATLAB commands.

Example 4.24. Consider the linear programming problem in Example 4.18. Use the problem-based method to describe it and then convert into a structured variable.

Solutions. The commands in Example 4.18 can be used to get model P:

```
>> P=optimproblem('ObjectiveSense','max'); % maximization
   x=optimvar('x',4,1,'LowerBound',-5);      % decision and bounds
   P.Objective=3*x(1)/4-150*x(2)+x(3)/50-6*x(4);
   P.Constraints.cons1=x(1)/4-60*x(2)-x(3)/50+9*x(4) <= 0;
   P.Constraints.cons2=-x(1)/2+90*x(2)+x(3)/50-3*x(4) >= 0;
   P.Constraints.cons3=x(3) <= 1;            % upper bound
```

With such a model, the command `showproblem()` can be used to display the optimization problem. It can be seen that the display here is very close to the original mathematical form. Therefore, it is easier to locate and fix a problem, if there are any.

```
>> showproblem(P), writeproblem(P,'c4myprob.txt')
```

The display of the problem is as follows. The format of display is slightly adjusted for typesetting purposes.

```
OptimizationProblem :
   max : 0.75*x(1, 1) - 150*x(2, 1) + 0.02*x(3, 1) - 6*x(4, 1)
   subject to cons1:
      0.25*x(1, 1) - 60*x(2, 1) - 0.02*x(3, 1) + 9*x(4, 1) <= 0
   subject to cons2:
      -0.5*x(1, 1) + 90*x(2, 1) + 0.02*x(3, 1) - 3*x(4, 1) >= 0
   subject to cons3:    x(3, 1) <= 1
   variable bounds:
      -5 <= x(1, 1)
      -5 <= x(2, 1)
      -5 <= x(3, 1)
      -5 <= x(4, 1)
```

It can be seen that the description of the problem is easy to understand. The user may then list the problem and the original mathematical format side-by-side, to compare whether there are differences.

Since `prob2struct()` can be used to convert optimization problems into ordinary linear programming problems in structured form, its `f` and `Aineq` members can be converted automatically. Note that under the default setting, `Aineq` and other members are provided in a sparse matrix format. Command `full()` can be used to convert it into a regular matrix form.

```
>> p=prob2struct(P), f=p.f, p.Aineq, A=full(p.Aineq)
```

The obtained results are as follows. It can be seen that they are the same as in Example 4.8, which were obtained manually. Therefore, the description method here is reliable:

$$\boldsymbol{f}=\begin{bmatrix}-0.75\\150\\-0.02\\6\end{bmatrix},\quad \boldsymbol{A}=\begin{bmatrix}0.25 & -60 & -0.02 & 9\\0.5 & -90 & -0.02 & 3\\0 & 0 & 1 & 0\end{bmatrix}.$$

It can be seen that the problem-based description method is flexible when handling small-scale problems. It is more straightforward than the matrix and vector description. Apart from that, even though nonlinear programming problems, which will be discussed later, are more involved, the linear part can be modeled in this way to establish the framework. We can manually convert it into the nonlinear programming framework, and add nonlinear items then. This will be discussed later through examples.

4.4 Quadratic programming

Quadratic programming is another type of simple constrained optimization problems. The objective function contains a quadratic form of $\boldsymbol{x}$, and the constraints are still linear equalities and inequalities of $\boldsymbol{x}$. In this section, the mathematical form of a quadratic programming problem is proposed. Then, the description method is presented, followed by demonstrative examples on how to solve quadratic programming problems.

4.4.1 Mathematical quadratic programming models

Definition 4.8. The mathematical form of a quadratic programming problem is

$$\min_{\boldsymbol{x} \text{ s.t. } \begin{cases}\boldsymbol{A}\boldsymbol{x}\leqslant\boldsymbol{B}\\ \boldsymbol{A}_{\rm eq}\boldsymbol{x}=\boldsymbol{B}_{\rm eq}\\ \boldsymbol{x}_{\rm m}\leqslant\boldsymbol{x}\leqslant\boldsymbol{x}_{\rm M}\end{cases}} \quad \boldsymbol{f}^{\rm T}\boldsymbol{x}+\frac{1}{2}\boldsymbol{x}^{\rm T}\boldsymbol{H}\boldsymbol{x}. \tag{4.4.1}$$

Compared with linear programming problems, it can be seen that the constraints are exactly the same. The difference is that, in the objective function, there is a quadratic term $\boldsymbol{x}^{\rm T}\boldsymbol{H}\boldsymbol{x}/2$, used to describe the weighting of x_i^2 and x_ix_j terms. Therefore, an $\boldsymbol{H}$ matrix should be constructed. It is called the Hessian matrix, named after German mathematical Ludwig Otto Hesse (1811–1874).

Definition 4.9. An alternative description of the quadratic term is

$$\frac{1}{2}(h_{11}x_1^2+h_{12}x_1x_2+\cdots+h_{1n}x_1x_n+h_{21}x_1x_2+h_{22}x_2^2+\cdots+h_{nn}x_n^2). \tag{4.4.2}$$

Theorem 4.3. *If matrix $\boldsymbol{H}$ is positive-definite, the quadratic programming problem is a convex problem.*

In other words, a convex quadratic programming problem is independent of the initial search point. If there are feasible solutions, the solution found by the search is the global optimum solution.

4.4.2 Direct solutions of quadratic programming problems

A solver `quadprog()` is provided in the Optimization Toolbox in MATLAB, with the following syntaxes:

[$\boldsymbol{x}$,$f_{\rm opt}$,flag,out]=quadprog(problem)

[$\boldsymbol{x}$,$f_{\rm opt}$,flag,out]=quadprog($\boldsymbol{H}$,$\boldsymbol{f}$,$\boldsymbol{A}$,$\boldsymbol{B}$)

[$\boldsymbol{x}$,$f_{\rm opt}$,flag,out]=quadprog($\boldsymbol{H}$,$\boldsymbol{f}$,$\boldsymbol{A}$,$\boldsymbol{B}$,$\boldsymbol{A}_{\rm eq}$,$\boldsymbol{B}_{\rm eq}$)

[$\boldsymbol{x}$,$f_{\rm opt}$,flag,out]=quadprog($\boldsymbol{H}$,$\boldsymbol{f}$,$\boldsymbol{A}$,$\boldsymbol{B}$,$\boldsymbol{A}_{\rm eq}$,$\boldsymbol{B}_{\rm eq}$,$\boldsymbol{x}_{\rm m}$,$\boldsymbol{x}_{\rm M}$)

[$\boldsymbol{x}$,$f_{\rm opt}$,flag,out]=quadprog($\boldsymbol{H}$,$\boldsymbol{f}$,$\boldsymbol{A}$,$\boldsymbol{B}$,$\boldsymbol{A}_{\rm eq}$,$\boldsymbol{B}_{\rm eq}$,$\boldsymbol{x}_{\rm m}$,$\boldsymbol{x}_{\rm M}$,options)

If the quadratic programming problem is described by a structured variable, its member H describes the H matrix, and the member `solver` is set to `'quadprog'`. Note that if the quadratic programming problem is nonconvex, the global optimum solution cannot be found in general, even a feasible solution may not be found with the solver.

Example 4.25. Solve the following quadratic programming problem with four variables:

$$\min_{\boldsymbol{x} \text{ s.t. } \begin{cases} x_1+x_2+x_3+x_4 \leqslant 5 \\ 3x_1+3x_2+2x_3+x_4 \leqslant 10 \\ x_1,x_2,x_3,x_4 \geqslant 0 \end{cases}} (x_1-1)^2+(x_2-2)^2+(x_3-3)^2+(x_4-4)^2.$$

Solutions. The problem should be expressed in the standard quadratic programming form. The objective function should be expanded first

$$f(x) = x_1^2+x_2^2+x_3^2+x_4^2-2x_1-4x_2-6x_3-8x_4+30.$$

Since the constant term in the objective function has no impact on the solution, it can be dropped. The $\boldsymbol{H}$ matrix and $\boldsymbol{f}^{\mathrm{T}}$ vector can then be written as

$$\boldsymbol{H} = \operatorname{diag}([2,2,2,2]), \quad \boldsymbol{f}^{\mathrm{T}} = [-2,-4,-6,-8].$$

The following MATLAB commands can be used in solving the problem:

```
>> f=[-2,-4,-6,-8]; H=diag([2,2,2,2]); % the objective function
   A=[1,1,1,1; 3,3,2,1]; B=[5;10];
   Aeq=[]; Beq=[]; xm=zeros(4,1);          % constraints
   [x,f_opt]=quadprog(H,f,A,B,Aeq,Beq,xm) % problem solution
```

The optimum solution can then be found as $\boldsymbol{x} = [0, 0.6667, 1.6667, 2.6667]^{\mathrm{T}}$, and the value of the objective function is −23.6667.

When the standard form of quadratic programming problem is used, one must be very careful with the 1/2 term in front of the $\boldsymbol{H}$ in (4.4.1). The diagonal elements in $\boldsymbol{H}$ matrix here should be 2 rather than 1, otherwise the problem is wrongly described. Besides, the objective function value obtained here is not the one in the original problem. The constant term was dropped off deliberately, and it should be added back. The optimal objective function should be 6.3333 for this example.

4.4.3 Problem-based quadratic programming problem description

Similar to linear programming problems, the quadratic programming problem can also be described by a problem-based format. In this section the description and solution of the problems and introduced.

Example 4.26. Solve the following quadratic programming problem:

$$\min \quad (x_1-1)^2+(x_2-2)^2+(x_3-3)^2+(x_4-4)^2.$$
$$\boldsymbol{x} \text{ s.t. } \begin{cases} x_1+x_2+x_3+x_4\leqslant 5 \\ 3x_1+3x_2+2x_3+x_4\leqslant 10 \\ x_1,x_2,x_3,x_4\geqslant 0 \end{cases}$$

Solutions. If the problem-based modeling technique is used, there is no need to derive the $\boldsymbol{H}$ matrix manually. The following commands can be used to describe the original problem. With command `solve()`, the problem can be solved, and the result is exactly the same as in Example 4.25:

```
>> P=optimproblem; x=optimvar('x',4,1,'LowerBound',0);
   P.Objective=(x(1)-1)^2+(x(2)-2)^2+(x(3)-3)^2+(x(4)-4)^2;
   P.Constraints.cons1=x(1)+x(2)+x(3)+x(4) <= 5;
   P.Constraints.cons2=3*x(1)+3*x(2)+2*x(3)+x(4) <= 10;
   sols=solve(P); x=sols.x
```

In fact, the objective function and constraints can also be modeled in vectorized form, such that the description is more concise. The result obtained is also the same.

```
>> P=optimproblem; x=optimvar('x',4,1,'LowerBound',0);
   P.Objective=sum((x-[1:4]').^2);
   P.Constraints.cons1=sum(x) <= 5;
   P.Constraints.cons2=[3 3 2 1]*x <= 10;
   sols=solve(P); x0=sols.x
```

Example 4.27. Solve the following quadratic programming problem:

$$\min \quad -2x_1+3x_2-4x_3+4x_1^2+2x_2^2+7x_3^2-2x_1x_2-2x_1x_3+3x_1x_3.$$
$$\boldsymbol{x} \text{ s.t. } \begin{cases} 2x_1+x_2+3x_3\geqslant 8 \\ x_1+2x_2+x_3\leqslant 7 \\ -3x_1+2x_2\leqslant -5 \\ x_1,x_2,x_3\geqslant 0 \end{cases}$$

Solutions. If one wants to solve the problem with `quadprog()` solver, some manual work is needed to convert it into the standard form in (4.4.1). For this particular problem, it may be hard to accurately write down the $\boldsymbol{H}$ matrix. The problem-based method can be used to describe this quadratic programming problem. The objective function and constraints can be expressed with expressions. No manual conversion work is needed. The quadratic programming problem can be described and solved, and the result obtained is $\boldsymbol{x}=[2.87,1.8051,0.5198]^{\mathrm{T}}$.

```
>> P=optimproblem;
   x=optimvar('x',3,1,'LowerBound',0);
```

```
P.Objective=-2*x(1)+3*x(2)-4*x(3)+4*x(1)^2+2*x(2)^2+...
        7*x(3)^2-2*x(1)*x(2)-2*x(1)*x(3)+3*x(1)*x(3);
P.Constraints.c1=2*x(1)+x(2)+3*x(3) >= 8;
P.Constraints.c2=x(1)+2*x(2)+x(3) >= 7;
P.Constraints.c3=-3*x(1)+2*x(2) <= -5;
sols=solve(P); x0=sols.x
```

In the solution process, it is pointed out that the automatically generated Hessian matrix is asymmetric. It is recommended to use $(\boldsymbol{H}+\boldsymbol{H}^{\mathrm{T}})/2$ to convert it into a symmetric one. Unfortunately, this cannot be done in the model. The model should be converted into a structured variable, and then the matrix can be converted. The conversion and solving commands are now changed to the following, and the result obtained is the same:

```
>> p=prob2struct(P); p.H=(p.H+p.H')/2; x1=quadprog(p)
```

Example 4.28. Solve the unconstrained quadratic optimization problem when $n = 20$:

$$f(x) = \sum_{i=1}^{n} i x_i^2 + \left(\sum_{i=1}^{n} x_i\right)^2.$$

Solutions. Even though there are no constraints, since the objective function is quadratic, the problem-based method can still be used in describing and solving the optimization problem. Since the representation in $\boldsymbol{H}$ matrix and $\boldsymbol{f}$ vector may not be simple, the problem-based method can be used, with the following commands, and the objective function can be modeled. The problem can then be solved, and the optimum solution is $x_i = 0$, with the error norm of 2.6668×10^{-15}.

```
>> n=20; P=optimproblem; x=optimvar('x',n,1);
   P.Objective=sum([[1:n]'].*x.^2)+sum(x)^2/n;
   sols=solve(P); x0=sols.x, norm(x0)
```

Example 4.29. In fact, the benchmark problems with analytical solutions set to 0 are not practical and meaningful, since normally random initial points may be generated around zeros. If a meaningful benchmark problem is to be constructed, letting $n = 100$, the quadratic programming problem can be rewritten in the following form. Describe and solve the new problem

$$f(x) = \sum_{i=1}^{n} i(x_i - i)^2 + \left(\sum_{i=1}^{n} (x_i - i)\right)^2.$$

Solutions. Since the objective function used here is quadratic, the above mentioned methods can be used to solve the problem directly. Of course, the difficulties are that it may be hard to write down $\boldsymbol{H}$ matrix and $\boldsymbol{f}$ vector. The problem-based method is a good choice in modeling this type of problems. Through simple manipulation, the original problem can be modeled and solved, with the global optimum solution at $x_i = i$. The error norm is 1.1662×10^{-13}.

```
>> n=100; P=optimproblem; x=optimvar('x',n,1);
   P.Objective=sum([[1:n]'].*(x-[1:n]').^2)+sum(x-[1:n]')^2/n;
   sols=solve(P); x0=sols.x, norm(x0-[1:n]')
```

If the user is interested in finding the $\boldsymbol{H}$ matrix and $\boldsymbol{f}$ vector, he/she can set $n = 7$ or consider another smaller positive number and convert the problem into a structured variable. Then, the corresponding matrices can be displayed.

```
>> n=7; P=optimproblem; x=optimvar('x',n,1);
   P.Objective=sum([[1:n]'].*(x-[1:n]').^2)+sum(x-[1:n]')^2/n;
   p=prob2struct(P); H=sym(p.H), f=sym(p.f)
```

The matrix and vector are

$$\boldsymbol{H} = \frac{1}{7}\begin{bmatrix} 16 & 2 & 2 & 2 & 2 & 2 & 2 \\ 2 & 30 & 2 & 2 & 2 & 2 & 2 \\ 2 & 2 & 44 & 2 & 2 & 2 & 2 \\ 2 & 2 & 2 & 58 & 2 & 2 & 2 \\ 2 & 2 & 2 & 2 & 72 & 2 & 2 \\ 2 & 2 & 2 & 2 & 2 & 86 & 2 \\ 2 & 2 & 2 & 2 & 2 & 2 & 100 \end{bmatrix}, \quad \boldsymbol{f} = \begin{bmatrix} -10 \\ -16 \\ -26 \\ -40 \\ -58 \\ -80 \\ -106 \end{bmatrix}.$$

Although it may be difficult to find manually the $\boldsymbol{H}$ matrix and $\boldsymbol{f}$ vector, the problem-based modeling technique can be adopted, and the quadratic programming problem can be described. Then, selecting an appropriate n, it can be converted into a structured variable, so that useful information can be displayed, and from these clues, the matrix can be finally set up.

Example 4.30. Solve the following quadratic programming test problem:[22]

$$\min \quad \boldsymbol{c}^{\mathrm{T}}\boldsymbol{x} + \boldsymbol{d}^{\mathrm{T}}\boldsymbol{y} - \frac{1}{2}\boldsymbol{x}^{\mathrm{T}}\boldsymbol{Q}\boldsymbol{x},$$
$$\boldsymbol{x} \text{ s.t. } \begin{cases} 2x_1+2x_2+y_6+y_7 \leqslant 10 \\ 2x_1+2x_3+y_6+y_8 \leqslant 10 \\ 2x_2+2x_3+y_7+y_8 \leqslant 10 \\ -8x_1+y_6 \leqslant 0 \\ -8x_2+y_7 \leqslant 0 \\ -8x_3+y_8 \leqslant 0 \\ -2x_4-y_1+y_6 \leqslant 0 \\ -2y_2-y_3+y_7 \leqslant 0 \\ -2y_4-y_5+y_8 \leqslant 0 \\ 0 \leqslant x_i \leqslant 1,\ i=1,2,3,4 \\ 0 \leqslant y_i \leqslant 1,\ i=1,2,3,4,5,9 \\ y_i \geqslant 0,\ i=6,7,8 \end{cases}$$

where $\boldsymbol{c} = [5,5,5,5]$, $\boldsymbol{d} = [-1,-1,-1,-1,-1,-1,-1,-1,-1]$, $\boldsymbol{Q} = 10\boldsymbol{I}$.

Solutions. It is hard to solve the problem with the regular `quadprog()` function, since in order to solve it, $\boldsymbol{x}$ and $\boldsymbol{y}$ vectors should be combined into a longer decision variable. Manual formulation is needed, which may be complicated and error-prone. Therefore, the problem-based method is a better choice. The problem can be entered into MATLAB, and solved directly.

```
>> P=optimproblem; c=5*ones(1,4); d=-1*ones(1,9); Q=10*eye(4);
   x=optimvar('x',4,1,'LowerBound',0,'UpperBound',1);
   y=optimvar('y',9,1,'LowerBound',0);
   P.Objective=c*x+d*y-x'*Q*x/2; clear cons1
   cons1=[2*x(1)+2*x(2)+y(6)+y(7)<=10;
          2*x(1)+2*x(3)+y(6)+y(8)<=10;
          2*x(2)+2*x(3)+y(7)+y(8)<=10;
          -8*x(1)+y(6)<=0; -8*x(2)+y(7)<=0; -8*x(3)+y(8)<=0;
          -2*x(4)-y(1)+y(6)<=0; -2*y(2)-y(3)+y(7)<=0;
          -2*y(4)-y(5)+y(8)<=0; y([1 2 3 4 5 9])<=1];
   P.Constraints.cons1=cons1;
   sols=solve(P); x0=sols.x, y0=sols.y
```

In the solution process, it is prompted that "The problem is nonconvex". In this case, the solution obtained may not be the global optimum solution. The results obtained with the previous commands are

$$\boldsymbol{x}_0 = [0.55, 0.55, 0.55, 0.55]^{\mathrm{T}}, \quad \boldsymbol{y}_0 = [0.55, 0.55, 0.55, 0.55, 0.55, 1, 1, 1, 1]^{\mathrm{T}}.$$

Since the problem is nonconvex, the solution is not globally optimal, in fact, the one given in [22] is better. This result cannot be obtained with solvers such as `quadprog()` and `solve()`:

$$\boldsymbol{x} = [1, 1, 1, 1]^{\mathrm{T}}, \quad \boldsymbol{y} = [1, 1, 1, 1, 1, 3, 3, 3, 1]^{\mathrm{T}}.$$

The global optimum solution of nonconvex problems will be further discussed in Chapter 5.

4.4.4 Quadratic programming problem with double subscripts

As a direct extension to the transportation problem in Definition 4.7, quadratic terms may be introduced, and a quadratic transportation problem can be defined, where double subscripts are involved. It can be shown that the quadratic programming problem may be concave. Ordinary solvers cannot ensure the global optimum solutions, sometimes even feasible solutions may not be found.

Definition 4.10. A concave-cost transportation problem is mathematically defined as

$$\min \quad \sum_{i=1}^{m}\sum_{j=1}^{n} c_{ij}x_{ij} + d_{ij}x_{ij}^2, \qquad (4.4.3)$$
$$\boldsymbol{X} \text{ s.t. } \begin{cases} \sum_{i=1}^{m} x_{ij}=b_j,\ j=1,2,\dots,n \\ \sum_{j=1}^{n} x_{ij}=a_i,\ i=1,2,\dots,m \\ x_{ij}\geqslant 0,\ i=1,2,\dots,m,\ j=1,2,\dots,n \end{cases}$$

with

$$d_{ij} \leqslant 0, \quad \sum_{i=1}^{m} a_i = \sum_{j=1}^{n} b_j. \qquad (4.4.4)$$

Example 4.31. Solve the concave-cost transportation problem in Example 4.10, if $n = 4$, $m = 6$, and $\boldsymbol{b} = [29, 41, 13, 21]^{\mathrm{T}}$, $\boldsymbol{a} = [8, 24, 20, 24, 16, 12]^{\mathrm{T}}$,

$$\boldsymbol{C} = \begin{bmatrix} 300 & 270 & 460 & 800 \\ 740 & 600 & 540 & 380 \\ 300 & 490 & 380 & 760 \\ 430 & 250 & 390 & 600 \\ 210 & 830 & 470 & 680 \\ 360 & 290 & 400 & 310 \end{bmatrix}, \quad \boldsymbol{D} = \begin{bmatrix} -7 & -4 & -6 & -8 \\ -12 & -9 & -14 & -7 \\ -13 & -12 & -8 & -4 \\ -7 & -9 & -16 & -8 \\ -4 & -10 & -21 & -13 \\ -17 & -9 & -8 & -4 \end{bmatrix}.$$

Solutions. According to the mathematical formula, it is easy to input the model into the computer. In fact, comparing the MATLAB commands and the mathematical formulas in Definition 4.10, it can be found that the concise statements match the original mathematical form suitably. Unfortunately, since the problem is known to be concave, the solution command may be tried, but the warning "The problem is nonconvex" is obtained, showing that $x_{ij} = 1$. Of course, the solution does not satisfy the constraints, meaning that it is not feasible. This problem will be solved and explored in Chapter 5.

```
>> n=4; m=6; b=[29,41,13,21]; a=[8,24,20,24,16,12]';
   C=[300,270,460,800; 740,600,540,380; 300,490,380,760;
      430,250,390,600; 210,830,470,680; 360,290,400,310];
   D=[-7,-4,-6,-8; -12,-9,-14,-7; -13,-12,-8,-4;
      -7,-9,-16,-8; -4,-10,-21,-13; -17,-9,-8,-4];
   P=optimproblem; x=optimvar('x',m,n,'LowerBound',0);
   P.Objective=sum(sum(C.*x+D.*x.^2));
   P.Constraints.c1=sum(x,1)==b; P.Constraints.c2=sum(x,2)==a;
   sols=solve(P); x0=sols.x
```

If the problem is converted into a structured variable, it is found that the Hessian matrix is diagonal, with negative diagonal elements. Therefore, the quadratic programming problem is nonconvex. The quadratic programming problem solvers cannot be used directly.

```
>> p=prob2struct(P); H=full(p.H)
```

4.5 Linear matrix inequalities

The linear matrix inequalities (LMIs) theory has attracted the attention of researchers in the control system community for over 30 years.[8] The concept of LMIs and its applications in automatic control systems was pioneered by Willems.[50] The essential task of the method is to convert control problems into linear programming problems. Since linear programming methods are mature, and problems are convex, LMIs solutions to control problems are practical.

In this section, the fundamental concepts and common forms of LMIs are introduced. Necessary conversions are presented. Then, MATLAB Robust Control Toolbox and the free YALMIP Toolbox are demonstrated through examples.

4.5.1 Description of linear matrix inequality problems

Definition 4.11. A linear matrix inequality is described as

$$\boldsymbol{F}(\boldsymbol{x}) = \boldsymbol{F}_0 + x_1\boldsymbol{F}_1 + \cdots + x_m\boldsymbol{F}_m < \mathbf{0}, \tag{4.5.1}$$

where $\boldsymbol{x} = [x_1, \dots, x_m]^\mathrm{T}$ are the coefficients of a polynomial, also known as a decision vector; $\boldsymbol{F}_i$ are real symmetric or complex Hermitian matrices. The whole $\boldsymbol{F}(\boldsymbol{x})$ satisfying LMI $\boldsymbol{F}(\boldsymbol{x}) < \mathbf{0}$ is referred to as a negative-definite matrix.

The solutions $\boldsymbol{x}$ of an LMI problem form a convex set, that is,

$$\boldsymbol{F}[\alpha\boldsymbol{x}_1 + (1-\alpha)\boldsymbol{x}_2] = \alpha\boldsymbol{F}(\boldsymbol{x}_1) + (1-\alpha)\boldsymbol{F}(\boldsymbol{x}_2) < \mathbf{0}, \tag{4.5.2}$$

where $\alpha > 0$, $1-\alpha > 0$. The solution set is also known as feasible solutions. The LMI can be used as the constraints of an LMI problem.

Definition 4.12. Assuming that there are two LMIs, $\boldsymbol{F}_1(\boldsymbol{x}) < \mathbf{0}$ and $\boldsymbol{F}_2(\boldsymbol{x}) < \mathbf{0}$, a single LMI can be formed as

$$\begin{bmatrix} \boldsymbol{F}_1(\boldsymbol{x}) & 0 \\ 0 & \boldsymbol{F}_2(\boldsymbol{x}) \end{bmatrix} < \mathbf{0}. \tag{4.5.3}$$

The two LMIs can be written as a single LMI. Similarly, multiple LMIs $\boldsymbol{F}_i(\boldsymbol{x}) < \boldsymbol{0}$, $i = 1, 2, \ldots, k$ can also be combined into a single LMI $\boldsymbol{F}(\boldsymbol{x}) < \boldsymbol{0}$, where

$$\boldsymbol{F}(\boldsymbol{x}) = \begin{bmatrix} \boldsymbol{F}_1(\boldsymbol{x}) & & & \\ & \boldsymbol{F}_2(\boldsymbol{x}) & & \\ & & \ddots & \\ & & & \boldsymbol{F}_k(\boldsymbol{x}) \end{bmatrix} < \boldsymbol{0}. \tag{4.5.4}$$

4.5.2 Lyapunov inequalities

To demonstrate the relationship between control problems and LMIs, Lyapunov stability assessment is presented. For linear systems, if, for a given positive-definite matrix $\boldsymbol{Q}$, Lyapunov equation

$$\boldsymbol{A}^{\mathrm{T}}\boldsymbol{X} + \boldsymbol{X}\boldsymbol{A} = -\boldsymbol{Q} \tag{4.5.5}$$

has a positive-definite solution $\boldsymbol{X}$, the system is stable. The above problem can be naturally extended to that of finding the solution of Lyapunov inequalities.

Definition 4.13. The mathematical form of Lyapunov inequality is as follows:

$$\boldsymbol{A}^{\mathrm{T}}\boldsymbol{X} + \boldsymbol{X}\boldsymbol{A} < \boldsymbol{0}. \tag{4.5.6}$$

Since $\boldsymbol{X}$ is a symmetric matrix with $n(n+1)/2$ elements, a vector $\boldsymbol{x}$ can be formulated to describe the matrix:

$$x_i = X_{i,1},\ i = 1, \ldots, n, \quad x_{n+i} = X_{i,2},\ i = 2, \ldots, n, \ldots \tag{4.5.7}$$

The requirements can be expressed as $x_{(2n-j+2)(j-1)/2+i} = X_{i,j}$, $j = 1, 2, \ldots, n$, $i = j, j+1, \ldots, n$, and from the subscripts i, j, the vector $\boldsymbol{x}$ can be matched. Based on this idea, the following MATLAB function can be written, which can be used to convert a Lyapunov equation into an LMI:

```
function F=lyap2lmi(A0)
if prod(size(A0))==1, n=A0; A=sym('a%d%d',n);
else, n=size(A0,1); A=A0; end
vec=0; for i=1:n, vec(i+1)=vec(i)+n-i+1; end
for k=1:n*(n+1)/2, %construct the needed inequalities with a loop
   X=zeros(n); i=find(vec>=k); i=i(1)-1; j=i+k-vec(i)-1;
   X(i,j)=1; X(j,i)=1; F(:,:,k)=A.'*X+X*A; %construct LMIs
end
```

Two syntaxes are designed. If $\boldsymbol{A}$ is a given matrix, command $\boldsymbol{F}$=`lyap2lmi(`$\boldsymbol{A}$`)` can be used and the returned $\boldsymbol{F}$ is a three-dimensional array, whose ith layer, i. e., $\boldsymbol{F}$`(:,:,`i`)`, is the needed matrix $\boldsymbol{F}_i$. If an $n \times n$ arbitrary $\boldsymbol{A}$ matrix is expected, command `F=lyap2lmi(n)` can be used, where $\boldsymbol{F}$ is still a three-dimensional array. If $x_i = 1$, while other components are 0, $\boldsymbol{F}_i$ matrix can be found.

Example 4.32. If $\boldsymbol{A} = \begin{bmatrix} 1 & 2 & 3 \\ 4 & 5 & 6 \\ 7 & 8 & 0 \end{bmatrix}$, use the LMI technique to express its Lyapunov inequality. If $\boldsymbol{A}$ is a 3×3 real matrix, write down the corresponding LMI.

Solutions. Input $\boldsymbol{A}$ matrix and use the following commands:

```
>> A=[1,2,3; 4,5,6; 7,8,0]; F=lyap2lmi(A) %generate the LMI
```

then, it is found that $\boldsymbol{F}_i$ matrices are the coefficient ones in the following inequality

$$x_1 \begin{bmatrix} 2 & 2 & 3 \\ 2 & 0 & 0 \\ 3 & 0 & 0 \end{bmatrix} + x_2 \begin{bmatrix} 8 & 6 & 6 \\ 6 & 4 & 3 \\ 6 & 3 & 0 \end{bmatrix} + x_3 \begin{bmatrix} 14 & 8 & 1 \\ 8 & 0 & 2 \\ 1 & 2 & 6 \end{bmatrix}$$
$$+ x_4 \begin{bmatrix} 0 & 4 & 0 \\ 4 & 10 & 6 \\ 0 & 6 & 0 \end{bmatrix} + x_5 \begin{bmatrix} 0 & 7 & 4 \\ 7 & 16 & 5 \\ 4 & 5 & 12 \end{bmatrix} + x_6 \begin{bmatrix} 0 & 0 & 7 \\ 0 & 0 & 8 \\ 7 & 8 & 0 \end{bmatrix} < \boldsymbol{0}.$$

To study an arbitrary 3×3 matrix, the following command can be used:

```
>> F=lyap2lmi(3) %if dimension is known, the LMI can be constructed
```

The following LMI is then obtained:

$$x_1 \begin{bmatrix} 2a_{11} & a_{12} & a_{13} \\ a_{12} & 0 & 0 \\ a_{13} & 0 & 0 \end{bmatrix} + x_2 \begin{bmatrix} 2a_{21} & a_{22}+a_{11} & a_{23} \\ a_{22}+a_{11} & 2a_{12} & a_{13} \\ a_{23} & a_{13} & 0 \end{bmatrix} + x_3 \begin{bmatrix} 2a_{31} & a_{32} & a_{33}+a_{11} \\ a_{32} & 0 & a_{12} \\ a_{33}+a_{11} & a_{12} & 2a_{13} \end{bmatrix}$$
$$+ x_4 \begin{bmatrix} 0 & a_{21} & 0 \\ a_{21} & 2a_{22} & a_{23} \\ 0 & a_{23} & 0 \end{bmatrix} + x_5 \begin{bmatrix} 0 & a_{31} & a_{21} \\ a_{31} & 2a_{32} & a_{33}+a_{22} \\ a_{21} & a_{33}+a_{22} & 2a_{23} \end{bmatrix} + x_6 \begin{bmatrix} 0 & 0 & a_{31} \\ 0 & 0 & a_{32} \\ a_{31} & a_{32} & 2a_{33} \end{bmatrix} < \boldsymbol{0}.$$

Some nonlinear inequalities can be converted into LMIs, where taking advantage of the Schur complement properties of the partitioned matrix[42] is a commonly used conversion method.

Theorem 4.4. *If an affine matrix function $\boldsymbol{F}(\boldsymbol{x})$ can be partitioned as*

$$\boldsymbol{F}(\boldsymbol{x}) = \left[\begin{array}{c|c} \boldsymbol{F}_{11}(\boldsymbol{x}) & \boldsymbol{F}_{12}(\boldsymbol{x}) \\ \hline \boldsymbol{F}_{21}(\boldsymbol{x}) & \boldsymbol{F}_{22}(\boldsymbol{x}) \end{array}\right], \tag{4.5.8}$$

where $\boldsymbol{F}_{11}(\boldsymbol{x})$ is a square matrix, the following LMIs are equivalent:

$$\boldsymbol{F}(\boldsymbol{x}) < \boldsymbol{0}, \tag{4.5.9}$$

$$F_{11}(x) < 0, F_{22}(x) - F_{21}(x)F_{11}^{-1}(x)F_{12}(x) < 0, \tag{4.5.10}$$

$$F_{22}(x) < 0, F_{11}(x) - F_{12}(x)F_{22}^{-1}(x)F_{21}(x) < 0. \tag{4.5.11}$$

For instance, for algebraic Riccati equations, slight changes can be made to convert them into a Riccati inequality

$$A^T X + XA + (XB - C)R^{-1}(XB - C^T)^T < 0, \tag{4.5.12}$$

where $R = R^T > 0$. It is obvious that there are quadratic terms, therefore, it is not an LMI. With Schur complement properties, it can be seen that the nonlinear inequality can be equivalently converted to

$$X > 0, \quad \begin{bmatrix} A^T X + XA & XB - C^T \\ B^T X - C & -R \end{bmatrix} < 0. \tag{4.5.13}$$

4.5.3 Classifications of LMI problems

LMI problems are usually classified into three types: feasible solution problems, linear objective function problems, and generalized eigenvalue problems:

(1) Feasible solution problems. The so-called feasible solution problem is the solution of the constraints

$$F(x) < 0. \tag{4.5.14}$$

The solution of the problem is equivalent to solving $F(x) < \sigma I$, where σ is the smallest possible value found with numerical methods. If a $\sigma < 0$ can be found, the original problem is feasible. Otherwise, it is considered that there is no feasible solution.

(2) Linear objective function optimization. Consider the following optimization problem:

$$\min_{x \text{ s.t. } F(x)<0} \quad c^T x. \tag{4.5.15}$$

Since the constraints are expressed as an LMI, and the objective function is a linear expression of the decision variables x, the problem is an ordinary linear programming problem.

(3) Generalized eigenvalue problem. An eigenvalue problem, $Ax = \lambda Bx$, can be converted to a generalized eigenvalue problem $A(x) < \lambda B(x)$. It can be classified as the following optimization problem:

$$\min_{\lambda, x \text{ s.t. } \begin{cases} A(x)<\lambda B(x) \\ B(x)>0 \\ C(x)<0 \end{cases}} \quad \lambda. \tag{4.5.16}$$

Besides, there are other types of constraints, classified as $C(x) < 0$. In fact, the constraints can be combined into a single LMI expression.

4.5.4 MATLAB solutions of LMI problems

In the Robust Control Toolbox, an LMI solver is provided. Unfortunately, the statements describing the LMIs are rather complicated. An example is used here to illustrate in details the uses of the LMI solver.

The following procedures are used to describe LMIs in MATLAB:

(1) Create an LMI model. A blank LMI model can be established with function `setlmis([])`. The blank model is then created in MATLAB workspace.

(2) Declare the decision variables. With function `lmivar()`, the decision variables can be declared using the syntax $\boldsymbol{P}$ = `lmivar(key,[`$\boldsymbol{n}_1$`,`$\boldsymbol{n}_2$`])`, where `key` indicates the data type of the decision matrix. If `key`=2, an ordinary $n_1 \times n_2$ matrix $\boldsymbol{P}$ is declared, while `key`=1 for an $n_1 \times n_1$ symmetric matrix is defined. If `key`=1, while $\boldsymbol{n}_1$ and $\boldsymbol{n}_2$ are both vectors, $\boldsymbol{P}$ is declared as a block-diagonal symmetric matrix. If `key`=3, $\boldsymbol{P}$ is a special matrix, which is not discussed in this book. The interested readers may refer to the Robust Control Toolbox manual[47] for details.

(3) Describe the partitioned LMIs. One may describe the LMIs by consecutive calls of the function `lmiterm()`, while the syntax is quite complicated

```
lmiterm([k,i,j,P],A,B,flag)
```

where k is the number of the LMIs. Since an LMI problem is usually composed of several LMIs, they should be numbered first. If an LMI $G_k(\boldsymbol{x}) > 0$ is to be described, k should be set to $-k$. In the partitioned matrix, a term in a block can be described by function `lmiterm()`, where i, j respectively represent the row and column numbers in the block; $\boldsymbol{P}$ contains the decision variables already declared, and $\boldsymbol{A}$ and $\boldsymbol{B}$ are given matrices, the term $\boldsymbol{APB}$ is then defined. If `flag` is assigned as `'s'`, a symmetric term of $\boldsymbol{APB} + (\boldsymbol{APB})^{\mathrm{T}}$ is defined. If the whole term is a constant matrix, $\boldsymbol{P}$ should be set to 0, with matrix $\boldsymbol{B}$ omitted.

(4) Finalizing the LMI model. When all the LMIs are specified by the function `lmiterm()`, G = `getlmis` can be called to finalize the G model.

(5) Solve the LMI problem. If the LMI model G is specified, the corresponding LMI optimization problems can be solved in one of the following three forms:

```
[t_min,x]=feasp(G,options,target), % feasible solution
[c_opt,x]=mincx(G,c,options,x_0,target), % linear objective function
[λ,x]=gevp(G,nlfc,options,λ_0,x_0,target), % generalized eigenvalues
```

The solution $\boldsymbol{x}$ thus obtained is a vector, and function `dec2mat()` can then be used to extract the matrix. The control variable `options` is expressed as a 5-element vector, whose first element specifies the error tolerance, with its default value of 10^{-5}.

Example 4.33. Consider the Riccati inequality $A^{\mathrm{T}}X + XA + XBR^{-1}B^{\mathrm{T}}X + Q < 0$, where

$$A = \begin{bmatrix} -2 & -2 & -1 \\ -3 & -1 & -1 \\ 1 & 0 & -4 \end{bmatrix}, \quad B = \begin{bmatrix} -1 & 0 \\ 0 & -1 \\ -1 & -1 \end{bmatrix}, \quad Q = \begin{bmatrix} -2 & 1 & -2 \\ 1 & -2 & -4 \\ -2 & -4 & -2 \end{bmatrix}, \quad R = I_2,$$

and find a feasible positive-definite solution X.

Solutions. It is obvious that the original nonlinear matrix inequality is not an LMI. With Schur complement, the Riccati inequality can be expressed by a partitioned LMI. Besides, for an expected positive-definite solution, the second LMI can be rewritten as

$$\left[\begin{array}{c:c} A^{\mathrm{T}}X + XA + Q & XB \\ \hdashline B^{\mathrm{T}}X & -R \end{array}\right] < 0.$$

The above inequality can be numbered as no. 1, and the positive-definite inequality is numbered as no. 2. Therefore, the value of k can be set to 1 and 2, respectively, when using the `lmiterm()` function. It is noted that the matrix X is a 3×3 symmetric matrix. Thus, a feasible solution to the original problem can be obtained with the following statements. It is seen that since the second inequality is $X > 0$, its number should be -2 instead of 2.

```
>> A=[-2,-2,-1; -3,-1,-1; 1,0,-4]; B=[-1,0; 0,-1; -1,-1];
   Q=[-2,1,-2; 1,-2,-4; -2,-4,-2]; R=eye(2); %enter the matrices
   setlmis([]);                   %create a blank LTI framework
   X=lmivar(1,[3 1]);             %declare X as a 3 × 3 symmetric matrix
   lmiterm([1 1 1 X],A',1,'s')    %(1,1)th block, 's' means A^T X + XA
   lmiterm([1 1 1 0],Q)           %(1,1)th, appended by constant matrix Q
   lmiterm([1 1 2 X],1,B)         %(1,2)th, meaning XB
   lmiterm([1 2 2 0],-1)          %(2,2)th, meaning -R
   lmiterm([-2,1,1,X],1,1)        %the second inequality meaning X > 0
   G=getlmis;                     %complete the LTI framework setting
   [tmin b]=feasp(G);             %solve the feasible problem
   X=dec2mat(G,b,X)               %extract the solution matrix X
```

It is found that $t_{\min} = -0.2427$, with the feasible solution found as

$$X = \begin{bmatrix} 1.0329 & 0.4647 & -0.23583 \\ 0.4647 & 0.77896 & -0.050684 \\ -0.23583 & -0.050684 & 1.4336 \end{bmatrix}.$$

It is worth mentioning that, due to possible problems in the new Robust Control Toolbox, if the command `lmiterm([1 2 1 X],B',1)` is used to describe the symmetric term in the first inequality, wrong results will be found. Thus, the symmetric terms should not be used again in describing the LMI problems.

4.5.5 Optimization solutions with YALMIP toolbox

An optimization toolbox, YALMIP Toolbox (yet another LMI package) is developed by Swedish scholar Dr Johan Löfberg.[34] LMI-based solution methods for optimization problems are provided. Sometimes it is simpler and more straightforward than the LMI solver supported in the Robust Control Toolbox. Many more optimization problems, other than LMI problems, can be solved directly with YALMIP Toolbox.[35]

Decision variables can easily be created in YALMIP, with the command `sdpvar()`. The syntax of the function is

$\boldsymbol{X}$=`sdpvar(`n`)`, % symmetric matrix

$\boldsymbol{X}$=`sdpvar(`n,m`)`, % rectangular matrix

$\boldsymbol{X}$=`sdpvar(`n,n`,'full')`, % standard square matrix

These commands can be extended, so that some of the special matrices can be defined. For instance, it can be combined with function `hankel()` such that a Hankel matrix is defined. Similarly, function `intvar()` and `binvar()` can be used to declare integer and binary decision matrices.

Square brackets can be used to join constraints together, so that the linear and quadratic expressions defined on variables of `sdpvar` can be combined together to construct the constraint set.

Of course, the objective function can be defined in a similar way, and then function `optimize()` can be used to solve various problems with

`optimize(`constraints, objective function`)`, % solve problem

After the solution process, `double(`$\boldsymbol{X}$`)` or `value(`$\boldsymbol{X}$`)` commands can be used to extract the solution matrices. The solution pattern is similar to the problem-based solution methods.

The current versions of YALMIP toolbox are much more than merely LMI solvers. Many other types of problems, including bilevel programming problems, which are difficult to handle with other tools, can be tackled. Some of the problem solutions will be discussed in the subsequent presentation.

Example 4.34. Solve the problem in Example 4.33 with YALMIP Toolbox.

Solutions. With YALMIP Toolbox, the following concise commands can be used to describe the LMI problem, and the solution obtained is the same, as obtained earlier. There are two things to be noted: first, $>$ and $<$ constraints are no longer supported in YALMIP Toolbox. Relationships such as $\geqslant$, $\leqslant$, and $=$ are supported. And second, if a feasible problem is to be solved, there is no need to specify the objective function. After the solution process, `double()` command should be used to extract the solutions.

```
>> A=[-2,-2,-1; -3,-1,-1; 1,0,-4]; B=[-1,0; 0,-1; -1,-1];
   Q=[-2,1,-2; 1,-2,-4; -2,-4,-2]; R=eye(2);     % given matrices
```

```
X=sdpvar(3);                                    % solution matrix
F=[[A'*X+X*A+Q, X*B; B'*X, -R]<=0, X>=0];       % describe LMIs
optimize(F); X0=double(X) % solve problem and extract solutions
```

Example 4.35. Solve the linear programming problem in Example 4.6 with YALMIP Toolbox functions.

Solutions. It is obvious that $\boldsymbol{x}$ is a 5×1 column vector. The following commands can be used to solve the problem, and it is immediately found that $\boldsymbol{x} = [19.785, 0, 3.32, 11.385, 2.57]^{\mathrm{T}}$. It is the same as obtained earlier.

```
>> x=sdpvar(5,1); A=[0 2 1 4 2; 3 4 5 -1 -1];
   B=[54; 62]; xm=[0,0,3.32,0.678,2.57]'; % matrix input
   F=[A*x<=B, x>=xm];                      % input LMIs
   optimize(F,-[2 1 4 3 1]*x); x=double(x) % solve the problem
```

Example 4.36. For a linear system $(\boldsymbol{A}, \boldsymbol{B}, \boldsymbol{C}, \boldsymbol{D})$, the $\mathcal{H}_\infty$ norm can be computed from the `norm()` function directly. Alternatively, LMI method can be used to compute the $\mathcal{H}_\infty$ norm, whose mathematical form is

$$\min_{\gamma,\boldsymbol{P} \text{ s.t.} \begin{cases} \begin{bmatrix} \boldsymbol{A}^{\mathrm{T}}\boldsymbol{P}+\boldsymbol{P}\boldsymbol{A} & \boldsymbol{P}\boldsymbol{B} & \boldsymbol{C}^{\mathrm{T}} \\ \boldsymbol{B}^{\mathrm{T}}\boldsymbol{P} & -\gamma\boldsymbol{I} & \boldsymbol{D}^{\mathrm{T}} \\ \boldsymbol{C} & \boldsymbol{D} & -\gamma\boldsymbol{I} \end{bmatrix} < \boldsymbol{0} \\ \boldsymbol{P} > \boldsymbol{0} \end{cases}} \gamma. \tag{4.5.17}$$

Compute the $\mathcal{H}_\infty$ norm of the linear system, with

$$\boldsymbol{A} = \begin{bmatrix} -4 & -3 & 0 & -1 \\ -3 & -7 & 0 & -3 \\ 0 & 0 & -13 & -1 \\ -1 & -3 & -1 & -10 \end{bmatrix}, \quad \boldsymbol{B} = \begin{bmatrix} 0 \\ -4 \\ 2 \\ 5 \end{bmatrix}, \quad \boldsymbol{C} = [0, 0, 4, 0], \quad \boldsymbol{D} = 0.$$

Solutions. With the following commands, the $\mathcal{H}_\infty$ description with YALMIP Toolbox can be issued and the solution obtained is 0.4640, which is the same as the result when using the `norm()` function.

```
>> A=[-4,-3,0,-1; -3,-7,0,-3; 0,0,-13,-1;
      -1,-3,-1,-10]; % given matrices
   B=[0; -4; 2; 5]; C=[0,0,4,0]; D=0; gam=sdpvar(1); P=sdpvar(4);
   F=[[A*P+P*A',P*B,C'; B'*P,-gam,D'; C,D,-gam]<0, P>0];
   sol=optimize(F,gam); double(gam), norm(ss(A,B,C,D),'inf')
```

YALMIP Toolbox can also be used in solving quadratic programming problems. Now let us consider the quadratic programming problem descriptions and solutions.

The kernel of the YALMIP Toolbox uses the low-level solvers in MATLAB Optimization Toolbox.

Example 4.37. Solve the unconstrained optimization problem in Example 4.28 with YALMIP.

Solutions. Since the objective function is quadratic, a similar method to the problem-based method in Example 4.28 can be used to describe the problem, then YALMIP can be used to solve the problem. The result obtained is the same as in Example 4.28, where the norm of $\boldsymbol{x}_0$ is 2.6668×10^{-15}, which is exactly the same as in the example, since the kernel in the solution process is the same.

```
>> n=20; x=sdpvar(n,1); opt=sum([[1:n]'].*x.^2)+sum(x)^2/n;
   optimize([],opt); x0=double(x), norm(x0)
```

Example 4.38. Solve the quadratic programming problem in Example 4.25 with YALMIP Toolbox functions.

Solutions. Similar to the problem-based description method, the problem can be described again, and the solution obtained is the same as that obtained in Example 4.25.

```
>> x=sdpvar(4,1); opt=sum((x-[1:4]').^2);
   const=[sum(x)<=5, [3 3 2 1]*x<=10, x>=0];
   optimize(const,opt); x0=double(x)
```

4.5.6 Trials on nonconvex problems

Traditional YALMIP Toolbox implemented some LMI-based algorithms. Therefore, nonconvex problems could not be solved. In the new versions, an algorithm call `bmibnb` can be used. Global solutions of some nonconvex quadratic programming problems can be found. The actual solution method is to set the solver directly to such an algorithm. Then, function `optimize()` can be called for solving such a problem with syntaxes

`options=sdpsettings('solver','bmibnb');`

`optimize(`constraints, objective function`,options)`

Example 4.39. Consider again Example 4.30, which was shown to be a nonconvex quadratic programming problem.

Solutions. Similar to the problem-based method, the problem can be input into MATLAB workspace. Then, function `optimize()` can be called for finding the solution. Again the warning "The problem is nonconvex" is given, indicating that the problem cannot be solved directly.

```
>> c=5*ones(1,4); d=-1*ones(1,9); Q=10*eye(4);
   x=sdpvar(4,1); y=sdpvar(9,1); opt=c*x+d*y-x'*Q*x/2;
   const=[2*x(1)+2*x(2)+y(6)+y(7)<=10;
          2*x(1)+2*x(3)+y(6)+y(8)<=10;
          2*x(2)+2*x(3)+y(7)+y(8)<=10;
          -8*x(1)+y(6)<=0; -8*x(2)+y(7)<=0; -8*x(3)+y(8)<=0;
          -2*x(4)-y(1)+y(6)<=0; -2*y(2)-y(3)+y(7)<=0;
          -2*y(4)-y(5)+y(8)<=0; y([1 2 3 4 5 9])<=1;
          x>=0; x<=1; y>=0];
   optimize(const,opt); x0=double(x), y0=double(y)
```

Now a global optimization method can be tried, so that the global optimum solution is found.

```
>> options=sdpsettings('solver','bmibnb');
   optimize(const,opt,options); x0=double(x), y0=double(y)
```

The global optimum solution of

$$\boldsymbol{x} = [1,1,1,1]^{\mathrm{T}}, \quad \boldsymbol{y} = [1,1,1,1,1,3,3,3,1]^{\mathrm{T}}$$

can be found, which is exactly the same as in [22], indicating that the global method is successful here.

4.5.7 Problems with quadratic constraints

The problem-based method cannot handle constraints with quadratic terms. Direct description in YALMIP Toolbox can be applied to present such constraints. Unfortunately, the programming problems with quadratic constraints are usually nonconvex, so the solvers in YALMIP Toolbox may not find the global optimum solutions.

Example 4.40. Describe the following problem with quadratic constraints using YALMIP and find its solutions:[26]

$$\min_{\boldsymbol{q},w,k \text{ s.t.} \begin{cases} q_3+9.625q_1w+16q_2w+16w^2+12-4q_1-q_2-78w=0 \\ 16q_1w+44-19q_1-8q_2-q_3-24w=0 \\ 2.25-0.25k\leqslant q_1\leqslant 2.25+0.25k \\ 1.5-0.5k\leqslant q_2\leqslant 1.5+0.5k \\ 1.5-1.5k\leqslant q_3\leqslant 1.5+1.5k \end{cases}} k.$$

Solutions. For the optimization with quadratic constraints, if the objective function is linear or quadratic, the solution can be found directly with YALMIP Toolbox, without bothering with manual conversions. It can be seen that the descriptions of the objective function and constraints are almost the same as those in mathematical formulas. This complicated problem can be described and solved directly with YALMIP Toolbox commands. The optimal solution found is $k = 1.1448$. For the time being, we are not able to judge whether it is global or not. This problem will further be explored in Chapter 5.

```
>> q=sdpvar(3,1); w=sdpvar(1,1); k=sdpvar(1,1);
   con=[q(3)+9.625*q(1)*w+16*q(2)*w+16*w^2+12-4*q(1)-q(2)-78*w==0;
        16*q(1)*w+44-19*q(1)-8*q(2)-q(3)-24*w==0;
        2.25-0.25*k <= q(1) <= 2.25+0.25*k;
        1.5-0.5*k   <= q(2) <= 1.5+0.5*k;
        1.5-1.5*k   <= q(3) <= 1.5+1.5*k];
   optimize(con,k), value(k), value(q), value(w)
```

Example 4.41. Solve the following problem with quadratic constraints:[28]

$$\min_{\boldsymbol{x} \text{ s.t. } \begin{cases} 8-x_1^2-x_2^2-x_3^2-x_4^2-x_1+x_2-x_3+x_4 \geqslant 0 \\ 10-x_1^2-2x_2^2-x_3^2-2x_4^2+x_1+x_4 \geqslant 0 \\ 5-2x_1^2-x_2^2-x_3^2-2x_1+x_2+x_4 \geqslant 0 \end{cases}} \quad x_1^2+x_2^2+2x_3^2+x_4^2-5x_1-5x_2-21x_3+7x_4.$$

Solutions. The following statements can be written directly to describe and solve the optimization problem. The result obtained is

$$\boldsymbol{x} = [0.0029, 0.9976, 1.9984, -1.0014]^{\mathrm{T}},$$

and the objective function value is -43.9928.

```
>> x=sdpvar(4,1);
   opt=x(1)^2+x(2)^2+2*x(3)^2+x(4)^2-5*x(1)-5*x(2)-21*x(3)+7*x(4);
   const=[8-sum(x.^2)-x(1)+x(2)-x(3)+x(4)>=0,
          10-x(1)^2-2*x(2)^2-x(3)^2-2*x(4)^2+x(1)+x(4)>=0,
          5-2*x(1)^2-x(2)^2-x(3)^2-2*x(1)+x(2)+x(4)>=0];
   optimize(const,opt), value(x), value(opt)
```

The global optimum solution given in Example [28] is $\boldsymbol{x} = [0, 1, 2, -1]$, and the objective function value is -44. It seems that the accuracy obtained is rather low. The default solver used is `lmilab`, and the precision control option under the algorithm is `reltol`. It is possible to modify the control options and solve the problem again. The new solution is global this time, and the actual error is 10^{-10}.

```
>> ops=sdpsettings('solver','lmilab','lmilab.reltol',eps);
   optimize(const,opt,ops), value(x), value(opt)
```

4.6 Exercises

4.1 Solve the following linear programming problems with graphical methods:

(1) max $2x_1 + x_2$;

$$\boldsymbol{x} \text{ s.t. } \begin{cases} 2x_1+x_2\leqslant 4 \\ 2x_1+3x_2\leqslant 3 \\ 4x_1+x_2\leqslant 5 \\ x_1+5x_2\leqslant 1 \\ x_1,x_2\geqslant 0 \end{cases}$$

(2) min $-2x_1 - x_2$.

$$\boldsymbol{x} \text{ s.t. } \begin{cases} x_1+x_2\leqslant 5 \\ 2x_1+3x_2\leqslant 12 \\ x_1\leqslant 4 \\ x_1,x_2\geqslant 0 \end{cases}$$

4.2 Solve the following optimization problem:

$$\min \quad -(x_1 + x_2 + x_3 + x_4 + x_5).$$

$$\boldsymbol{x} \text{ s.t. } \begin{cases} -\sum_{i=1}^{5}(9+i)x_i+50\,000\geqslant 0 \\ x_i\geqslant 0, i=1,2,3,4,5 \end{cases}$$

4.3 Solve the linear programming problem

$$\min \quad 10x_1 - 57x_2 + 9x_3 - 24x_4.$$

$$\boldsymbol{x} \text{ s.t. } \begin{cases} 0.5x_1-5.5x_2-2.5x_3+9x_4\leqslant 0 \\ 0.5x_1-1.5x_2-0.5x_3+x_4\leqslant 0 \\ x_1\leqslant 1 \\ x_1,x_2,x_3,x_4\geqslant 0 \end{cases}$$

4.4 Solve the linear programming problem

$$\max \quad v.$$

$$\boldsymbol{x},v \text{ s.t. } \begin{cases} -x_2+2x_3+v\leqslant 0 \\ 3x_1-4x_3+v\leqslant 0 \\ -5x_1+6x_2+v\leqslant 0 \\ x_1+x_2+x_3=1 \\ x_1,x_2,x_3\geqslant 0 \end{cases}$$

4.5 Solve the following linear programming problems:

(1) min $-3x_1 + 4x_2 - 2x_3 + 5x_4$;

$$\boldsymbol{x} \text{ s.t. } \begin{cases} 4x_1-x_2+2x_3-x_4=-2 \\ x_1+x_2-x_3+2x_4\leqslant 14 \\ 2x_1-3x_2-x_3-x_4\geqslant -2 \\ x_{1,2,3}\geqslant -1,\ x_4 \text{unconstrained} \end{cases}$$

(2) min $x_6 + x_7$.

$$\boldsymbol{x} \text{ s.t. } \begin{cases} x_1+x_2+x_3+x_4=4 \\ -2x_1+x_2-x_3-x_6+x_7=1 \\ 3x_2+x_3+x_5+x_7=9 \\ x_{1,2,\cdots,7}\geqslant 0 \end{cases}$$

4.6 Solve the linear programming problem

$$\min \quad x_{11} + 8x_{13} + 9x_{14} + 2x_{23} + 7x_{24} + 3x_{34}.$$
$$x \text{ s.t. } \begin{cases} x_{12}+x_{13}+x_{14}\geqslant 1 \\ -x_{12}+x_{23}+x_{24}=0 \\ -x_{13}-x_{23}+x_{34}=0 \\ x_{14}+x_{24}+x_{34}\leqslant 0 \\ x_{12},x_{13},\cdots,x_{34}\geqslant 0 \end{cases}$$

4.7 Solve the linear programming problem

$$\max \quad -3x_1 - x_2 + x_3 + 2x_4 - x_5 + x_6 - x_7 - 4x_8.$$
$$x \text{ s.t. } \begin{cases} x_1+4x_3+x_4-5x_5-2x_6+3x_7-6x_8=7 \\ x_2-3x_3-x_4+4x_5+x_6-2x_7+5x_8=-3 \\ 0\leqslant x_1\leqslant 8,\ 0\leqslant x_2\leqslant 6,\ 0\leqslant x_3\leqslant 10,\ 0\leqslant x_4\leqslant 15 \\ 0\leqslant x_5\leqslant 2,\ 0\leqslant x_6\leqslant 10,\ 0\leqslant x_7\leqslant 4,\ 0\leqslant x_8\leqslant 3 \end{cases}$$

4.8 Solve the following transportation problems and interpret the results:

(1)

suppliers	destination				amount
S1	3	7	6	4	5
S2	2	4	3	2	2
S3	4	3	8	5	3
D	3	3	2	2	

(2)

suppliers	costs				export
S1	464	513	654	867	75
S2	352	416	690	791	125
S3	995	682	388	685	100
D	80	65	70	85	

4.9 Solve the following quadratic programming problems, and explain the results with graphical methods:

(1) $$\min \quad 2x_1^2 - 4x_1x_2 + 4x_2^2 - 6x_1 - 3x_2;$$
$$x \text{ s.t. } \begin{cases} x_1+x_2\leqslant 3 \\ 4x_1+x_2\leqslant 9 \\ x_{1,2}\geqslant 0 \end{cases}$$

(2) $$\min \quad (x_1-1)^2 + (x_2-2)^2.$$
$$x \text{ s.t. } \begin{cases} -x_1+x_2=1 \\ x_1+x_2\leqslant 2 \\ x_{1,2}\geqslant 0 \end{cases}$$

4.10 Solve the Finkbeiner–Kall quadratic programming problem:[4]

$$\min \quad \frac{1}{2}x_1^2 + \frac{1}{2}x_2^2 + 3x_1 + 7x_2 + x_4.$$
$$x \text{ s.t. } \begin{cases} x_1+2x_2+x_3=8 \\ x_1+2x_2+x_4=5 \\ x_{1,2,3,4}\geqslant 0 \end{cases}$$

4.11 Consider the problem in Example 4.29. Find the general descriptions of matrix $\boldsymbol{H}$ and vector $\boldsymbol{f}$.

4.12 Solve the nonconvex quadratic programming problem

$$\min \quad \boldsymbol{c}^{\mathrm{T}}\boldsymbol{x} - \frac{1}{2}\boldsymbol{x}^{\mathrm{T}}\boldsymbol{Q}\boldsymbol{x},$$
$$x \text{ s.t. } \begin{cases} \boldsymbol{A}\boldsymbol{x}\leqslant \boldsymbol{b} \\ 0\leqslant x_i\leqslant 10,\ i=1,2,\ldots,10 \end{cases}$$

where $\boldsymbol{c}^{\mathrm{T}} = [48, 42, 48, 45, 44, 41, 47, 42, 45, 46]$, $\boldsymbol{Q} = 10\boldsymbol{I}$, and

$$\boldsymbol{A} = \begin{bmatrix} -2 & -6 & -1 & 0 & -3 & -3 & -2 & -6 & -2 & -2 \\ 6 & -5 & 8 & -3 & 0 & 1 & 3 & 8 & 9 & -3 \\ -5 & 6 & 5 & 3 & 8 & -8 & 9 & 2 & 0 & -9 \\ 9 & 5 & 0 & -9 & 1 & -8 & 3 & -9 & -9 & -3 \\ -8 & 7 & -4 & -5 & -9 & 1 & -7 & -1 & 3 & -2 \end{bmatrix}, \quad \boldsymbol{b} = \begin{bmatrix} -4 \\ 22 \\ -6 \\ -23 \\ -12 \end{bmatrix}.$$

4.13 Input the following quadratic programming problem[22] into MATLAB workspace:

$$\min_{\boldsymbol{x} \text{ s.t. } \begin{cases} \boldsymbol{AX} \leqslant \boldsymbol{b}, \text{ where } \boldsymbol{X} = [\boldsymbol{x}; \boldsymbol{y}] \\ \boldsymbol{0} \leqslant \boldsymbol{X} \leqslant \boldsymbol{1} \end{cases}} \quad \boldsymbol{c}^{\mathrm{T}}\boldsymbol{x} + \boldsymbol{d}^{\mathrm{T}}\boldsymbol{y} - \frac{1}{2}\boldsymbol{x}^{\mathrm{T}}\boldsymbol{Q}\boldsymbol{x},$$

where

$$\boldsymbol{A} = \begin{bmatrix} -2 & -6 & -1 & 0 & -3 & -3 & -2 & -6 & -2 & -2 \\ 6 & -5 & 8 & -3 & 0 & 1 & 3 & 8 & 9 & -3 \\ -5 & 6 & 5 & 3 & 8 & -8 & 9 & 2 & 0 & -9 \\ 9 & 5 & 0 & -9 & 1 & -8 & 3 & -9 & -9 & -3 \\ -8 & 7 & -4 & -5 & -9 & 1 & -7 & -1 & 3 & -2 \\ -7 & -5 & -2 & 0 & -6 & -6 & -7 & -6 & 7 & 7 \\ 1 & -3 & -3 & -4 & -1 & 0 & -4 & 1 & 6 & 0 \\ 1 & -2 & 6 & 9 & 0 & -7 & 9 & -9 & -6 & 4 \\ -4 & 6 & 7 & 2 & 2 & 0 & 6 & 6 & -7 & 4 \\ 1 & 1 & 1 & 1 & 1 & 1 & 1 & 1 & 1 & 1 \\ -1 & -1 & -1 & -1 & -1 & -1 & -1 & -1 & -1 & -1 \end{bmatrix}, \quad \boldsymbol{b} = \begin{bmatrix} -4 \\ 22 \\ -6 \\ -23 \\ -12 \\ -3 \\ 1 \\ 12 \\ 15 \\ 9 \\ -1 \end{bmatrix},$$

and it is known that $\boldsymbol{x}$ has 3 elements, $\boldsymbol{y}$ has 7 elements, $\boldsymbol{Q} = 10\boldsymbol{I}$, and

$$\boldsymbol{d} = [10, 10, 10]^{\mathrm{T}}, \quad \boldsymbol{c} = [-20, -80, -20, -50, -60, -90, 0]^{\mathrm{T}}.$$

4.14 Solve the following problem with both the Robust Control Toolbox and YALMIP Toolbox:

$$\min_{\boldsymbol{X} \text{ s.t. } \begin{cases} \begin{bmatrix} \boldsymbol{A}^{\mathrm{T}}\boldsymbol{X} + \boldsymbol{X}\boldsymbol{A} + \boldsymbol{Q} & \boldsymbol{X}\boldsymbol{B} \\ \boldsymbol{B}^{\mathrm{T}}\boldsymbol{X} & -\boldsymbol{I} \end{bmatrix} < \boldsymbol{0} \\ \boldsymbol{X} < \boldsymbol{0} \end{cases}} \quad \mathrm{trace}(\boldsymbol{X}),$$

where

$$\boldsymbol{A} = \begin{bmatrix} -1 & -2 & 1 \\ 3 & 2 & 1 \\ 1 & -2 & -1 \end{bmatrix}, \quad \boldsymbol{B} = \begin{bmatrix} 1 \\ 0 \\ 1 \end{bmatrix}, \quad \boldsymbol{Q} = \begin{bmatrix} 1 & -1 & 0 \\ -1 & -3 & -12 \\ 0 & -12 & -36 \end{bmatrix}.$$

4.15 Solve the following linear matrix inequalities problem:

$$\begin{cases} \boldsymbol{P}^{-1} > \mathbf{0}, \text{ or equivalently, } \boldsymbol{P} > \mathbf{0}, \\ \boldsymbol{A}_1\boldsymbol{P} + \boldsymbol{P}\boldsymbol{A}_1^{\mathrm{T}} + \boldsymbol{B}_1\boldsymbol{Y} + \boldsymbol{Y}^{\mathrm{T}}\boldsymbol{B}_1^{\mathrm{T}} < \mathbf{0}, \\ \boldsymbol{A}_2\boldsymbol{P} + \boldsymbol{P}\boldsymbol{A}_2^{\mathrm{T}} + \boldsymbol{B}_2\boldsymbol{Y} + \boldsymbol{Y}^{\mathrm{T}}\boldsymbol{B}_2^{\mathrm{T}} < \mathbf{0}, \end{cases}$$

where

$$\boldsymbol{A}_1 = \begin{bmatrix} -1 & 2 & -2 \\ -1 & -2 & 1 \\ -1 & -1 & 0 \end{bmatrix}, \quad \boldsymbol{B}_1 = \begin{bmatrix} -2 \\ 1 \\ -1 \end{bmatrix}, \quad \boldsymbol{A}_2 = \begin{bmatrix} 0 & 2 & 2 \\ 2 & 0 & 2 \\ 2 & 0 & 1 \end{bmatrix}, \quad \boldsymbol{B}_2 = \begin{bmatrix} -1 \\ -2 \\ -1 \end{bmatrix}.$$

4.16 Represent the following optimization problem with quadratic constraints[23] with YALMIP Toolbox and find the solutions:

$$\min_{\boldsymbol{x} \text{ s.t. } \begin{cases} -1+0.0025(x_4+x_6)\leqslant 0 \\ -1+0.0025(-x_4+x_5+x_7)\leqslant 0 \\ -1+0.01(-x_5+x_8)\leqslant 0 \\ 100x_1-x_1x_6+8\,333.33252x_4-83\,333.333\leqslant 0 \\ x_2x_4-x_2x_7-1\,250x_4+1\,250x_5\leqslant 0 \\ x_3x_5-x_3x_8-2\,500x_5+1\,250\,000\leqslant 0 \\ 100\leqslant x_1\leqslant 10\,000 \\ 1\,000\leqslant x_2,x_3\leqslant 10\,000 \\ 10\leqslant x_4,x_5,x_6,x_7,x_8\leqslant 1\,000 \end{cases}} x_1 + x_2 + x_3.$$

In [23], a solution is provided as follows. Are we able to get a better solution with YALMIP Toolbox?

$$\boldsymbol{x}^* = [579.19, 1\,360.13, 5\,109.92, 182.01, 295.6, 271.99, 286.4, 395.6]^{\mathrm{T}}, \quad f_0 = 7\,079.25.$$

5 Nonlinear programming

Linear and quadratic programming problems discussed in Chapter 4 are special cases of nonlinear programming problems. Since they are widely used, and especially because they are convex problems, global optimum solutions can be found, and the user needs not worry about local minimum solutions. If a solution can be found, it is the global optimum solution. For quadratic programming problems, if the quadratic matrix is positive-definite, the quadratic programming problems are also known as convex problems. Of course, it has been demonstrated in Chapter 4 that if the quadratic term is not positive-definite, quadratic problem solvers may not be able to get the global optimum solutions. Sometimes, even feasible solutions cannot be found. There is a need to introduce better solvers for solving these problems.

Besides, linear and quadratic programming problems are limited, for instance, because the constraints are only linear expressions, not nonlinear ones. The objective function should also be extended to the nonlinear terms. Therefore, here nonlinear optimization problem concepts and solution methods are introduced. Nonlinear optimization problems is also referred to as nonlinear programming problems.

In Section 5.1, a mathematical model of nonlinear programming problems is presented, and the constraints are further classified into linear and nonlinear constraints. The concept and geometric interpretations of feasible regions in simple problems are illustrated. A low-level MATLAB solution methods are discussed. Since the low-level algorithm is not really practical in solving complicated problems, in Section 5.2, we discuss a universal and powerful solver which is provided in MATLAB Optimization Toolbox. Some techniques are introduced through examples. For instance, sometimes the solver may fail to find an actual optimum solution, then a loop structure can be used to find the expected one. Besides, some complicated nonlinear programming problems are explored. Since traditional nonlinear programming solvers are not able to find global optimum solutions, in Section 5.3, a general-purpose search method is presented, aiming at finding global optimum solutions to ordinary nonlinear programming problems. Nonconvex quadratic and nonlinear programming problems are explored. In Section 5.4, the definition and solutions of bilevel programming problems are formulated. In Section 5.5, applications of nonlinear programming problems, such as semiinfinite programming problems and several others, are presented.

5.1 Introduction to nonlinear programming

In this section, ordinary nonlinear programming problems are explored. The graphical method is introduced first to demonstrate the evaluation of feasible regions, and then solutions of nonlinear programming problems with two decision variables are demonstrated. An idea and its MATLAB implementation when solving a nonlinear problem is presented.

https://doi.org/10.1515/9783110667011-005

5.1.1 Mathematical models of nonlinear programming problems

A general mathematical description on nonlinear programming problems is introduced in Definition 4.1. From the easy manipulation point of view, the linear constraints can be separated from the nonlinear ones, so that the general form of nonlinear programming problems as follows can be established.

Definition 5.1. The mathematical form of constrained nonlinear programming problems is

$$\min_{\boldsymbol{x} \text{ s.t. } \begin{cases} \boldsymbol{A}\boldsymbol{x}\leqslant\boldsymbol{B} \\ \boldsymbol{A}_{\rm eq}\boldsymbol{x}=\boldsymbol{B}_{\rm eq} \\ \boldsymbol{x}_{\rm m}\leqslant\boldsymbol{x}\leqslant\boldsymbol{x}_{\rm M} \\ \boldsymbol{C}(\boldsymbol{x})\leqslant\boldsymbol{0} \\ \boldsymbol{C}_{\rm eq}(\boldsymbol{x})=\boldsymbol{0} \end{cases}} f(\boldsymbol{x}), \tag{5.1.1}$$

where $\boldsymbol{x} = [x_1, x_2, \dots, x_n]^{\rm T}$ is decision vector, and the objective function $f(\boldsymbol{x})$ is a scalar function.

5.1.2 Feasible regions and graphical methods

In this section, the concept of a feasible region of a nonlinear programming problem is introduced. Then, the graphical method is demonstrated by a simple nonlinear programming problem.

Definition 5.2. The set of all $\boldsymbol{x}$ satisfying the constraints in (5.1.1) is referred to as the feasible region of a nonlinear programming problem.

Similar to equation solution problems, problems with one or two independent variables can be solved graphically. The feasible region and graphical solutions of problems with two decision variables are demonstrated through examples.

Example 5.1. Use the graphical method to study the following optimization problem with two decision variables:

$$\max_{\boldsymbol{x} \text{ s.t. } \begin{cases} 9\geqslant x_1^2+x_2^2 \\ x_1+x_2\leqslant 1 \end{cases}} -x_1^2 - x_2.$$

Solutions. Selecting $-3 \leqslant x_1, x_2 \leqslant 3$, generate mesh grids in the x_1–x_2 plane. Find the function values at the mesh grids points, and solve the unconstrained optimization problem from the three-dimensional data.

```
>> [x1,x2]=meshgrid(-3:0.01:3); % generate mesh grids
   z=-x1.^2-x2;                 % compute objective function values
```

If constraints are introduced, the values of the objective function at points where the constraints do not hold can be removed, that is, we find the subscripts of such points and set the function values there to NaN. The problem can be solved with the following commands:

```
>> i=find(x1.^2+x2.^2>9); z(i)=NaN; % find the points x1^2+x2^2>9, set to NaN
   i=find(x1+x2>1); z(i)=NaN;        % find the points x1+x2>1, set to NaN
   surfc(x1,x2,z); shading flat;     % draw surface plot
```

The surface plot in Figure 5.1 can be obtained. Since function surfc() is used, the contours on the x_1–x_2 plane are also superimposed. It can be seen that the maximum value of the remaining plot is the solution of the original problem, labeled on the plot.

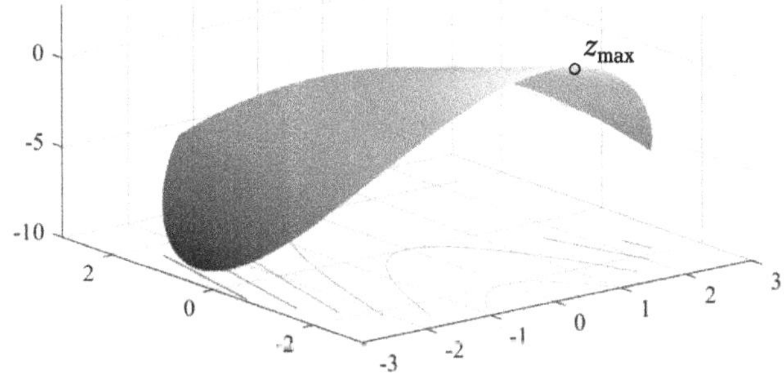

Figure 5.1: Surface plot of the feasible solutions.

It can be seen mathematically that the first constraint forces the solution to be in the interior of a circle, centered at the origin and with a radius of 3. The second constraint is the part under a straight line. Solving the simultaneous inequalities, the feasible region can easily be found.

In fact, if one wants to observe the plot downwards, function view() can be used to set the view point. Implicit function drawing commands can be used to superimpose the bounds of the two constraints, as shown in Figure 5.2.

```
>> view(0,90), hold on; syms x1 x2;
   fimplicit(x1+x2==1), fimplicit(x1^2+x2^2==9), hold off
```

The shaded area is the feasible region of the corresponding optimization problem. All the points satisfying the constraints appear in the feasible region. The point with the maximum objective function value is the global optimum solution of the problem. It

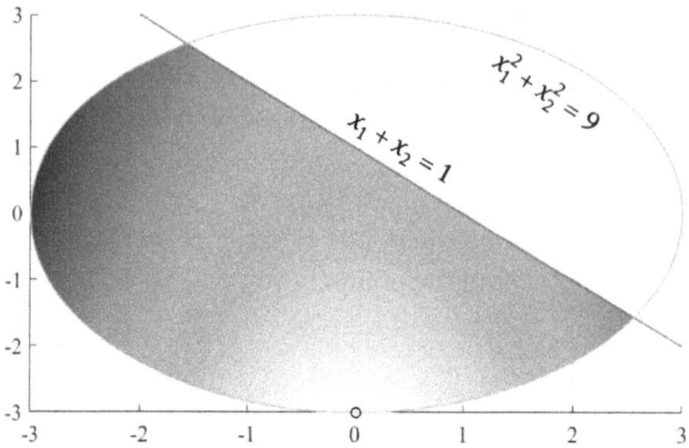

Figure 5.2: Illustration of a feasible region.

can be found from the figure that the solution is $x_1 = 0$ and $x_2 = -3$. The function value of 3 can be obtained with `max(z(:))` command.

Example 5.2. Consider the following Marsha test function. Use the graphical method to solve the optimization problem:

$$\min_{x,y \text{ s. t. } (x+5)^2+(y+5)^2<25} \quad e^{(1-\cos x)^2}\sin y + e^{(1-\sin y)^2}\cos x + (x-y)^2.$$

Solutions. It can be seen that the feasible region is defined as the interior part of a circle, centered at $(-5,-5)$, with radius 5. Mesh grids can be generated in the region $-10 \leqslant x, y \leqslant 0$, and function values at the mesh grids can be computed. With function `find()`, the mesh grid points that do not satisfy the constraints can be found, and the function values there can be set to `NaN`. The surface in the feasible region can be obtained as shown in Figure 5.3.

```
>> [x,y]=meshgrid(-10:0.01:0);
   z=exp((1-cos(x)).^2).*sin(y)+exp((1-sin(y)).^2).*cos(x)+(x-y).^2;
   i=find((x+5).^2+(y+5).^2>25); z(i)=NaN;
   surfc(x,y,z), shading flat
```

For problems with one or two decision variables, of course, graphical methods can be used. For ordinary optimization problems, graphical methods are not suitable. Numerical methods should be used to handle the complicated problems. There is no method to test whether a solution found is global or not.

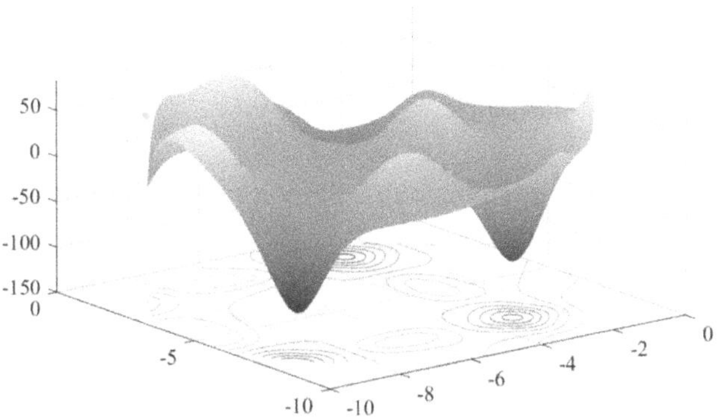

Figure 5.3: Graphical interpretation of Marsha function problem.

5.1.3 Examples of numerical methods

In Chapter 3, solutions of unconstrained optimization problems are studied. Of course, the solver `fminunc()` cannot be used to handle constrained problems. There should be an alternative method, to set the values of the objective function at points which do not satisfy the constraints to a very large value, and then convert the problem into an unconstrained one. In this way, function `fminunc()` can be tried to solve constrained optimization problems.

The method of modifying objective function values manually is referred to as punish function method. The selection of the punish function is not unique. The idea considered above is the most straightforward one. An example will be used to demonstrate the use of an unconstrained optimization solver in handling unconstrained problems.

Example 5.3. Use the idea in the punish function method presented above to solve the constrained optimization problem in Example 5.2.

Solutions. Since in the original problem the decision variables are x and y, while in the solver `fminunc()` a vectorized decision variable is expected, it is natural to let $x_1 = x$ and $x_2 = y$, and modify the original objective function manually as

$$f(\boldsymbol{x}) = \mathrm{e}^{(1-\cos x_1)^2}\sin x_2 + \mathrm{e}^{(1-\sin x_2)^2}\cos x_1 + (x_1 - x_2)^2.$$

If $(x_1+5)^2 + (x_2+5)^2 \geqslant 25$, the value of the objective function can be set to a very larger value, e. g., 10^5. Then, a forbidden area can be set up in the infeasible region. With such an idea, the objective function with penalties can be written as follows:

```
function y=c5mmar1(x)
if sum((x+5).^2)>=25, y=1e5;
else
```

```
    y=exp((1-cos(x(1)))^2)*sin(x(2))+...
      exp((1-sin(x(2)))^2)*cos(x(1))+(x(1)-x(2))^2;
end
```

For the problem, since there are local optimum points, the solver `fminunc()` is not suitable. The global function recommended in Chapter 3, `fminunc_global()`, can be used to solve the problem. The following commands can be tried, and the global optimum solution is found at $\boldsymbol{x}_1 = [-3.1302, -1.5821]^{\mathrm{T}}$. The objective function value is $f_0 = -106.7645$. If the command is called repeatedly, no better solution is found, so the solution can be regarded as the global optimum solution:

```
>> x1=fminunc_global(@c5mmar1,-10,0,2,100), f0=c5mmar1(x1)
```

From the above solution process, it seems that such a method can be used in handling nonlinear programming problems. In fact, this is not true. For inequality constraints, such a conversion can be made. Although the efficiency is low, at least it can be tried. However, if there are equality constraints, the method cannot be used at all. Effective and feasible algorithms should be used to solve the problems, not by merely setting forbidden regions.

If there exist equality constraints, Lagrange multiplier method can be used to convert a nonlinear programming problem into an unconstrained one. The method is named after French mathematician Joseph-Louis Lagrange (1736–1813). The basic idea of the method can be demonstrated through an example.

For a nonlinear programming problem with one equality constraint

$$\min_{\boldsymbol{x} \text{ s. t. } g(\boldsymbol{x})=0} f(\boldsymbol{x}),$$

a Lagrange multiplier $\lambda > 0$ can be introduced to construct a new objective function

$$H(\boldsymbol{x}) = f(\boldsymbol{x}) + \lambda g(\boldsymbol{x}), \tag{5.1.2}$$

such that the original nonlinear programming problem can be converted into an unconstrained one for $H(\boldsymbol{x})$. If there are many constraints, more Lagrange multipliers can be used. For unconstrained problems, many simultaneous equations can be set up:

$$\frac{\partial H}{\partial x_1} = 0, \quad \frac{\partial H}{\partial x_2} = 0, \quad \ldots, \quad \frac{\partial H}{\partial x_n} = 0, \quad \frac{\partial H}{\partial \lambda} = 0. \tag{5.1.3}$$

With the equation solution method, plus the second-order derivative information, the solution can be found and shown to be the minimum or maximum solution of the original problem.

Example 5.4. Solve the following nonlinear programming problem with Lagrange multiplier method:

$$\min_{x_1,x_2 \text{ s. t. } x_1^2+x_2^2=4} \quad \frac{3}{2}x_2^2 + 4x_1^2 + 5x_1x_2.$$

Solutions. A Lagrange multiplier $\lambda > 0$ should be selected, such that the new objective function $\boldsymbol{H}(\boldsymbol{x})$ can be constructed. Next we find the first-order derivative of the function with respect to each variable, that is, the gradient of the objective function, and then function `vpasolve()` can be used to find the quasianalytical solution of the problem:

```
>> syms x1 x2; syms lam positive
   f=3*x2^2/2+4*x1^2+5*x1*x2; g=x1^2+x2^2-4;
   H=f+lam*g; J=jacobian(H,[x1,x2,lam]);
   [x0,y0,lam0]=vpasolve(J,[x1,x2,lam])
```

It can be seen that the solution is $\lambda = 0.045085$, and $\boldsymbol{x}_a = [1.05146, -1.7013]$, $\boldsymbol{x}_b = -\boldsymbol{x}_a$. Substituting both points into the original objective function, it is found that $f(\boldsymbol{x}_a) = f(\boldsymbol{x}_b) = -0.1803$. They are both global optimum solutions of the original problem.

```
>> fa=subs(f,{x1,x2},{x0(1),y0(1)})
   fb=subs(f,{x1,x2},{x0(2),y0(2)})
```

Of course, if the derivatives of the original function are not known, numerical methods can be adopted to solve the optimization problems. The related methods are not further demonstrated here, since powerful and flexible solvers are provided in MATLAB Optimization Toolbox. In this chapter, we concentrate on the direct solutions of nonlinear programming problems.

5.2 Direct solutions of nonlinear programming problems

Many low-level nonlinear programming problem algorithms have been introduced previously. However, if the problems are complicated, their handling using these methods is rather complicated. Better solvers are needed to solve a more general nonlinear programming problem. In this section, a direct solver is proposed, then examples are shown how to use the solver to handle complicated problems.

5.2.1 Direct solution using MATLAB

Consider the nonlinear programming problem in Definition 5.1. For such a standard form, a solver `fmincon()` is provided in MATLAB Optimization Toolbox, dedicated to solving optimization problems under different constraints. The syntaxes of the solver are

$[\boldsymbol{x},f_0,\texttt{flag},\texttt{out}]=\texttt{fmincon(problem)}$

$[\boldsymbol{x},f_0,\texttt{flag},\texttt{out}]=\texttt{fmincon}(\texttt{F},\boldsymbol{x}_0,\boldsymbol{A},\boldsymbol{B})$

$[\boldsymbol{x},f_0,\texttt{flag},\texttt{out}]=\texttt{fmincon}(\texttt{F},\boldsymbol{x}_0,\boldsymbol{A},\boldsymbol{B},\boldsymbol{A}_{\rm eq},\boldsymbol{B}_{\rm eq})$

$[\boldsymbol{x},f_0,\texttt{flag},\texttt{out}]=\texttt{fmincon}(\texttt{F},\boldsymbol{x}_0,\boldsymbol{A},\boldsymbol{B},\boldsymbol{A}_{\rm eq},\boldsymbol{B}_{\rm eq},\boldsymbol{x}_{\rm m},\boldsymbol{x}_{\rm M})$

$[\boldsymbol{x},f_0,\texttt{flag},\texttt{out}]=\texttt{fmincon}(\texttt{F},\boldsymbol{x}_0,\boldsymbol{A},\boldsymbol{B},\boldsymbol{A}_{\rm eq},\boldsymbol{B}_{\rm eq},\boldsymbol{x}_{\rm m},\boldsymbol{x}_{\rm M},\texttt{C})$

$[\boldsymbol{x},f_0,\texttt{flag},\texttt{out}]=\texttt{fmincon}(\texttt{F},\boldsymbol{x}_0,\boldsymbol{A},\boldsymbol{B},\boldsymbol{A}_{\rm eq},\boldsymbol{B}_{\rm eq},\boldsymbol{x}_{\rm m},\boldsymbol{x}_{\rm M},\texttt{C},\texttt{ff})$

$[\boldsymbol{x},f_0,\texttt{flag},\texttt{c}]=\texttt{fmincon}(\texttt{F},\boldsymbol{x}_0,\boldsymbol{A},\boldsymbol{B},\boldsymbol{A}_{\rm eq},\boldsymbol{B}_{\rm eq},\boldsymbol{x}_{\rm m},\boldsymbol{x}_{\rm M},\texttt{C},\texttt{ff},p_1,p_2,\ldots)$

where `F` is the function handle of the objective function, provided as either a MATLAB or anonymous function. The argument $\boldsymbol{x}_0$ is the initial search point. If a certain constraint does not exist, empty matrices should be used. The argument `C` is the handle of MATLAB function to describe nonlinear constraints. The function should have two returned arguments, $\boldsymbol{c}$ and $\boldsymbol{c}_{\rm eq}$. The former describes the nonlinear inequalities and the latter is for nonlinear equalities, both in vectorized form. Since two returned arguments are expected, anonymous functions cannot be used. Argument `ff` contains the control options.

After the function call, the result is returned in vector $\boldsymbol{x}$, and the objective function value is returned in f_0. If the returned argument `flag` is positive, the solution process is successful.

If additional parameters are needed, they should be described in the same way in the objective function and constraints.

A structured variable can be used to describe nonlinear programming problems. The members are listed in Table 5.1. They can be set directly. If there are no constraints,

Table 5.1: Members in nonlinear programming problem structured variables.

member names	member description
`Objective`	objective function handle
`Aineq, bineq`	linear inequality constraints $\boldsymbol{A}$ and $\boldsymbol{b}$
`Aeq, beq`	linear equality constraints $\boldsymbol{A}_{\rm eq}$ and $\boldsymbol{b}_{\rm eq}$, where if the constraints are not used, they can be set to empty matrices, or not used at all
`ub,lb`	upper and lower bounds of decision variables $\boldsymbol{x}_{\rm M}$, $\boldsymbol{x}_{\rm m}$
`options`	control options. As before, the user may set the control parameters first, then assign the result to the `options` member
`solver`	must be set to `'fmincon'`
`nonlcon`	nonlinear constraint handle of MATLAB function, where two returned arguments $\boldsymbol{c}$ and $\boldsymbol{c}_{\rm eq}$ are needed
`x0`	initial search point $\boldsymbol{x}_0$

the members may not be set. If a structured variable is used to describe optimization problems, additional parameters are not supported, or additional parameters can be written in the relevant MATLAB functions.

Example 5.5. Solve the following constrained nonlinear programming problem:

$$\min_{\boldsymbol{x} \text{ s.t. } \begin{cases} x_1^2+x_2^2+x_3^2-25=0 \\ 8x_1+14x_2+7x_3-56=0 \\ x_1,x_2,x_3\geqslant 0 \end{cases}} 1000 - x_1^2 - 2x_2^2 - x_3^2 - x_1x_2 - x_1x_3.$$

Solutions. Analyzing the given nonlinear programming problem, it is found that the constraints have nonlinear equalities. Therefore, the quadratic programming solver cannot be used. A nonlinear programming solver must be adopted. From the mathematical description of the problem, the objective function can be written as the anonymous function

```
>> f=@(x)1000-x(1)*x(1)-2*x(2)*x(2)-x(3)*x(3)-x(1)*x(2)-x(1)*x(3);
```

Meanwhile since the two constraints are equalities, the inequality constraint should be set to an empty matrix. Then, the following commands can be used to describe the nonlinear constraints:

```
function [c,ceq]=opt_con1(x)
c=[];     %no nonlinear inequalities, hence an empty matrix
ceq=[x(1)*x(1)+x(2)*x(2)+x(3)*x(3)-25;
     8*x(1)+14*x(2)+7*x(3)-56];
```

The nonlinear constraints can be modified as `c` and `ceq`, where the former describes inequality constraints, while the latter the equality ones. If a certain item does not exist, it must be set to an empty matrix.

When the nonlinear equality is described this way, then $\boldsymbol{A}$, $\boldsymbol{B}$, $\boldsymbol{A}_{\text{eq}}$, and $\boldsymbol{B}_{\text{eq}}$ are empty matrices. Besides, the lower bound of the decision variable is $\boldsymbol{x}_{\text{m}} = [0,0,0]^{\text{T}}$. The initial point can be set to $\boldsymbol{x}_0 = [1,1,1]^{\text{T}}$. The solver `fmincon()` can be called to solve the optimization problem.

```
>> ff=optimset; ff.Display='iter'; %control options
   ff.TolFun=eps; ff.TolX=eps; ff.TolCon=eps;
   x0=[1;1;1]; xm=[0;0;0]; xM=[];
   A=[]; B=[]; Aeq=[]; Beq=[]; %constraints
   [x,f_opt,flag,d]=fmincon(f,x0,A,B,Aeq,Beq,xm,xM,@opt_con1,ff)
```

The optimum solution obtained is $\boldsymbol{x} = [3.5121, 0.2170, 3.5522]^{\text{T}}$, with the objective function value of $f_{\text{opt}} = 961.7151$. From the components in d it can be seen that 15 iterations are made, and the objective function is called 113 times. The intermediate iteration results are obtained as shown below.

```
                                            First-order      Norm of
 Iter F-count          f(x)  Feasibility   optimality          step
    0       4  9.940000e+02    2.700e+01    4.070e-01
    1       8  9.862197e+02    2.383e+01    1.355e+00     1.898e+00
    2      12  9.854621e+02    2.307e+01    1.407e+00     1.026e-01
    3      16  9.509871e+02    1.323e+01    4.615e+00     3.637e+00
    4      20  9.570458e+02    4.116e+00    3.569e+00     2.029e+00
    5      24  9.611754e+02    4.729e-01    1.203e+00     6.877e-01
    6      28  9.615389e+02    1.606e-01    4.060e-01     4.007e-01
  ...
   13      56  9.617152e+02    1.208e-13    9.684e-07     3.351e-07
   14      76  9.617152e+02    0.000e+00    8.081e-07     3.272e-11
```

Nonlinear programming problems can be described and solved with the following statements. The results obtained are still the same. It can be seen that the structured variable description is more concise, and the solution process is more straightforward.

```
>> clear P; P.objective=f; P.nonlcon=@opt_con1;
   P.x0=x0; P.lb=xm; P.options=ff; %structured variable description
   P.solver='fmincon'; [x,f_opt,c,d]=fmincon(P) %solution
```

The second constraint is a linear equality, it can be removed from the constraint function, so that the constraints can be simplified as

```
function [c,ceq]=opt_con2(x) %new nonlinear constraints
ceq=x(1)*x(1)+x(2)*x(2)+x(3)*x(3)-25; c=[];
```

Linear constraints can be described by relevant matrices. The following commands can be used to solve the problem, and the solution is exactly the same as that obtained earlier.

```
>> x0=[1;1;1]; Aeq=[8,14,7]; Beq=56; %use matrix for linear equality
   [x,f_opt,c,d]=fmincon(f,x0,A,B,Aeq,Beq,xm,xM,@opt_con2,ff)
```

Example 5.6. Solve again the Marsha test problem in Example 5.2:

$$\min_{x,y \text{ s.t. } (x+5)^2+(y+5)^2<25} \mathrm{e}^{(1-\cos x)^2}\sin y + \mathrm{e}(1-\sin y)^2\cos x + (x-y)^2.$$

Solutions. Since the decision variables in the original problem are x and y, not the expected $\boldsymbol{x}$ vector, it is necessary to introduce a vectorized decision variable $\boldsymbol{x}$ by setting

$x_1 = x$ and $x_2 = y$. Then, the problem can be manually rewritten as

$$\min_{\boldsymbol{x} \text{ s.t. } (x_1+5)^2+(x_2+5)^2-25<0} \quad e^{(1-\cos x_1)^2}\sin x_2 + e(1-\sin x_2)^2\cos x_1 + (x_1-x_2)^2.$$

Since the constraints are nonlinear, a MATLAB function should be written to describe the constraints. There is an inequality constraint and no equality ones, so the returned argument `ce` must be set to an empty matrix.

```
function [c,ce]=c5mmarsha(x)
ce=[]; c=(x(1)+5)^2+(x(2)+5)^2-25;
```

With the constraints description, the other linear constraints can be set to empty matrices. With the following commands, the problem can be solved, and the result is $\boldsymbol{x} = [-3.1302, -1.5821]$ while $f_0 = -106.7645$.

```
>> f=@(x)exp((1-cos(x(1)))^2)*sin(x(2))+...
        exp((1-sin(x(2)))^2)*cos(x(1))+(x(1)-x(2))^2;
   x0=[-10; 10]; A=[]; B=[]; Aeq=[]; Beq=[]; xm=[]; xM=[];
   [x f0 flag d]=fmincon(f,x0,A,B,Aeq,Beq,xm,xM,@c5mmarsha)
```

In fact, since many constraints are empty, the problem is more suitable for using structured variables. The following commands can be used to solve the problem, and the result obtained is the same as obtained above.

```
>> clear P; P.solver='fmincon'; P.options=optimset;
   P.Objective=f; P.x0=x0; P.nonlcon=@c5mmarsha;
   [x f0 flag d]=fmincon(P)
```

Example 5.7. Solve the following optimization problem with Townsend function:

$$\min_{x,y \text{ s.t. } \begin{cases} x^2+y^2-(2\cos t-\cos 2t/2-\cos 3t/4-\cos 4t/8)^2-4\sin^2 t\leqslant 0 \\ \text{where } t=\text{atan2}(x,y)\end{cases}} \quad -\cos^2((x-0.1)y) - x\sin(3x+y).$$

Solutions. The decision variables here are x and y, not the expected decision variable vector. One may let $x_1 = x$ and $x_2 = y$, such that the problem can be written as

$$\min_{\boldsymbol{x} \text{ s.t. } \begin{cases} x_1^2+x_2^2-(2\cos t-\cos 2t/2-\cos 3t/4-\cos 4t/8)^2-4\sin^2 t\leqslant 0 \\ \text{where } t=\text{atan2}(x_1,x_2)\end{cases}} \quad -\cos^2((x_1-0.1)x_2) - x_1\sin(3x_1+x_2).$$

In describing the nonlinear constraints, the intermediate variable t and then the nonlinear constraint `c` can be computed. The equality constant `ce` should be set to an empty matrix. The following commands can be used in the MATLAB function to describe nonlinear constraints:

```
function [c,ce]=c5mtown(x)
ce=[]; t=atan2(x(1),x(2));
c=x(1)^2+x(2)^2-4*sin(t)^2-(2*cos(t)-cos(2*t)/2 ...
    -cos(3*t)/4-cos(4*t)/8)^2;
```

With this description, the following commands can be used to solve the nonlinear programming problem. For instance, if the initial vector is selected as $\boldsymbol{x}_0 = [1,1]^\mathrm{T}$, the problem can be solved, and the optimum solution is $\boldsymbol{x} = [0.7525, -0.3235]^\mathrm{T}$, with the value of objective function being $f_0 = -1.6595$.

```
>> f=@(x)-(cos((x(1)-0.1)*x(2)))^2-x(1)*sin(3*x(1)+x(2));
   x0=[1; 1]; A=[]; B=[]; Aeq=[]; Beq=[]; xm=[]; xM=[];
   [x f0 flag d]=fmincon(f,x0,A,B,Aeq,Beq,xm,xM,@c5mtown)
```

If another initial value $\boldsymbol{x}_0 = [-10, 10]^\mathrm{T}$ is selected, the solution obtained is $\boldsymbol{x} = [-2.0206, -0.0561]^\mathrm{T}$, while the objective function value is $f_0 = -2.0240$. It is obvious that this is better than the previously found result.

```
>> f=@(x)-(cos((x(1)-0.1)*x(2)))^2-x(1)*sin(3*x(1)+x(2));
   x0=[-10; 10]; A=[]; B=[]; Aeq=[]; Beq=[]; xm=[]; xM=[];
   [x f0 flag d]=fmincon(f,x0,A,B,Aeq,Beq,xm,xM,@c5mtown)
```

If the graphical method is used to find the global optimum of the original problem, the following commands should be used. The mesh grid data can be generated and contours of the objective function can be obtained as shown in Figure 5.4. With an appropriate viewpoint setting, the global optimum solution can be found. It can be seen that the solution is a global one. The original function may have several valleys, so if the initial value is not selected properly, the global solution may not be found.

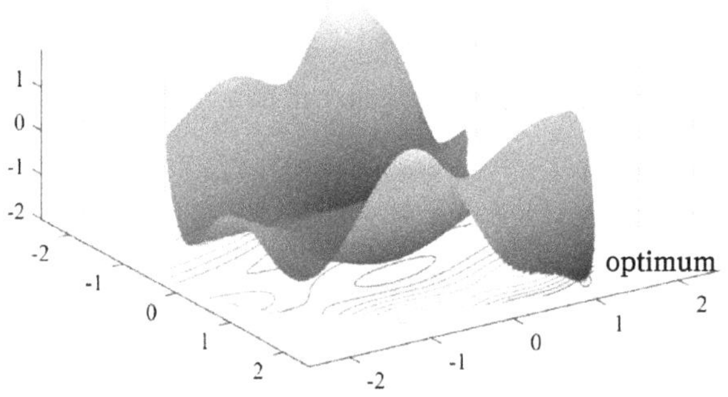

Figure 5.4: Townsend function optimization.

```
>> [x,y]=meshgrid(-2.5:0.01:2.5); t=atan2(x,y);
   z=-(cos((x-0.1).*y)).^2-x.*sin(3*x+y);
   ii=find(x.^2+y.^2-4*sin(t).^2 ...
       -(2*cos(t)-cos(2*t)/2-cos(3*t)/4-cos(4*t)/8).^2>0);
   z(ii)=NaN; surfc(x,y,z)
```

5.2.2 Handling of earlier termination phenomenon

When solving nonlinear programming problems, the solution process may sometimes be unsuccessful. For instance, if the `flag` returned is negative or zero, the phenomenon may sometimes be neglected. In fact, the reason for nonpositive `flag` values may be that the maximum allowed number of iterations or function calls is set too small, so that a premature termination in the optimization process happens. The result obtained may not be an optimum solution to the original problem.

How to avoid this phenomenon? Normally, it is possible to modify the control options, and set `MaxIter` or `MaxFunEvals` members to larger ones, for instance, 10 000, as it was done in Chapter 3. Also loops can be used to search for the optimum solutions. If `flag` is zero or negative, the result obtained can be used as the initial search point, until the value of `flag` becomes positive. The loop can then be terminated.

With such a method, normally optimum solutions can be found for nonlinear programming problems. It should be noted that the optimum value may not be a global optimum for the original problem. Global optimum solution methods will be further explored later.

Example 5.8. Solve the following nonlinear programming problem:

$$\min_{\boldsymbol{x} \text{ s.t. } \begin{cases} x_1+x_2\leqslant 0 \\ -x_1x_2+x_1+x_2\geqslant 1.5 \\ x_1x_2\geqslant -10 \\ -10\leqslant x_1,x_2\leqslant 10 \end{cases}} \quad e^{x_1}(4x_1^2+2x_2^2+4x_1x_2+2x_2+1).$$

Solutions. The constraints can be described with the following commands. Since the original problem has no nonlinear equalities, they should be set to an empty matrix. Besides, the first constraint is a linear one, but we do not have to bother with it. It can also be described in the nonlinear constraints. The last two constraints have $\geqslant$ inequalities, so they should be converted to the standard $\leqslant 0$ form. The constraint function can be written as follows:

```
function [c,ce]=c5exmcon(x)
ce=[]; % nonlinear constraints, with equality ones set to an empty matrix
c=[x(1)+x(2); x(1)*x(2)-x(1)-x(2)+1.5; -10-x(1)*x(2)];
```

Therefore, the following commands can be used to solve the original optimization problem directly. A structured variable is used here to describe the whole nonlinear programming problem. The solution can then be found directly.

```
>> clear P; P.nonlcon=@c5exmcon;
   P.solver='fmincon'; P.options=optimset;
   P.objective=@(x)exp(x(1))*(4*x(1)^2+2*x(2)^2+...
                4*x(1)*x(2)+2*x(2)+1);
   ff=optimset; ff.TolX=eps; ff.TolFun=eps; P.options=ff;
   P.lb=[-10; -10]; P.ub=-P.lb; P.x0=[0;0];  %structured
   [x,f0,flag]=fmincon(P) %direct solution
```

The "optimum solution" obtained is $\boldsymbol{x} = [0.4195, 0.4195]^{\mathrm{T}}$, with the objective function value of $f_1 = 5.4737$. By observing the results, it can be seen that `flag` is 0, which means that the solution may not be an optimum one. A warning "`fmincon` stopped because it exceeded the function evaluation limit" is also displayed, indicating that the restriction on the maximum number of objective function calls is violated. Abnormal termination of the solver is experienced. This also prompts us that when solving problems with MATLAB, we should not only pay attention to the results, but also notice the warning or error messages, if any. If a warning or error is displayed, alternative methods should be consider to solve the problem again. The result $\boldsymbol{x}$ can be used as an initial search point to solve the problem again. For the results, the warning message should also be detected, or we can see whether `flag` is positive or not. If not, the result should be used as a new initial search point to solve again. A loop structure is suitable for solving such problems. If the value of `flag` is positive, the loop may be terminated. With the following commands, the solution obtained is $\boldsymbol{x} = [1.1825, -1.7398]^{\mathrm{T}}$, while the objective function value is 3.0608 and the number of iteration steps is $i = 4$:

```
>> i=1;
   while 1
      P.x0=x; [x,f0,flag]=fmincon(P);
      if flag>0, break; end, i=i+1; %if solution found
   end
```

It should be noted that the result may still not be the global optimum. Another initial value can be selected as $\boldsymbol{x}_0 = [-10, -10]$, and we should see whether better solutions can be found.

5.2.3 Gradient information

Gradient information has been demonstrated in Section 3.2, for its applications in unconstrained problems. In fact, for constrained problems, if necessary, gradients can also be used, to speed up the search process, or improve the accuracy. In this section the gradient information in nonlinear programming problems is presented.

Example 5.9. Consider again the optimization problem in Example 5.5. Use gradient information to solve the problem again, and compare with the previous method.

Solutions. From the given objective function $f(x)$, the gradient can be found with the following commands:

```
>> syms x1 x2 x3;
   f=1000-x1*x1-2*x2*x2-x3*x3-x1*x2-x1*x3; % symbolic expression
   J=jacobian(f,[x1,x2,x3])                % compute gradients
```

The mathematical form can be written as

$$J = \left[\frac{\partial f}{\partial x_1}, \frac{\partial f}{\partial x_2}, \frac{\partial f}{\partial x_3}\right]^{\mathrm{T}} = \begin{bmatrix} -2x_1 - x_2 - x_3 \\ -4x_2 - x_1 \\ -2x_3 - x_1 \end{bmatrix}.$$

Compared with the method in Section 3.2, the gradient information should be described in the objective function file, as the second returned argument. The revised objective function file is then

```
function [y,Gy]=opt_fun2(x) %objective function & gradient
y=1000-x(1)*x(1)-2*x(2)*x(2)-x(3)*x(3)-x(1)*x(2)-x(1)*x(3);
Gy=[-2*x(1)-x(2)-x(3); -4*x(2)-x(1); -2*x(3)-x(1)]; %gradient
```

where Gy represents the gradient vector.

In the solution process, if gradient information is needed, the member GradObj in the control option should be set to 'on', then the solver can be called to solve the problem

```
>> x0=[1;1;1]; xm=[0;0;0]; xM=[];
   A=[]; B=[]; Aeq=[]; Beq=[]; ff=optimset; ff.GradObj='on';
   ff.TolFun=eps; ff.TolX=eps; ff.TolCon=eps;
   [x,f0,flag,cc]=fmincon(@opt_fun2,x0,A,B,Aeq,Beq,...
         xm,xM,@opt_con1,ff)
```

If a structured variable is used to model the problem, the following commands can be issued, and the same results are found.

```
>> clear P; P.x0=x0; P.lb=xm; P.options=ff;
   P.objective=@opt_fun2; P.nonlcon=@opt_con1;
   P.solver='fmincon'; x=fmincon(P) %structured variable
```

It can be seen that if the partial derivatives of the objective function are known, only 14 iterations are needed, with the number of calls to the objective function being 86, so that the original problem can be solved. It is faster than the previous one, where 113 iterations were carried out. While if the time of evaluating and programming is considered, the use of gradient information may not be a good choice. Note that if the gradient information is known, the `GradObj` option should be set to `'on'`, otherwise it may not be recognized.

5.2.4 Solving problems with multiple decision vectors

It can be seen from the standard form of nonlinear programming problems that the decision variable should be given in the $\boldsymbol{x}$ vector. While in actual problems, if there is more than one decision variable vector, a single decision variable vector should be defined. The original problem should be manually manipulated such that the standard form of nonlinear programming problems can be set up. Then, the solver `fmincon()` can be called to solve the problem. Next an example is given to show the conversion and solution process of such problems.

Example 5.10. Consider again the problem in Example 4.40. For convenience of presentation, the problem is given again:[26]

$$\min_{\boldsymbol{q},w,k \text{ s.t. } \begin{cases} q_3+9.625q_1w+16q_2w+16w^2+12-4q_1-q_2-78w=0 \\ 16q_1w+44-19q_1-8q_2-q_3-24w=0 \\ 2.25-0.25k\leqslant q_1\leqslant 2.25+0.25k \\ 1.5-0.5k\leqslant q_2\leqslant 1.5+0.5k \\ 1.5-1.5k\leqslant q_3\leqslant 1.5+1.5k \end{cases}} k.$$

Solutions. It can be seen that the decision variables for this problem are $\boldsymbol{q}$, w, and k. Variable substitutions should be made, by letting $x_1 = q_1$, $x_2 = q_2$, $x_3 = q_3$, $x_4 = w$, $x_5 = k$. The inequalities should be further handled such that the problem can be converted manually into the form

$$\min_{\boldsymbol{x} \text{ s.t. } \begin{cases} x_3+9.625x_1x_4+16x_2x_4+16x_4^2+12-4x_1-x_2-78x_4=0 \\ 16x_1x_4+44-19x_1-8x_2-x_3-24x_4=0 \\ -0.25x_5-x_1\leqslant -2.25 \\ x_1-0.25x_5\leqslant 2.25 \\ -0.5x_5-x_2\leqslant -1.5 \\ x_2-0.5x_5\leqslant 1.5 \\ -1.5x_5-x_3\leqslant -1.5 \\ x_3-1.5x_5\leqslant 1.5 \end{cases}} x_5.$$

From the conversion results it can be seen that there are two nonlinear equalities, and no nonlinear equalities. The following function can be established to describe nonlinear constraints:

```
function [c,ce]=c5mnls(x)
c=[];          % nonlinear constraints with no nonlinear equalities
ce=[x(3)+9.625*x(1)*x(4)+16*x(2)*x(4)+16*x(4)^2+12-...
        4*x(1)-x(2)-78*x(4);
   16*x(1)*x(4)+44-19*x(1)-8*x(2)-x(3)-24*x(4)];
```

The linear constraints can be expressed as nonlinear inequalities in matrix form $\boldsymbol{A}\boldsymbol{x} \leqslant \boldsymbol{b}$, where

$$\boldsymbol{A} = \begin{bmatrix} -1 & 0 & 0 & 0 & -0.25 \\ 1 & 0 & 0 & 0 & -0.25 \\ 0 & -1 & 0 & 0 & -0.5 \\ 0 & 1 & 0 & 0 & -0.5 \\ 0 & 0 & -1 & 0 & -1.5 \\ 0 & 0 & 1 & 0 & -1.5 \end{bmatrix}, \quad \boldsymbol{b} = \begin{bmatrix} -2.25 \\ 2.25 \\ -1.5 \\ 1.5 \\ -1.5 \\ 1.5 \end{bmatrix}.$$

Since there are neither linear equalities nor lower and upper bounds of decision variables, the constraints can be described by empty matrices. A structured variable can be used to describe the whole nonlinear programming problem. An initial search point can be generated to solve it. The solutions obtained is $\boldsymbol{x} = [1.9638, 0.9276, -0.2172, 0.0695, 1.1448]$, the same as that obtained in Example 4.40. The `flag` returned is 1, indicating that the solution process is successful.

```
>> clear P; P.objective=@(x)x(5);
   P.nonlcon=@c5mnls; P.solver='fmincon'; % structured variable
   P.Aineq=[-1,0,0,0,-0.25; 1,0,0,0,-0.25;
            0,-1,0,0,-0.5; 0,1,0,0,-0.5;
            0,0,-1,0,-1.5; 0,0,1,0,-1.5];
   P.Bineq=[-2.25; 2.25; -1.5; 1.5; -1.5; 1.5];
   P.options=optimoptions;
   P.x0=rand(5,1); [x,f0,flag]=fmincon(P) % solve the problem
```

It should be pointed out that local optimum solutions may be found in this way. Several random initial values can be tried. It can be seen that after some trials, a solution with the best optimum objective value of 0.8175 can be found. The result shows at least that the above result and that obtained in Example 4.40 are not the global optimum solution. In the next section, a global optimization solver is presented.

5.2.5 Complicated nonlinear programming problems

In many practical problems, the objective function and constraints are not always given in exactly the same manner as in (5.1.1). Sometimes the user must summarize and convert the problem into the standard form. In this section, an illustrative example is used to show how to solve a complicated optimization problem.

Example 5.11. Assume that in a minimization problem, the objective function is[12]

$$f(\boldsymbol{x}) = 0.7854x_1x_2^2(3.3333x_3^2 + 14.9334x_3 - 43.0934) - 1.508x_1(x_6^2 + x_7^2)$$
$$+ 7.477(x_6^3 + x_7^3) + 0.7854(x_4x_6^2 + x_5x_7^2).$$

The corresponding constraints are

$$x_1x_2^2x_3 \geqslant 27, \quad x_1x_2^2x_3^2 \geqslant 397.5, \quad x_2x_3x_6^4/x_4^3 \geqslant 1.93, \quad x_2x_3x_7^4/x_5^3 \geqslant 1.93 \tag{5.2.1}$$

$$A_1/B_1 \leqslant 1100, \text{ where } A_1 = \sqrt{(745x_4/(x_2x_3))^2 + 16.91\times 10^6}, \quad B_1 = 0.1x_6^3, \tag{5.2.2}$$

$$A_2/B_2 \leqslant 850, \text{ where } A_2 = \sqrt{(745x_5/(x_2x_3))^2 + 157.5\times 10^6}, \quad B_2 = 0.1x_7^3, \tag{5.2.3}$$

$$x_2x_3 \leqslant 40, \quad 5 \leqslant x_1/x_2 \leqslant 12, \quad 1.5x_6 + 1.9 \leqslant x_4, \quad 1.1x_7 + 1.9 \leqslant x_5, \tag{5.2.4}$$

$$2.6 \leqslant x_1 \leqslant 3.6, \quad 0.7 \leqslant x_2 \leqslant 0.8, \quad 17 \leqslant x_3 \leqslant 28, \tag{5.2.5}$$

$$7.3 \leqslant x_4, x_5 \leqslant 8.3, \quad 2.9 \leqslant x_6 \leqslant 3.9, \quad 5 \leqslant x_7 \leqslant 5.5. \tag{5.2.6}$$

Solutions. The form of the objective function is simple, and can be expressed directly with an anonymous function:

```
>> f=@(x)0.7854*x(1)*x(2)^2*(3.3333*x(3)^2+14.9334*x(3)-43.0934)...
      -1.508*x(1)*(x(6)^2+x(7)^2)+7.477*(x(6)^3+x(7)^3)...
      +0.7854*(x(4)*x(6)^2+x(5)*x(7)^2);
```

Equations (5.2.1)–(5.2.4) are used to describe nonlinear inequalities. Although the latter two constraints in (5.2.4) are linear, they are not distinguished here, and written together as other nonlinear ones. Besides, there are no nonlinear equality constraints. Before computing (5.2.2) and (5.2.3), matrices A_1, B_1, A_2, and B_2 should be computed first. The following nonlinear constraints can be written. Note that before writing each inequality, one must make sure that they are converted to the $\leqslant 0$ form.

```
function [c,ceq]=c5mcpl(x)
ceq=[];    % since no equality constraints, use empty matrix
A1=sqrt((745*x(4)/x(2)/x(3))^2+16.91e6); B1=0.1*x(6)^3;
A2=sqrt((745*x(5)/x(2)/x(3))^2+157.5e6); B2=0.1*x(7)^3;
c=[-x(1)*x(2)^2*x(3)+27; -x(1)*x(2)^2*x(3)^2+397.5;
   -x(2)*x(6)^4*x(3)/x(4)^3+1.93; A1/B1-1100;
   -x(2)*x(7)^4*x(3)/x(5)^3+1.93; A2/B2-850;
   x(2)*x(3)-40; -x(1)/x(2)+5; x(1)/x(2)-12;
   1.5*x(6)+1.9-x(4); 1.1*x(7)+1.9-x(5)];
```

With (5.2.5) and (5.2.6), the lower and upper bounds of the decision variables can be constructed as

$$\boldsymbol{x}_{\mathrm{m}} = [2.6, 0.7, 17, 17, 7.3, 2.9, 5]^{\mathrm{T}}, \quad \boldsymbol{x}_{\mathrm{M}} = [3.6, 0.8, 28, 8.3, 8.3, 3.9, 5.5]^{\mathrm{T}}.$$

With the following preparations, the following statements can be used in solving this optimization problem directly:

```
>> A=[]; B=[]; Aeq=[]; Beq=[];
   xm=[2.6,0.7,17,7.3,7.3,2.9,5];
   xM=[3.6,0.8,28,8.3,8.3,3.9,5.5];
   [x,f0]=fmincon(f,rand(7,1),A,B,Aeq,Beq,xm,xM,@c5mcp1)
```

The result obtained is $\boldsymbol{x} = [3.5, 0.7, 17, 7.3, 7.7153, 3.3505, 5.2867]$, with the optimal objective function value being $f_0 = 2994.4$. The result is the same as in [12].

If a structured variable used to study this optimization problem, the following commands can be issued. With 8 iterations and 72 objective function calls, the problem can be solved. The result is successfully obtained, within 0.036 seconds.

```
>> clear P; P.Objective=f; P.nonlcon=@c5mcp1;
   P.lb=xm; P.ub=xM; P.solver='fmincon'; P.options=optimset;
   P.x0=rand(7,1); tic, [x,f0,flag,cc]=fmincon(P), toc
```

5.3 Trials with global nonlinear programming solver

A powerful solver for nonlinear programming problems is provided in MATLAB Optimization Toolbox. The seemingly complicated nonlinear programming problems can be solved easily with the provided tools, since the standard form of nonlinear programming problems can be entered into MATLAB environment, so as to find the solutions directly. The biggest problem of the solver is that sometimes it may depend too much upon the use of initial search point. If the initial point is not selected properly, the global optimum solutions cannot be found. There are to date no well-accepted general-purpose methods in initial point selection, so trials must be made when finding global optimum solutions.

An attempt is made in this section. Based on the idea, a general-purpose global nonlinear programming problem solver is provided. Besides, further attempts are made to two classes of problems – global optimum solutions of nonconvex quadratic programming and concave-cost transportation problems.

5.3.1 Trials on global optimum solutions

It has been demonstrated that the classical search method may lead to local minimum solutions. Considering the idea introduced in Section 3.3, a loop structure can be used to assign the randomly generated initial points to the `fmincon()` solver, and we can call `fmincon()` each time to find an optimum solution. Each time a solution is found, it is compared with the recorded one, to see whether a better solution is found. If it is, we record the obtained one. In this case, it is quite likely that a global optimum solution is found.

With such an idea, a general-purpose nonlinear programming problem global solver can be written, for finding global optimum solutions.

```
function [x,f0,flag]=fmincon_global(f,a,b,n,N,varargin)
x0=rand(n,1); k0=0;
if strcmp(class(f),'struct'), k0=1; end
if k0==1, f.x0=x0; [x f0,flag]=fmincon(f); %structured variable
else
   [x f0,flag]=fmincon(f,x0,varargin{:});   %non structured variable
end
if flag==0, f0=1e10; end
for i=1:N
   x0=a(:)+(b(:)-a(:)).*rand(n,1) %try different initial points
   if k0==1, f.x0=x0; [x1 f1 flag]=fmincon(f); %structured variable
   else, [x1 f1 flag]=fmincon(f,x0,varargin{:}); end %normal one
   if flag>0 & f1<f0, x=x1; f0=f1; end %if a better solution is found
end
```

The general syntax of the function is

[$\boldsymbol{x}$,f_0]=`fmincon_global(fun,`a,b,n,N,other arguments)

where `fun` is a structured variable or the handle of the objective function. Arguments a and b are used to describe the decision variable bounds. If $\boldsymbol{x}_{\rm m}$ and $\boldsymbol{x}_{\rm M}$ are finite vectors, a and b can be set respectively to $\boldsymbol{x}_{\rm m}$ and $\boldsymbol{x}_{\rm M}$. Argument n is the number of decision variables; N is the number of calls of low-level function `fmincon()`. Generally, $N = 5 \sim 10$ are sufficient. If `fun` is the handle of the objective function, then "other arguments" may be set to other parameters for describing the constraints. The orders of the arguments are the same as those used in the solver `fmincon()`, that is, all the subsequent arguments after `F` and $\boldsymbol{x}_0$ in the `fmincon()` function call. It is quite likely that the returned argument $\boldsymbol{x}$ is the global optimum solution. The argument f_0 is the optimum objective function.

In the general case, the possible syntaxes are listed below. If a certain constraint does not exist, empty matrices can be used:

[$\boldsymbol{x}$,f_0]=`fmincon_global(problem,`a,b,n,N)

[$\boldsymbol{x}$,f_0]=`fmincon_global(`f,a,b,n,N,$\boldsymbol{A}$,$\boldsymbol{B}$,$\boldsymbol{A}_{\rm eq}$,$\boldsymbol{B}_{\rm eq}$,$\boldsymbol{x}_{\rm m}$,$\boldsymbol{x}_{\rm M}$)

[$\boldsymbol{x}$,f_0]=`fmincon_global(`f,a,b,n,N,$\boldsymbol{A}$,$\boldsymbol{B}$,$\boldsymbol{A}_{\rm eq}$,$\boldsymbol{B}_{\rm eq}$,$\boldsymbol{x}_{\rm m}$,$\boldsymbol{x}_{\rm M}$,`nfun)`

[$\boldsymbol{x}$,f_0]=`fmincon_global(`f,a,b,n,N,$\boldsymbol{A}$,$\boldsymbol{B}$,$\boldsymbol{A}_{\rm eq}$,$\boldsymbol{B}_{\rm eq}$,$\boldsymbol{x}_{\rm m}$,$\boldsymbol{x}_{\rm M}$,`nfun,ff)`

Example 5.12. Try to find the global optimum solution in Example 5.10 with the new `fmincon_global()` function.

Solutions. If the low-level `fmincon()` function is called 10 times in each run, the following commands can be used to solve the problem. It can be found that every time the solver is called, the global optimum solution is $\boldsymbol{x} = [2.4544, 1.9088, 2.7263, 1.3510, 0.8175]^{\mathrm{T}}$, where the fifth decision variable, x_5, is the objective function. The total elapsed time is around 0.342 seconds.

```
>> clear P; P.objective=@(x)x(5);
   P.nonlcon=@c5mnls; P.solver='fmincon';
   P.Aineq=[-1,0,0,0,-0.25; 1,0,0,0,-0.25;
            0,-1,0,0,-0.5; 0,1,0,0,-0.5;
            0,0,-1,0,-1.5; 0,0,1,0,-1.5]; %structured variable
   P.Bineq=[-2.25; 2.25; -1.5; 1.5; -1.5; 1.5];
   P.options=optimset; P.x0=rand(5,1);
   tic, [x,f0]=fmincon_global(P,-10,10,5,10), toc %find global solutions
```

If the solver is called 100 times, it is quite likely to find the global optimum solution. For this example, if the solver is called 100 times, every time the global optimum solution is found. The total elapsed time is 38.7 seconds.

```
>> tic, X=[];  %start stop watch to measure the time elapse
   for i=1:100 %run the solver 100 times and assess the success rate
     [x,f0]=fmincon_global(P,-10,10,5,10);
     X=[X; x'];                          %record results
   end, toc                              %display the total time elapse
```

In another syntax, where the original problem is expressed by the objective function f, constraints $\boldsymbol{A}$, $\boldsymbol{B}$ and other constraints, the result is exactly the same as that obtained with the structured variable syntax.

```
>> f=@(x)x(5); Aeq=[]; Beq=[]; xm=[]; xM=[];
   A=[-1,0,0,0,-0.25; 1,0,0,0,-0.25; 0,-1,0,0,-0.5;
      0,1,0,0,-0.5; 0,0,-1,0,-1.5; 0,0,1,0,-1.5];
   B=[-2.25; 2.25; -1.5; 1.5; -1.5; 1.5];
   X=[];   %start the stop watch and record the total elapsed time
   for i=1:100
     [x,f0]=fmincon_global(f,0,5,5,10,A,B,Aeq,Beq,xm,xM,@c6exnls);
     X=[X; x']; %record the obtained results
   end, toc
```

5.3.2 Nonconvex quadratic programming problems

Many nonconvex quadratic programming benchmark problems are provided in [22]. These problems cannot be solved with the `quadprog()` function discussed in Chapter 4. Therefore, trials on these problems can be made with the general-purpose nonlinear programming problem solvers. In this section, examples are given to show the global optimum solutions.

Example 5.13. Solve the nonconvex quadratic programming problem studied in Example 4.30. For the convenience of presentation, the mathematical model is given again:

$$\min \quad \boldsymbol{c}^{\mathrm{T}}\boldsymbol{x}+\boldsymbol{d}^{\mathrm{T}}\boldsymbol{y}-\frac{1}{2}\boldsymbol{x}^{\mathrm{T}}\boldsymbol{Q}\boldsymbol{x},$$
$$\boldsymbol{x} \text{ s. t.} \begin{cases} 2x_1+2x_2+y_6+y_7\leqslant 10\\ 2x_1+2x_3+y_6+y_8\leqslant 10\\ 2x_2+2x_3+y_7+y_8\leqslant 10\\ -8x_1+y_6\leqslant 0\\ -8x_2+y_7\leqslant 0\\ -8x_3+y_8\leqslant 0\\ -2x_4-y_1+y_6\leqslant 0\\ -2y_2-y_3+y_7\leqslant 0\\ -2y_4-y_5+y_8\leqslant 0\\ 0\leqslant x_i\leqslant 1,\ i=1,2,3,4\\ 0\leqslant y_i\leqslant 1,\ i=1,2,3,4,5,9\\ y_i\geqslant 0,\ i=6,7,8 \end{cases}$$

where $\boldsymbol{c}=[5,5,5,5]$, $\boldsymbol{d}=[-1,-1,-1,-1,-1,-1,-1,-1,-1]$, $\boldsymbol{Q}=10\boldsymbol{I}$.

Solutions. The problem is rather complicated, since two decision vectors, $\boldsymbol{x}$ and $\boldsymbol{y}$, are used. The conversional method needs some manual handling to change the problem into one with a single decision vector. The problem-based description method in Example 4.30 should be adopted. The problem can be sent to the computer first. Since the problem is no longer convex, the quadratic programming solver fails in solving it. New solvers must be tried for such a problem. For instance, function `fmincon()` and the global solver `fmincon_global()` can be tried.

If the two functions are tried, the problem itself must be entered into MATLAB environment. It is rather complicated, therefore, the method in Example 4.30 can be adopted. Then, `prob2struct()` function can be used to convert the problem into a structured variable p. In the converted problem, the first 4 elements in the decision vector come from vector $\boldsymbol{x}$, while the subsequent 9 are from the original vector $\boldsymbol{y}$.

```
>> P=optimproblem; c=5*ones(1,4); d=-1*ones(1,9); Q=10*eye(4);
   x=optimvar('x',4,1,'Lower',0,'Upper',1);
   y=optimvar('y',9,1,'Lower',0); % set swo decision vectors
   cons1=[2*x(1)+2*x(2)+y(6)+y(7)<=10;   % constraints
          2*x(1)+2*x(3)+y(6)+y(8)<=10;
          2*x(2)+2*x(3)+y(7)+y(8)<=10;
```

```
        -8*x(1)+y(6)<=0; -8*x(2)+y(7)<=0; -8*x(3)+y(8)<=0;
        -2*x(4)-y(1)+y(6)<=0; -2*y(2)-y(3)+y(7)<=0;
        -2*y(4)-y(5)+y(8)<=0; y([1 2 3 4 5 9])<=1];
P.Constraints.cons1=cons1; %same statements in Example 4.30
p=prob2struct(P);          %convert directly to structured variable
```

Besides, some of the members should be modified manually. For instance, the member `solver` in p should be changed from `'quadprog'` to `'fmincon'`. Besides, the decision variable should be set to the anonymous function handle. The error tolerance should be set again, and so on. With the following commands, the global optimum solution can be found:

```
>> p.solver='fmincon'; ff=optimset; ff.TolX=eps; ff.TolFun=eps;
   f=@(x)c*x(1:4)+d*x(5:13)-0.5*x(1:4)'*Q*x(1:4);
   p.Objective=f; p.options=ff; p.x0=100*rand(13,1); %precision
   x0=fmincon_global(p,-100,100,13,10); norm(x0-round(x0))
```

The solution obtained is $\boldsymbol{x}_0 = [1,1,1,1,1,1,1,1,1,3,3,3,1]^{\mathrm{T}}$, with the objective function value of −15, which is exactly the same as that in [22]. The norm in the error vector is 1.9978×10^{-10}.

Example 5.14. Solve the nonconvex quadratic programming problem[22]

$$\min_{\boldsymbol{x} \text{ s.t.} \begin{cases} \boldsymbol{A}_1\boldsymbol{x}+\boldsymbol{A}_2\boldsymbol{y}\leqslant\boldsymbol{b} \\ \boldsymbol{x},\boldsymbol{y}\geqslant\boldsymbol{0} \end{cases}} \quad -\frac{1}{2}\sum_{i=1}^{10}\lambda_i(x_i-\alpha_i)^2+\frac{1}{2}\sum_{i=1}^{10}\mu_i(y_i-\beta_i)^2,$$

where the matrices and vectors in the constraints are:

$$\boldsymbol{\lambda} = [63,15,44,91,45,50,89,58,86,82]^{\mathrm{T}},$$

$$\boldsymbol{\mu} = [42,98,48,91,11,63,61,61,38,26]^{\mathrm{T}},$$

$$\boldsymbol{\alpha} = [-19,-27,-23,-53,-42,26,-33,-23,41,19]^{\mathrm{T}},$$

$$\boldsymbol{\beta} = [-52,-3,81,30,-85,68,27,-81,97,-73]^{\mathrm{T}},$$

$$\boldsymbol{A}_1 = \begin{bmatrix} 3 & 5 & 5 & 6 & 4 & 4 & 5 & 6 & 4 & 4 \\ 5 & 4 & 5 & 4 & 1 & 4 & 4 & 2 & 5 & 2 \\ 1 & 5 & 2 & 4 & 7 & 3 & 1 & 5 & 7 & 6 \\ 3 & 2 & 6 & 3 & 2 & 1 & 6 & 1 & 7 & 3 \\ 6 & 6 & 6 & 4 & 5 & 2 & 2 & 4 & 3 & 2 \\ 5 & 5 & 2 & 1 & 3 & 5 & 5 & 7 & 4 & 3 \\ 3 & 6 & 6 & 3 & 1 & 6 & 1 & 6 & 7 & 1 \\ 1 & 2 & 1 & 7 & 8 & 7 & 6 & 5 & 8 & 7 \\ 8 & 5 & 2 & 5 & 3 & 8 & 1 & 3 & 3 & 5 \\ 1 & 1 & 1 & 1 & 1 & 1 & 1 & 1 & 1 & 1 \end{bmatrix}, \quad \boldsymbol{A}_2 = \begin{bmatrix} 8 & 2 & 4 & 1 & 1 & 1 & 2 & 1 & 7 & 3 \\ 3 & 6 & 1 & 7 & 7 & 5 & 8 & 7 & 2 & 1 \\ 1 & 7 & 2 & 4 & 7 & 5 & 3 & 4 & 1 & 2 \\ 7 & 7 & 8 & 2 & 3 & 4 & 5 & 8 & 1 & 2 \\ 7 & 5 & 3 & 6 & 7 & 5 & 8 & 4 & 6 & 3 \\ 4 & 1 & 7 & 3 & 8 & 3 & 1 & 6 & 2 & 8 \\ 4 & 3 & 1 & 4 & 3 & 6 & 4 & 6 & 5 & 4 \\ 2 & 3 & 5 & 5 & 4 & 5 & 4 & 2 & 2 & 8 \\ 4 & 5 & 5 & 6 & 1 & 7 & 1 & 2 & 2 & 4 \\ 1 & 1 & 1 & 1 & 1 & 1 & 1 & 1 & 1 & 1 \end{bmatrix}, \quad \boldsymbol{b} = \begin{bmatrix} 380 \\ 415 \\ 385 \\ 405 \\ 470 \\ 415 \\ 400 \\ 460 \\ 400 \\ 200 \end{bmatrix}.$$

Solutions. Direct modeling from objective function to find $\boldsymbol{H}$ matrix and $\boldsymbol{f}$ vector is not a simple thing. Therefore, a problem-based description of the problem must be used to describe the entire quadratic programming problem. Then, it must be converted into a structured variable.

```
>> A1=[3,5,5,6,4,4,5,6,4,4; 5,4,5,4,1,4,4,2,5,2;
       1,5,2,4,7,3,1,5,7,6; 3,2,6,3,2,1,6,1,7,3;
       6,6,6,4,5,2,2,4,3,2; 5,5,2,1,3,5,5,7,4,3;
       3,6,6,3,1,6,1,6,7,1; 1,2,1,7,8,7,6,5,8,7;
       8,5,2,5,3,8,1,3,3,5; 1,1,1,1,1,1,1,1,1,1];
   A2=[8,2,4,1,1,1,2,1,7,3; 3,6,1,7,7,5,8,7,2,1;
       1,7,2,4,7,5,3,4,1,2; 7,7,8,2,3,4,5,8,1,2;
       7,5,3,6,7,5,8,4,6,3; 4,1,7,3,8,3,1,6,2,8;
       4,3,1,4,3,6,4,6,5,4; 2,3,5,5,4,5,4,2,2,8;
       4,5,5,6,1,7,1,2,2,4; 1,1,1,1,1,1,1,1,1,1];
   b=[380 415 385 405 470 415 400 460 400 200]';
   alpha=[-19,-27,-23,-53,-42,26,-33,-23,41,19]';
   beta=[-52,-3,81,30,-85,68,27,-81,97,-73]';
   lambda=[63,15,44,91,45,50,89,58,86,82]';
   mu=[42,98,48,91,11,63,61,61,38,26]';
   P=optimproblem;
   x=optimvar('x',10,1,'Lower',0); y=optimvar('y',10,1,'Lower',0);
   P.Constraints.cons=A1*x+A2*y<=b;
   P.Objective=(sum([-lambda.*(x-alpha).^2; mu.*(y-beta).^2]))/2;
   p=prob2struct(P);              % directly convert to structured variable
```

When the structured variable is created, the `Objective` and `solver` members must be set. The global optimum problem solver can then be used, and the solution found is $x_4 = 62.6087$, $y_6 = 4.3478$, with the other entries being zeros. The result is the same as in [22].

```
>> f=@(x)0.5*x(:)'*p.H*x(:)+[p.f]'*x(:); p.Objective=f;
   p.solver='fmincon'; x=fmincon_global(p,0,100,20,10)
   x0=x(1:10)', y0=x(11:20)'
```

5.3.3 Concave-cost transportation problem

The mathematical model of a concave-cost transportation problem is described in Definition 4.10. The problem itself is a double subscript problem. Through proper conversions, it can be transformed into an ordinary quadratic programming problem. Since matrix $\boldsymbol{H}$ is not positive-definite, the problem cannot be solved by the quadratic programming solver. In this section, the general-purpose global optimum problem solver is used to tackle the problem.

Example 5.15. Use the global optimum solver to solve again the concave-cost transportation problem in Example 4.31.

Solutions. Since the problem is concave, the solver in Chapter 4 failed to find a global optimum solution, not even a feasible solution could be found. If the problem is converted into an ordinary nonlinear programming problem, conventional solvers are sensitive to the initial search point. Unless one is really lucky, the global optimum solution cannot be found. The solver `fmincon_global()` can be tried to solve the problem.

If functions like `fmincon()` can be applied to nonlinear programming problems, they must be entered into MATLAB environment. Inputting everything once again is not a wise choice. The problem-based description should be completed first, and then function `prob2struct()` can be used to convert it into quadratic programming problem for the structured variable p:

```
>> n=4; m=6; b=[29,41,13,21]; a=[8,24,20,24,16,12];
   C=[300,270,460,800; 740,600,540,380; 300,490,380,760;
      430,250,390,600; 210,830,470,680; 360,290,400,310];
   D=[-7,-4,-6,-8; -12,-9,-14,-7; -13,-12,-8,-4;
      -7,-9,-16,-8; -4,-10,-21,-13; -17,-9,-8,-4];
   P=optimproblem; X=optimvar('X',m,n,'LowerBound',0);
   P.Constraints.c1=sum(X)==b; P.Constraints.c2=sum(X')==a;
   p=prob2struct(P);
```

Of course, this problem is nonconvex. For it, the member `solver` in p should be changed from `'quadprog'` to `'fmincon'`. An anonymous function can be written instead for the objective function, and assigned to the member `Objective`. The constraints of the problem remain the same. A new member `x0` can be set. It can be any $nm \times 1$ vector to hold the space. In order to get high-precision solutions, a strict error tolerance such as `eps` can be set. The following statements can be used to solve the problem directly:

```
>> p.solver='fmincon'; %converting the problem into structured variable
   f=@(x)sum(C(:).*x+D(:).*x.^2); p.Objective=f; %objective function
   ff=optimset; ff.TolX=eps; ff.TolFun=eps; p.options=ff; %precision
   p.x0=100*rand(m*n,1); x0=fmincon_global(p,-10,10,n*m,50)
   X0=reshape(x0,m,n), f0=f(x0), norm(x0-round(x0))
```

The global solution matrix obtained is given below, the same as in [22]. The objective function value is $f_0 = 15\,639$, and the norm of error in $\boldsymbol{X}_0$ is 4.5309×10^{-15} where

$$\boldsymbol{X}_0^{\mathrm{T}} = \begin{bmatrix} 6 & 2 & 0 & 0 \\ 0 & 3 & 0 & 21 \\ 20 & 0 & 0 & 0 \\ 0 & 24 & 0 & 0 \\ 3 & 0 & 13 & 0 \\ 0 & 12 & 0 & 0 \end{bmatrix}.$$

5.3.4 Testing of the global optimum problem solver

When globally optimizing in nonconvex domains, if initial values are not properly selected, local optimum solutions may be found. In order to find global optimum solutions, various algorithms are presented, while the quality of the algorithms may be completely different. Also they are usually quite time demanding. Here a test example is presented to show the benefit of the global optimum solver provided in this section.

Example 5.16. A test problem is introduced in [11], where the objective function is

$$f(x) = l(x_1x_2 + x_3x_4 + x_5x_6 + x_7x_8 + x_9x_{10}),$$

and the constraints are

$$\frac{6Pl}{x_9x_{10}^2} - \sigma_{\max} \leqslant 0, \quad \frac{6P(2l)}{x_7x_8^2} - \sigma_{\max} \leqslant 0,$$

$$\frac{6P(3l)}{x_5x_6^2} - \sigma_{\max} \leqslant 0, \quad \frac{6P(4l)}{x_3x_4^2} - \sigma_{\max} \leqslant 0, \quad \frac{6P(5l)}{x_1x_2^2} - \sigma_{\max} \leqslant 0,$$

$$\frac{Pl^3}{E}\left(\frac{244}{x_1x_2^3} + \frac{148}{x_3x_4^3} + \frac{76}{x_5x_6^3} + \frac{28}{x_7x_8^3} + \frac{4}{x_9x_{10}^3}\right) - \delta_{\max} \leqslant 0,$$

$$\frac{x_2}{x_1} - 20 \leqslant 0, \quad \frac{x_4}{x_3} - 20 \leqslant 0, \quad \frac{x_6}{x_5} - 20 \leqslant 0, \quad \frac{x_8}{x_7} - 20 \leqslant 0, \quad \frac{x_{10}}{x_9} - 20 \leqslant 0.$$

The upper and lower bounds of the decision variables are

$$1 \leqslant x_{1,7,9} \leqslant 5, \quad 30 \leqslant x_{2,8,10} \leqslant 65, \quad 2.4 \leqslant x_{3,5} \leqslant 3.1, \quad 45 \leqslant x_{4,6} \leqslant 60,$$

where $L = 500$, $l = 100$, $P = 50\,000$, $\delta_{\max} = 2.7$, $\sigma_{\max} = 14\,000$, $E = 2 \times 10^7$. Find the global minimum solution for the nonlinear programming problem, and compare with the results provided in [11].

Solutions. In [11], many different values of the objective function are listed: 63 113.61, 63 631.55, and 74 125.97. What will be the objective function with the global solver provided in the book?

The following function can be used to model nonlinear constraints:

```
function [c,ceq]=c5mglo1(x)
P=50000; del=2.7; L=500; l=100; sig=14000; E=2e7; ceq=[];
c=[6*P*l/x(9)/x(10)^2-sig;
   6*P*(2*l)/x(7)/x(8)^2-sig; 6*P*(3*l)/x(5)/x(6)^2-sig;
   6*P*(4*l)/x(3)/x(4)^2-sig; 6*P*(5*l)/x(1)/x(2)^2-sig;
   P*l^3/E*(244/x(1)/x(2)^3+148/x(3)/x(4)^3+76/x(5)/x(6)^3+...
       28/x(7)/x(8)^3+4/x(9)/x(10)^3)-del;
   x(2)/x(1)-20; x(4)/x(3)-20; x(6)/x(5)-20;
   x(8)/x(7)-20; x(10)/x(9)-20];
```

With such a function, the following commands can be used to solve the problem directly and find the global minimum solution. The total elapsed time is 0.32 seconds. The solution obtained is $x_1 = 2.992$, $x_2 = 59.8408$, $x_3 = 2.7776$, $x_4 = 55.5513$, $x_5 = 2.5236$, $x_6 = 50.4717$, $x_7 = 2.2046$, $x_8 = 44.0911$, $x_9 = 1.7498$, and $x_{10} = 34.9951$.

```
>> l=100; ff=optimset; ff.TolX=eps; ff.TolFun=eps;
   f=@(x)l*(x(1)*x(2)+x(3)*x(4)+x(5)*x(6)+x(7)*x(8)+x(9)*x(10));
   xm=[1,30,2.4,45,2.4,45,1,30,1,30]'; n=10; N=10;
   xM=[5,65,3.1,60,3.1,60,5,65,5,65]'; A=[]; B=[]; tic
   [x,fv]=fmincon_global(f,xm,xM,n,N,A,B,[],[],xm,xM,@c5mglo1,ff)
   toc, d1=max(c5mglo1(x)), d2=max(x-xM), d3=max(xm-x)
```

The value of the objective function is 61 914.789, which is significantly lower than those mentioned in the above reference. The maximum values of the constraints are $d_1 = -6.6791 \times 10^{-13}$, $d_2 = -0.3224$, and $d_3 = -0.1236$, indicating that the solution process is successful. If the solver is called 100 times, every time the global optimum solution is found. The success rate is 100 %.

5.3.5 Handling piecewise objective functions

If the objective function or constraints are piecewise functions, or they are described in other complicated forms, the straightforward method can also be used directly to solve the problem.

Example 5.17. Assume that the objective function is $f(\boldsymbol{x}) = f_1(\boldsymbol{x}) + f_2(\boldsymbol{x})$, where the two piecewise subfunctions are:

$$f_1(\boldsymbol{x}) = \begin{cases} 30x_1, & 0 \leqslant x_1 < 300, \\ 31x_1, & 300 \leqslant x_1 \leqslant 400, \end{cases} \qquad f_2(\boldsymbol{x}) = \begin{cases} 28x_2, & 0 \leqslant x_2 < 100, \\ 29x_2, & 100 \leqslant x_2 < 200, \\ 30x_2, & 200 \leqslant x_2 < 1\,000. \end{cases}$$

The constraints are

$$\begin{cases} 300 - x_1 - x_3x_4\cos(b - x_6)/a + cdx_3^2/a = 0, \\ -x_2 - x_3x_4\cos(b + x_6)/a + cdx_4^2/a = 0, \\ -x_5 - x_3x_4\sin(b + x_6)/a + cex_4^2/a = 0, \\ 200 - x_3x_4\sin(b - x_6)/a + cex_3^2/a = 0, \end{cases}$$

where $a = 131.078$, $b = 1.48577$, $c = 0.90798$, $d = \cos 1.47588$, $e = \sin 1.47588$, and the bounds for the decision variables are

$$0 \leqslant x_1 \leqslant 400, \quad 0 \leqslant x_2 \leqslant 1\,000, \quad 340 \leqslant x_3, x_4 \leqslant 420,$$
$$-1\,000 \leqslant x_5 \leqslant 10\,000, \quad 0 \leqslant x_6 \leqslant 0.5236.$$

Solutions. From the given constraints, the MATLAB function for the nonlinear constraints can be written as follows:

```
function [cc,ceq]=c5mtest(x)
cc=[];            % pay attention to the variable name here
a=131.078; b=1.48577; c=0.90798; d=cos(1.47588); e=sin(1.47588);
ceq=[300-x(1)-x(3)*x(4)*cos(b-x(6))/a+c*d*x(3)^2/a;
    -x(2)-x(3)*x(4)*cos(b+x(6))/a+c*d*x(4)^2/a;
    -x(5)-x(3)*x(4)*sin(b+x(6))/a+c*e*x(4)^2/a;
    200-x(3)*x(4)*sin(b-x(6))/a+c*e*x(3)^2/a];
```

Note that the inequality constraints are modeled with the variable cc, not c, since c was used to represent a constant. They must be separated, otherwise the problem is wrongly described, which may lead to wrong results.

Now the following function is described as a piecewise function. A structured variable can be used to describe the original problem. Random initial values can be used to solve the following nonlinear programming problem:

```
>> f1=@(x)30*x(1).*(x(1)<300)+31*x(1).*(x(1)>=300);
   f2=@(x)28*x(2).*(x(2)<100)+29*x(2).*(100<=x(2)&x(2)<200)+...
          30*x(2).*(x(2)>=200);
   f=@(x)f1(x)+f2(x); P.Objective=f; P.nonlcon=@c5mtest;
   P.lb=[0,0,340,340,-1000,0];
   P.ub=[400,1000,420,420,10000,0.5236];
   P.options=optimset; P.solver='fmincon'; P.x0=100*rand(6,1);
   [x0,fx,flag]=fmincon(P)
```

When the above codes are executed, the following results are found. The value of the objective function is slightly larger with that listed in [28]. It should be studied further whether the solution is a global optimum solution or not:

$$x = [106.5627, 200, 373.6757, 419.9997, 22.0580, 0.1554]^{\mathrm{T}}, \quad f_x = 8\,996.9.$$

Now the solver `fmincon_global()` is executed 100 times. Each time the number of runs is set to $N = 10$. The rate of finding a global optimum solution is 55 %, with a total elapsed time of 95.01 seconds. If a global solution is found, it can be seen that the norm of the equality constraints is about 6.2112×10^{-12}.

```
>> X=[]; N=10; tic
   for i=1:100
      [x fx]=fmincon_global(P,P.lb,P.ub,6,N); X=[X; x' fx];
   end, toc
   C=length(find(X(:,7)<8926.2)), [c,ce]=c5mtest(x), norm(ce)
```

The optimum solution of the problem and objective function are:

$$x = [204.2024, 100, 383.0571, 419.8597, -11.2665, 0.07221]^{\mathrm{T}}, \quad f_x = 8\,926.1.$$

If one selects $N = 20$, the success rate of finding a global solution is increased to 69 %, with the elapsed time of 157.17 seconds. If $N = 30$, the success rate is as high as 91 %, with the elapsed time being 264.42 seconds, i. e., with an average of 2.64 seconds per run.

Now observe the results given in [28]. They can be validated with the following commands:

```
>> x0=[107.8199,196.3186,373.8707,420,213.0713,0.1533];
   fx=P.Objective(x0), [cc,ceq]=c5mtest(x0)
```

The objective function value obtained is $f_x = 8\,927.8$, the same as provided in [28], slightly larger than the above mentioned global optimum solution. If the result in [28] is substituted into the constraints, the values of the four constraints are $\boldsymbol{c}_{\mathrm{eq}} = [1.1359, 1.2134, -191.8102, -0.1994]^{\mathrm{T}}$, far away from the expected values of 0. It can be concluded that since the equalities are not satisfied, the solutions given in [28] are wrong!

5.4 Bilevel programming problems

If the result of a nonlinear programming problem is used in the constraints of another optimization problem, this programming problem is referred to as a bilevel programming problem, constituting a class of important programming problems. A bilevel programming problem was first proposed by German economist Heinrich Freiherr von Stackelberg (1905–1946) and used for solving game problems.

Definition 5.3. The mathematical model of a bilevel programming is

$$\min_{\boldsymbol{x},\boldsymbol{y} \text{ s.t. } \begin{cases} \boldsymbol{G}(\boldsymbol{x},\boldsymbol{y})\leqslant\boldsymbol{0} \\ \boldsymbol{H}(\boldsymbol{x},\boldsymbol{y})=\boldsymbol{0} \\ \min\limits_{\boldsymbol{y} \text{ s.t. } \begin{cases} \boldsymbol{g}(\boldsymbol{x},\boldsymbol{y})\leqslant\boldsymbol{0} \\ \boldsymbol{h}(\boldsymbol{x},\boldsymbol{y})=\boldsymbol{0} \end{cases}} f(\boldsymbol{x},\boldsymbol{y}) \end{cases}} F(\boldsymbol{x},\boldsymbol{y}). \tag{5.4.1}$$

It can be seen from the mathematical form that the problem can be regarded as a game of two players. The first is called the leader, and the other is called the follower. The leader makes a decision upon his own objective function and constraints. After that, the follower makes a decision according to the current situation and his objective function and constraints.

In this section, bilevel linear and quadratic programming problems are mainly considered. Based on the solvers provided in the YALMIP Toolbox, the bilevel programming problems can be solved directly.

5.4.1 Bilevel linear programming problems

The processing of ordinary bilevel programs is rather complicated. Here only bilevel linear programming problems are considered, and they can be converted into those of ordinary linear programming. Then, the YALMIP Toolbox is used in solving related problems.

Definition 5.4. The most widely used bilevel linear programming problem is given by

$$\min_{\boldsymbol{x},\boldsymbol{y} \text{ s.t. } \begin{cases} \boldsymbol{A}_1\boldsymbol{x}+\boldsymbol{B}_1\boldsymbol{y}\leqslant\boldsymbol{f}_1 \\ \boldsymbol{C}_1\boldsymbol{x}+\boldsymbol{D}_1\boldsymbol{y}=\boldsymbol{g}_1 \\ \boldsymbol{x}_\mathrm{m}\leqslant\boldsymbol{x}\leqslant\boldsymbol{x}_\mathrm{M},\ \boldsymbol{y}_\mathrm{m}\leqslant\boldsymbol{y}\leqslant\boldsymbol{y}_\mathrm{M} \\ \min\limits_{\boldsymbol{y} \text{ s.t. } \begin{cases} \boldsymbol{A}_2\boldsymbol{x}+\boldsymbol{B}_2\boldsymbol{y}\leqslant\boldsymbol{f}_2 \\ \boldsymbol{C}_2\boldsymbol{x}+\boldsymbol{D}_2\boldsymbol{y}=\boldsymbol{g}_2 \\ \boldsymbol{x}_\mathrm{m}\leqslant\boldsymbol{x}\leqslant\boldsymbol{x}_\mathrm{M},\ \boldsymbol{y}_\mathrm{m}\leqslant\boldsymbol{y}\leqslant\boldsymbol{y}_\mathrm{M} \end{cases}} \boldsymbol{c}_2^\mathrm{T}\boldsymbol{x}+\boldsymbol{d}_2^\mathrm{T}\boldsymbol{y} \end{cases}} \boldsymbol{c}_1^\mathrm{T}\boldsymbol{x}+\boldsymbol{d}_1^\mathrm{T}\boldsymbol{y}. \tag{5.4.2}$$

There are no bilevel linear programming solvers provided in MATLAB Optimization Toolbox. The conversion to conventional linear programming problems is presented here. Karush–Kuhn–Tucker algorithm can be used to carry out the conversion.

Consider the inner linear programming first. If there are m inequality constraints, a vector $\boldsymbol{s}$ can be defined such that $\boldsymbol{s} \geqslant \boldsymbol{0}$, where $\boldsymbol{s}$ has also m components. Besides, the outer objective function can be reconstructed with multipliers $\boldsymbol{\lambda}, \boldsymbol{\mu}$, such that the original problem is converted into a conventional linear programming problem, solvable with solvers such as `linprog()`.

Theorem 5.1. Bilevel linear programming problems can be converted into the following linear programming problem:

$$\min_{x,y \text{ s.t. } \begin{cases} A_1x+B_1y\leqslant f_1 \\ C_1x+D_1y=g_1 \\ d_2+\lambda^T B_2+\mu^T D_2=0 \\ A_2x+B_2y+s=f_2 \\ C_2x+D_2y=g_2 \\ \lambda^T s=0 \\ \lambda,s\geqslant 0 \\ x_m\leqslant x\leqslant x_M,\ y_m\leqslant y\leqslant y_M \end{cases}} c_1^T x + d_1^T y. \tag{5.4.3}$$

Since the conversion process to the original linear programming problem is rather complicated, low-level solutions are not presented here. Later examples will be used to demonstrate the solutions of bilevel linear programming problems with suitable tools.

5.4.2 Bilevel quadratic programming problem

Bilevel quadratic programming problems also are a class of commonly used programming problems. In this section, definitions and conversions are presented.

Definition 5.5. The mathematical model of a bilevel quadratic programming problem is

$$\min_{x,y \text{ s.t. } \begin{cases} A_1x+B_1y\leqslant f_1 \\ C_1x+D_1y=g_1 \\ x_m\leqslant x\leqslant x_M,\ y_m\leqslant y\leqslant y_M \\ \min\limits_{y \text{ s.t. } \begin{cases} A_2x+B_2y\leqslant f_2 \\ C_2x+D_2y=g_2 \\ x_m\leqslant x\leqslant x_M,\ y_m\leqslant y\leqslant y_M \end{cases}} c_2^T x + d_2^T y + x^T Q_{21} y + y^T Q_{22} y \end{cases}} c_1^T x + d_1^T y + x^T Q_{11} x + x^T Q_{12} y + y^T Q_{13} y. \tag{5.4.4}$$

Theorem 5.2. A bilevel quadratic programming problem can be converted into the following quadratic programming problem:

$$\min_{x,y \text{ s.t. } \begin{cases} A_1x+B_1y\leqslant f_1 \\ C_1x+D_1y=g_1 \\ 2y^T Q_{22}+x^T Q_{21}+c_2+\lambda^T B_2+\mu^T D_2=0 \\ d_2+\lambda^T B_2+\mu^T D_2=0 \\ A_2x+B_2y+s=f_2 \\ C_2x+D_2y=g_2 \\ \lambda^T s=0 \\ \lambda,s\geqslant 0 \\ x_m\leqslant x\leqslant x_M,\ y_m\leqslant y\leqslant y_M \end{cases}} c_1^T x + d_1^T y + x^T Q_{11} x + x^T Q_{12} y + y^T Q_{13} y. \tag{5.4.5}$$

5.4.3 Bilevel program solutions with YALMIP Toolbox

Dedicated functions are provided in YALMIP Toolbox to solve bilevel programming problems. For linear and quadratic bilevel programs, `solvebilevel()` function provided in YALMIP Toolbox can be used to solve the problems directly. Before the solution process, four variables, namely `CI` (for inner constraints), `OI` (for inner objective function), `CO` (outer constraints), and `OO` (outer objective function), should be expressed. Then, the following command can be used to solve bilevel programming problems:

`solvebilevel(CO,OO,CI,OI,`***y***`)`

The current solvers in YALMIP Toolbox can be used to solve bilevel linear and quadratic programming problems. Nonlinear bilevel programming problems cannot be solved with such a function. Examples are used here to demonstrate the direct solutions of bilevel programs with YALMIP Toolbox.

Example 5.18. Solve the following bilevel linear programming problem:[23]

$$\min_{\boldsymbol{x},\boldsymbol{y}\text{ s.t. }\begin{cases}\boldsymbol{x}\geqslant\boldsymbol{0},\,\boldsymbol{y}\geqslant\boldsymbol{0}\\ \min\limits_{\boldsymbol{y}\text{ s.t. }\begin{cases}\boldsymbol{H}_1\boldsymbol{y}+\boldsymbol{H}_2\boldsymbol{x}\leqslant\boldsymbol{b}\\ \boldsymbol{y}\geqslant\boldsymbol{0}\end{cases}} x_1+2x_2+y_1+y_2+2y_3\end{cases}} -8x_1-4x_2+4y_1-40y_2-4y_3,$$

where

$$\boldsymbol{H}_1=\begin{bmatrix}-1 & 1 & 1 & 1 & 0 & 0\\ -1 & 2 & -0.5 & 0 & 1 & 0\\ 2 & -1 & -0.5 & 0 & 0 & 1\end{bmatrix},\quad \boldsymbol{H}_2=\begin{bmatrix}0 & 0\\ 2 & 0\\ 0 & 2\end{bmatrix},\quad \boldsymbol{b}=\begin{bmatrix}1\\ 1\\ 1\end{bmatrix}.$$

Solutions. It can be seen from the sizes of matrices $\boldsymbol{H}_1$ and $\boldsymbol{H}_2$, that the decision variable $\boldsymbol{y}$ is a 6×1 column vector, while $\boldsymbol{x}$ is a 2×1 column vector. It is not hard to write down the two objective functions and two constraints. The solver `solvebilevel()` can be used to solve the original problem directly:

```
>> x=sdpvar(2,1); y=sdpvar(6,1);
   H1=[-1,1,1,1,0,0; -1,2,-0.5,0,1,0; 2,-1,-0.5,0,0,1];
   H2=[0,0; 2,0; 0,2]; b=[1; 1; 1];
   CI=[y>=0, H1*y+H2*x==b];
   OI=[y(1)+y(2)+2*y(3)+x(1)+2*x(2)];
   CO=[x>=0, y>=0]; OO=-8*x(1)-4*x(2)+4*y(1)-40*y(2)-4*y(3);
   solvebilevel(CO,OO,CI,OI,y), value(x), value(y), value(OO)
```

The optimum solutions obtained are $\boldsymbol{x}=[0,0.9]^{\mathrm{T}}$ and $\boldsymbol{y}=[0,0.6,0.4,0,0,0]^{\mathrm{T}}$, which are the same as those in [23].

Example 5.19. Solve the following bilevel quadratic programming problem:[23]

$$\min_{\boldsymbol{x},\boldsymbol{y}\ \text{s.t.}\left\{\begin{array}{l} x_1+x_2+y_1-2y_2-40\leqslant 0\\ 0\leqslant x_1,x_2\leqslant 50,\ -10\leqslant y_1,y_2\leqslant 20\\ \min\limits_{\boldsymbol{y}\ \text{s.t.}\left\{\begin{array}{l}-x_1+2y_1\leqslant -10\\ -x_2+2y_2\leqslant -10\\ -10\leqslant y_1,y_2\leqslant 20\end{array}\right.} (y_1-x_1+20)^2+(y_2-x_2+20)^2\end{array}\right.} 2x_1+2x_2-3y_1-3y_2-60.$$

Solutions. Function `sdpvar()` can be used to describe the two decision variables $\boldsymbol{x}$ and $\boldsymbol{y}$. Then, the four quantities, `CI`, `OI`, `CO`, and `OO`, can be described by the expressions in YALMIP Toolbox. Although there are quadratic terms, it does not matter. With the descriptions, function `solvebilevel()` can be called to solve the original problem directly:

```
>> x=sdpvar(2,1); y=sdpvar(2,1);
   CI=[-x(1)+2*y(1)<=-10, -x(2)+2*y(2)<=-10, -10<=y<=20];
   OI=(y(1)-x(1)+20)^2+(y(2)-x(2)+20)^2;
   CO=[x(1)+x(2)+y(1)-2*y(2)-40<=0, 0<=x<=50, -10<=y<=20];
   OO=2*x(1)+2*x(2)-3*y(1)-3*y(2)-60;
   solvebilevel(CO,OO,CI,OI,y), value(x), value(y), value(OO)
```

It is known from the results that $\boldsymbol{x} = [0,30]^{\mathrm{T}}$, $\boldsymbol{y} = [-10,10]^{\mathrm{T}}$, which are different from those provided in [23], namely $\boldsymbol{x} = [0,0]^{\mathrm{T}}$, $\boldsymbol{y} = [10,-10]^{\mathrm{T}}$. When computing the value of the objective function at these points, it can be seen that it is the same, and the constraints are all satisfied. It can be concluded that this problem does not have a unique solution. Both found solutions solve the original problem.

5.5 Nonlinear programming applications

Nonlinear programming is important in economics, science, and engineering. In this section, several examples are provided to describe optimization techniques.

5.5.1 Maximum inner polygon inside a circle

In [21], there is an inner polygon problem inside a unit circle, as shown in Figure 5.5. For convenience, a semicircle is considered here. Assume that there are n inner points inside the semicircle. A polygon can be drawn, and the total area enclosed by the polygon can be computed from

$$J = \frac{1}{2}\sum_{i=1}^{n-1} r_{i+1}r_i \sin(\theta_{i+1}-\theta_i). \tag{5.5.1}$$

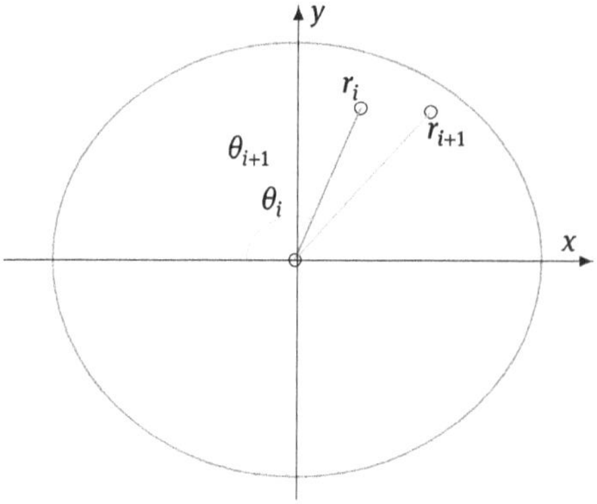

Figure 5.5: Illustration of inner polynomial in a circle.

For some selected points, the angle of the latter point is larger than that of the former. This can be described in mathematical terms as

$$\theta_i < \theta_{i+1}, \quad 1 \leqslant i < n. \tag{5.5.2}$$

For the inner points in the semicircle, the following constraints can be established:

$$0 < r_i \leqslant 1, \quad 0 \leqslant \theta_i \leqslant \pi, \quad 1 \leqslant i \leqslant n, \tag{5.5.3}$$

and at the terminal points, $\theta_1 = 0$, $\theta_n = \pi$.

Selecting decision variables $x_i = r_i$, $x_{n+i} = \theta_i$, $i = 1, 2, \ldots, n$, the original standard form can be written as

$$\min_{\boldsymbol{x} \text{ s.t. } \begin{cases} x_{n+1}=0,\ x_{2n}=\pi \\ x_{n+i}-x_{n+i+1}\leqslant 0,\ i=1,2,\ldots,n-1 \\ 0\leqslant x_{n+i}\leqslant\pi,\ i=1,2,\ldots,n \\ 0\leqslant x_i\leqslant 1,\ i=1,2,\ldots,n \end{cases}} \quad -\frac{1}{2}\sum_{i=1}^{n-1} x_{i+1}x_i \sin(x_{n+i+1} - x_{n+i}). \tag{5.5.4}$$

Example 5.20. Selecting $n = 11$, find the maximum area of the inner polygons and compute the approximate value of π.

Solutions. Since all the following constraints are linear, if one does not want to write down the corresponding linear matrices, nonlinear constraints can be expressed in the following MATLAB function:

```
function [c,ceq]=c5mpi1(x)
n=length(x)/2; ceq=[x(n+1); x(end)-pi]; c=[];
for i=n+1:2*n-1, c=[c; x(i)-x(i+1)]; end
```

Of course, considering the computational load, the linear inequality constraints for $x_{n+i} - x_{n+1+i}$ can be expressed as

$$\boldsymbol{A} = \begin{bmatrix} \boldsymbol{0}_{(n-1)\times n} & \begin{matrix} 1 & -1 & 0 & \cdots & 0 \\ 0 & 1 & -1 & \cdots & 0 \\ \vdots & \vdots & \vdots & \ddots & \vdots \\ 0 & 0 & 0 & \cdots & -1 \end{matrix} \end{bmatrix}, \quad \boldsymbol{B} = \boldsymbol{0}_{(n-1)\times 1}.$$

The equalities $x_{n+1} = 0$ and $x_{2n} = \pi$ can be expressed in matrix form as

$$\boldsymbol{A}_{\mathrm{eq}} = \begin{bmatrix} \boldsymbol{0}_{2\times n} & \begin{matrix} 1 & 0 & \cdots & 0 \\ 0 & 0 & \cdots & 1 \end{matrix} \end{bmatrix}, \quad \boldsymbol{B}_{\mathrm{eq}} = \begin{bmatrix} 0 \\ \pi \end{bmatrix}.$$

If one selects $n = 11$, the following commands can be used to generate the objective function in an anonymous function. Also, the solution to the problem can be found from

```
>> n=11; A=[zeros(n) eye(n)-diag(ones(n-1,1),1)];
   A(end,:)=[]; B=zeros(n-1,1);
   f=@(x)-sum(x(1:n-1).*x(2:n).*sin(x(n+2:2*n)-x(n+1:2*n-1)))/2;
   xm=zeros(2*n,1); xM=[ones(n,1); pi*ones(n,1)];
   Aeq=zeros(2,2*n); Aeq(1,n+1)=1; Aeq(2,end)=1; Beq=[0; pi];
   x=fmincon_global(f,0,pi,2*n,10,A,B,Aeq,Beq,xm,xM); 2*f(x)
   x1=x(1:n).*cos(x(n+1:2*n)); y1=x(1:n).*sin(x(n+1:2*n));
```

If the value of n is increased, the maximization problem may need much more time, and the quality becomes poorer. The two sets of inner points for different values of n can obtained as shown in Figure 5.6. It can be seen that when n is too large, the inner polygon becomes worse. Therefore, this method is not suitable for solving such problems.

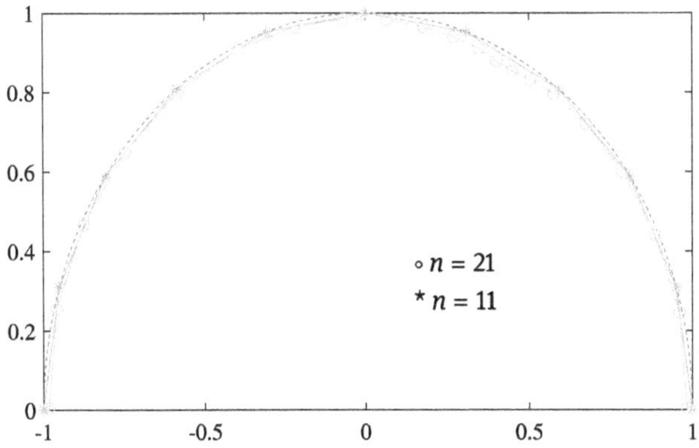

Figure 5.6: Demonstration in the semicircle.

```
>> n=21;
   f=@(x)-sum(x(1:n-1).*x(2:n).*sin(x(n+2:2*n)-x(n+1:2*n-1)))/2;
   xm=zeros(2*n,1); xM=[ones(n,1); pi*ones(n,1)];
   A=[zeros(n) eye(n)-diag(ones(n-1,1),1)]; A(end,:)=[];
   B=zeros(n-1,1); Aeq=zeros(2,2*n); Aeq(1,n+1)=1; Aeq(2,end)=1;
   Beq=[0; pi]; ff=optimset; ff.MaxFunEvals=10000;
   x=fmincon_global(f,0,pi,2*n,30,A,B,Aeq,Beq,xm,xM,'',ff); 2*f(x)
   x2=x(1:n).*cos(x(n+1:2*n)); y2=x(1:n).*sin(x(n+1:2*n));
   t=0:pi/1000:pi; plot(cos(t),sin(t),'--',x1,y1,'*-',x2,y2,'o-')
```

It can be seen from the polygon that the values of r_i should be set to 1. The decision vector can be set as $x_i = \theta_{i+1}$. The area of the inner polygon can be written as

$$\min_{\boldsymbol{x} \text{ s.t. } \begin{cases} x_1=0,\ x_n=\pi \\ x_i-x_{i+1}\leqslant 0,\ i=1,2,\dots,n-1 \\ 0\leqslant x_i\leqslant \pi,\ i=1,2,\dots,n \end{cases}} \quad -\frac{1}{2}\sum_{i=1}^{n-1} \sin(x_{i+1}-x_i). \tag{5.5.5}$$

Example 5.21. Solve the new optimization problem to find the largest area of the inner polygon.

Solutions. The following statements can be used to describe the objective function and constraints. Then, the problem can be solved directly, and the solutions are shown graphically in Figure 5.7. It can be seen that the points are not quite evenly distributed.

```
>> n=501; f=@(x)-sum(sin(diff(x)))/2;
   A=eye(n)-diag(ones(n-1,1),1); A(end,:)=[];  B=zeros(n-1,1);
   Aeq=[1 zeros(1,n-1); zeros(1,n-1),1]; Beq=[0; pi];
   xm=zeros(n,1); xM=pi*ones(n,1); x0=rand(n,1);
```

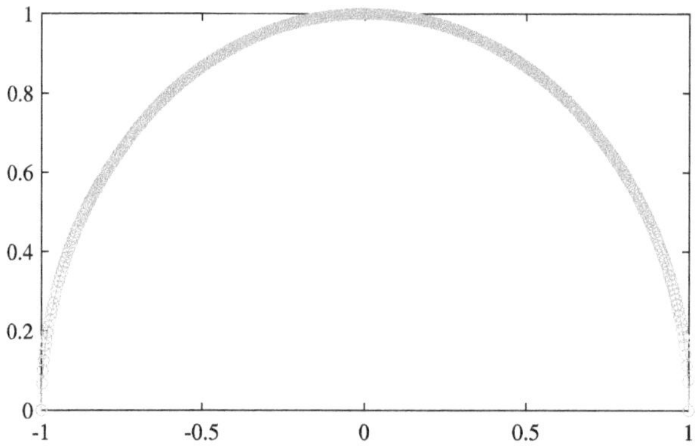

Figure 5.7: Demonstration of the polygons.

```
x=fmincon_global(f,0,pi,n,20,A,B,Aeq,Beq,xm,xM); 2*f(x)
x2=cos(x); y2=sin(x); plot(x2,y2,'o-')
```

If the unit circle is divided evenly into n parts, it can be seen that the area of each is $S = \sin(\pi/n)$. The value of π can be found as $S = 3.141592653589793$.

```
>> n=1000000000; S=n*sin(pi/n)
```

5.5.2 Semiinfinite programming problems

Semiinfinite programming problems form a special class of optimization problems. In this section, the definition and physical explanation of semiinfinite programming are given. Then, examples are used to demonstrate solutions of one- and multidimensional semiinfinite programming problems.

Definition 5.6. The mathematical model of a semiinfinite programming problem is

$$\min_{\boldsymbol{x} \text{ s.t. } \begin{cases} \boldsymbol{G}(\boldsymbol{x})\leqslant \boldsymbol{0} \\ K_i(\boldsymbol{x},w_i)\leqslant 0,\ i=1,\ldots,m \end{cases}} f(\boldsymbol{x}). \tag{5.5.6}$$

The physical interpretation of the semiinfinite programming problem is that, when the constraint $\boldsymbol{G}(\boldsymbol{x}) \leqslant 0$ is satisfied, and also for any w_i the constraints $K_i(\boldsymbol{x}, w_i) \leqslant 0$ are satisfied, then we seek an $\boldsymbol{x}$ such that the objective function $f(\boldsymbol{x})$ is minimized.

Function `fseminf()` is provided in MATLAB Optimization Toolbox for solving such problems. The syntax of the function is

$[\boldsymbol{x}, f_0, \texttt{flag}, \texttt{c}] = \texttt{fseminf}(\texttt{Fun}, \boldsymbol{x}_0, m, \texttt{SFun}, \boldsymbol{A}, \boldsymbol{B}, \boldsymbol{A}_{\rm eq}, \boldsymbol{B}_{\rm eq}, \boldsymbol{x}_{\rm m}, \boldsymbol{x}_{\rm M}, \texttt{OPT})$

where `SFun` is used to describe the constraints $\boldsymbol{G}(\boldsymbol{x}) \leqslant \boldsymbol{0}$. Also it may include the special function $K_i(\cdot)$, satisfying all the constraints.

In fact, a semiinfinite programming problem can be easily converted into an ordinary nonlinear programming problem, by introducing extra constraints – for all the allowed values of w_i, the constraints $K_j(\boldsymbol{x}, w_i) \leqslant 0$ are satisfied. Therefore, the solvers such as `fmincon()` can be used to solve these problems. In this book, function `fseminf()` is not recommended. A new idea and tool can be used to solve semiinfinite programming problems.

Example 5.22. Solve the one-dimensional semiinfinite programming problem:[41]

$$\min_{\boldsymbol{x} \text{ s.t. } \begin{cases} K(\boldsymbol{x},t)=x_1\cos t+x_2\sin t-1\leqslant 0 \\ 0\leqslant t\leqslant \pi \end{cases}} (x_1+x_2-2)^2+(x_1-x_2)^2+30\big[\min(0,x_1-x_2)\big]^2.$$

Solutions. Since t is defined in the interval $(0, \pi)$, some samples can be generated. For instance, if N samples $t_1, t_2, \ldots, t_N$ are generated, to ensure each sample of t satisfies

$K(\boldsymbol{x}, t) \leqslant 0$, it can be equivalently converted to a set of extra constraints $K(\boldsymbol{x}, t_i) \leqslant 0$, namely

$$\begin{cases} x_1 \cos t_1 + x_2 \sin t_1 - 1 \leqslant 0, \\ x_1 \cos t_2 + x_2 \sin t_2 - 1 \leqslant 0, \\ \quad \vdots \\ x_1 \cos t_m + x_2 \sin t_m - 1 \leqslant 0. \end{cases}$$

Therefore, the following function can be established to convert the original semi-infinite programming problem into a conventional nonlinear programming problem:

```
function [c,ceq]=c5msinf1(x)
ceq=[]; N=100000;  % no equality constraints, hence returns an empty matrix
t=[linspace(0,pi,N)]'; c=x(1)*cos(t)+x(2)*sin(t)-1;
```

An exaggerated number $N = 100\,000$ is selected. If an anonymous function is used to describe the objective function, the following statements can be employed to solve the semiinfinite programming problem:

```
>> f=@(x)(x(1)+x(2)-2)^2+(x(1)-x(2))^2+30*min(0,x(1)-x(2))^2;
   A=[]; B=[]; Aeq=[]; Beq=[]; xm=[]; xM=[]; x0=rand(2,1); tic
   [x,f0,flag,cc]=fmincon(f,x0,A,B,Aeq,Beq,xm,xM,@c5msinf1), toc
```

The optimum solution found is $\boldsymbol{x} = [0.7073, 0.7069]^{\mathrm{T}}$, and the objective function value is $f_0 = 0.3432$, which is close to the result listed in [41], indicating that the solution process is successful. The number of iterations is 25 and elapsed time is 2.936 seconds. Since there is only one parameter t, the time needed is acceptable. If there are more parameters, which is known as a multidimensional problem to be discussed later, one should strive to reduce the computational load. Since the above selected N is large, the converted problem is equivalent to a problem with 100 000 inequality constraints. Therefore, the solution process is slow.

If one wants to reduce the required time, an obvious method is to find the maximum value of the 100 000 inequality constraints and use the new constraint $K_{\mathrm{M}}(\boldsymbol{x}, t_i) \leqslant 0$. Therefore, the 100 000 constraints can be reduced to one inequality constraint. The new constraint function can be constructed:

```
function [c,ceq]=c5msinf2(x)
ceq=[]; N=100000;  % no equality constraints
t=[linspace(0,pi,N)]'; c=x(1)*cos(t)+x(2)*sin(t)-1; c=max(c);
```

Now the problem can be solved again. The number of iterations is reduced to 17, and more importantly, the elapsed time is reduced to 0.052 seconds. The solution obtained is more accurate, $\boldsymbol{x} = [0.7071, 0.7071]^{\mathrm{T}}$, the same as given in [41], the value of the objective function is $f_0 = 0.3431$, smaller than in the previous result.

```
>> tic, [x,f0,flag]=fmincon(f,x0,A,B,Aeq,Beq,xm,xM,@c5msinf2), toc
```

In order to demonstrate the physical meaning of the semiinfinite programming problem, with the optimum solution of $\boldsymbol{x}$, the $K(\boldsymbol{x},t)$ curve can be drawn, as shown in Figure 5.8. It can be seen that the curve is always located beneath the $y = 0$ line, meaning that for all the values of t, the extra inequality constraints are also satisfied.

```
>> t=linspace(0,pi,1000); c=x(1)*cos(t)+x(2)*sin(t)-1;
   [x1,i1]=max(c); i1=i1(1); plot(t,c,t(i1),c(i1),'o')
```

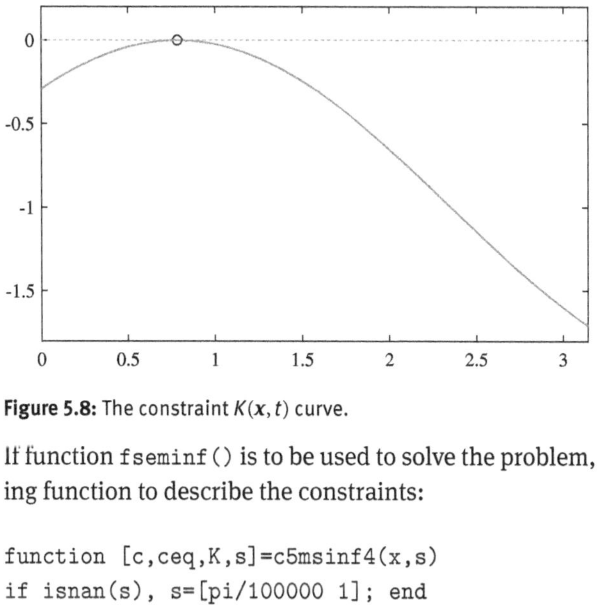

Figure 5.8: The constraint $K(\boldsymbol{x},t)$ curve.

If function `fseminf()` is to be used to solve the problem, one should write the following function to describe the constraints:

```
function [c,ceq,K,s]=c5msinf4(x,s)
if isnan(s), s=[pi/100000 1]; end
t=0:s(1):pi; c=[]; ceq=[]; K=x(1)*cos(t)+x(2)*sin(t)-1;
```

The following statements can be used to solve the semiinfinite programming problem. Some runs can be made, and the expected solution $\boldsymbol{x} = [0.7071, 0.7071]^{\mathrm{T}}$ is found. The elapsed time of each run is about 0.27 seconds, higher than the time required by the above method. The “1” in the statement means that there is one returned variable, K, in the `c5msinf4()` function.

```
>> f=@(x,s)(x(1)+x(2)-2)^2+(x(1)-x(2))^2+30*min(0,x(1)-x(2))^2;
   tic, [x,f0]=fseminf(f,rand(2,1),1,@c5msinf4), toc
```

Example 5.23. Now let us consider a six-dimensional semiinfinite programming problem (also known as Problem S, $p = 6$):[41]

$$\max_{\boldsymbol{x} \text{ s.t. } \begin{cases} K(\boldsymbol{x},\boldsymbol{t})\leqslant 0 \\ 0\leqslant \boldsymbol{t}\leqslant 1 \end{cases}} \quad x_1x_2 + x_2x_3 + x_3x_4,$$

where

$$\begin{aligned} K(\boldsymbol{x},\boldsymbol{t}) = {} & 2(x_1^2 + x_2^2 + x_3^2 + x_4^2) - 6 - 2p + \sin(t_1 - x_1 - x_4) + \sin(t_2 - x_2 - x_3) \\ & + \sin(t_3 - x_1) + \sin(2t_4 - x_2) + \sin(t_5 - x_3) + \sin(2t_6 - x_4). \end{aligned}$$

The words "six-dimensional problem" mean that the $\boldsymbol{t}$ vector has 6 components, $t_1, t_2, \dots, t_6$. Solve the problem.

Solutions. Function `ndgrid()` can be used to generate t_i mesh grid samples, then the mesh grid data can be manipulated, and $K(\boldsymbol{x},\boldsymbol{t}) \leqslant 0$ can be appended to the constraints so that the problem can be converted into a conventional nonlinear programming problem. For multidimensional problems, the maximum value of the inequalities should be extracted.

```
functions [c,ceq]=c5msinf3(x)
N=10; p=6; ceq=[];     % no equality constraints
[t1,t2,t3,t4,t5,t6]=ndgrid(linspace(0,1,N)); % mesh grid samples
c=2*sum(x.^2)-6-2*p+sin(t1-x(1)-x(4))+sin(t2-x(2)-x(3))+...
   sin(t3-x(1))+sin(2*t4-x(2))+sin(t5-x(3))+sin(2*t6-x(4));
c=max(c(:));   % this operation is essential
```

The following commands can be used to solve the problem. After 4.31 seconds of waiting, the optimum solution is obtained as

$$\boldsymbol{x} = [1.3863, -1.6141, 1.6886, -0.7446]^{\mathrm{T}},$$

with the objective function value of $f_0 = -6.2204$. It is significantly better than that provided in [41], which is $\boldsymbol{x}_1 = [0.960921, -1.456291, 1.581476, -0.90587]^{\mathrm{T}}$, with $f_1 = -5.1351$, indicating that the method here is feasible, and better solutions can be found.

```
>> f=@(x)x(1)*x(2)+x(2)*x(3)+x(3)*x(4);
   A=[]; B=[]; Aeq=[]; Beq=[]; xm=[]; xM=[]; x0=rand(4,1); tic
   [x,f0,flag,cc]=fmincon(f,x0,A,B,Aeq,Beq,xm,xM,@c5msinf3), toc
```

If the number of segments is increased to $N = 15$, the same results can be obtained. The elapsed time is increased significantly to 61.94 seconds. If $N = 15$, the number of equivalent inequalities is increased to $15^6 = 11\,390\,625$. For the large-scale problem, if the value of K_{M} is not found, the computational load is far too large. For multidimensional problems, selecting a large N is not possible.

If function `fseminf()` is used, the following constraint function should be written. Note that the maximum number of the inequalities should not be used, otherwise the solution process may not converge.

```
function [c,ceq,K,s]=c5msinf5(x,s)
N=10; p=6; ceq=[]; c=[]; %mesh grid samples
[t1,t2,t3,t4,t5,t6]=ndgrid(linspace(0,1,N));
K=2*sum(x.^2)-6-2*p+sin(t1-x(1)-x(4))+sin(t2-x(2)-x(3))+...
    sin(t3-x(1))+sin(2*t4-x(2))+sin(t5-x(3))+sin(2*t6-x(4));
K=K(:);
```

One may set $N = 10$ to solve the problem. To total elapsed time is 79.13 seconds. The result obtained is $\boldsymbol{x} = [1.4094, -1.6228, 1.6628, -0.7104]^{\mathrm{T}}$, $g_0 = -6.1667$. The result with function `fseminf()` is much worse than for the previous method. Therefore, it is not recommended to use such a solver for semiinfinite programming problems.

```
>> tic, [x,f0]=fseminf(f,rand(4,1),1,@c5msinf5), toc
```

5.5.3 Pooling and blending problem

Bilinear programming problems form a class of convex problems. In this section, the general form of a bilinear programming problem is presented, followed by an example introducing pooling and blending optimization problems.

Definition 5.7. If function $f(\boldsymbol{x},\boldsymbol{y})$ can be expressed in the following form, the function is referred to as a bilinear function:

$$f(\boldsymbol{x},\boldsymbol{y}) = \boldsymbol{a}^{\mathrm{T}}\boldsymbol{x} + \boldsymbol{x}^{\mathrm{T}}\boldsymbol{Q}\boldsymbol{y} + \boldsymbol{b}^{\mathrm{T}}\boldsymbol{y}, \tag{5.5.7}$$

where $\boldsymbol{a}$ and $\boldsymbol{b}$ are vectors, $\boldsymbol{Q}$ is a matrix. If $\boldsymbol{x}$ and $\boldsymbol{y}$ are of different dimensions, $\boldsymbol{Q}$ is a rectangular matrix.

Definition 5.8. The mathematical model of a bilinear programming problem is

$$\max_{\boldsymbol{x} \text{ s.t. } \begin{cases} \boldsymbol{x}^{\mathrm{T}}\boldsymbol{A}_i\boldsymbol{y}+\boldsymbol{c}_i^{\mathrm{T}}\boldsymbol{x}+\boldsymbol{d}_i^{\mathrm{T}}\boldsymbol{y}\leqslant \boldsymbol{b}_i,\ i=1,2,\dots,p \\ \boldsymbol{x}^{\mathrm{T}}\boldsymbol{A}_i\boldsymbol{y}+\boldsymbol{c}_i^{\mathrm{T}}\boldsymbol{x}+\boldsymbol{d}_i^{\mathrm{T}}\boldsymbol{y}= \boldsymbol{b}_i,\ i=p+1,\dots,p+q \end{cases}} \boldsymbol{x}^{\mathrm{T}}\boldsymbol{A}_0\boldsymbol{y} + \boldsymbol{c}_0^{\mathrm{T}}\boldsymbol{x} + \boldsymbol{d}_0^{\mathrm{T}}\boldsymbol{y}. \tag{5.5.8}$$

In chemical and petrol industry, bilinear programming problems are commonly encountered. The optimal petrol blending problems are always solved to maximize the product profits. Here, an example is given showing how to use bilinear problems to reach related objectives.

The mathematical model and illustration shown in Figure 5.9 is proposed by Haverly,[23] where a and b are the two channels flowing into the pool, c is the flow of

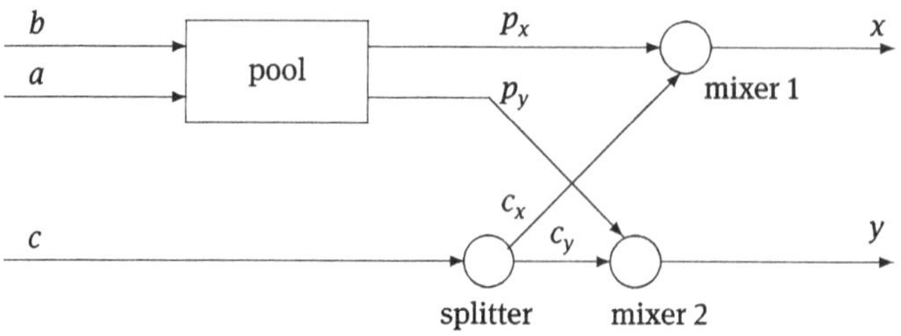

Figure 5.9: Pooling and blending illustration.

the splitter. The sulphur contents of the three input channels are respectively 3 %, 2 %, and 1 %. After the blending process, the flows of the two output channels are respectively x and y, with sulphur contents of 2.5 % and 1.5 %, respectively; the variables p_x, p_y, c_x, and c_y are the flows in the figure. Based on the given conditions, considering the price information, the following mathematical model can be established to maximize the profit:

$$\max \quad 9x + 15y - 6a - c_1 b - 10(c_x + c_y), \tag{5.5.9}$$
$$\boldsymbol{v} \text{ s. t.} \begin{cases} p_x+p_y-a-b=0 \\ x-p_x-c_x=0 \\ y-p_y-c_y=0 \\ pp_x+2c_x-2.5x\leqslant 0 \\ pp_y+2c_y-1.5y\leqslant 0 \\ pp_x+pp_y-3a-b=0 \\ 0\leqslant x\leqslant c_2,\ 0\leqslant y\leqslant 200 \\ 0\leqslant a,b,c_x,c_y,p,p_x,p_y\leqslant 500 \end{cases}$$

where the decision variables are $\boldsymbol{v} = [x, y, a, b, c_x, c_y, p, p_x, p_y]^{\mathrm{T}}$, and c_1, c_2 are given quantities.

Example 5.24. Consider the above constrained optimization model. If $c_1 = 13$ and $c_2 = 600$ (the third scheme in [23]), solve the optimization problem.

Solutions. To solve the problem in a simple manner, YALMIP Toolbox can be used to describe and solve the problem. Note that a maximization problem is to be solved here.

```
>> x=sdpvar(1,1); y=sdpvar(1,1); a=sdpvar(1,1);
   b=sdpvar(1,1); cx=sdpvar(1,1); cy=sdpvar(1,1);
   p=sdpvar(1,1); px=sdpvar(1,1); py=sdpvar(1,1);
   c1=13; c2=600;
   const=[px+py-a-b==0, x-px-cx==0, y-py-cy==0,
          p*px+2*cx-2.5*x<=0, p*py+2*cy-1.5*y<=0,
          p*px+p*py-3*a-b==0, 0<=x<=c2, 0<=y<=200,
          0<=a<=500,0<=b<=500, 0<=cx<=500, 0<=cy<=500,
          0<=p<=500, 0<=px<=500, 0<=py<=500]; % constraints
```

```
obj=9*x+15*y-6*a-c1*b-10*(cx+cy);
optimize(const,-obj); value(x), value(y), value(a), value(b)
value(cx), value(cy), value(p), value(px), value(py)
```

The results obtained are $y = 200$, $a = 50$, $b = 150$, $p = 1.5$, and $p_y = 200$, the other decision variables are all 0. The objective function value is 750. It can be seen that the results are the same as those provided in [23].

Example 5.25. Using YALMIP Toolbox directly is, of course, a good choice, since the modeling is straightforward and flexible. However, there are limitations, since YALMIP solvers may not be satisfactory in finding global optimum solutions. Solve again the problem in Example 5.24 with ordinary solvers provided as MATLAB Optimization Toolbox functions.

Solutions. If one defines the decision variables $x_1 = x$, $x_2 = y$, $x_3 = a$, $x_4 = b$, $x_5 = c_x$, $x_6 = c_y$, $x_7 = p$, $x_8 = p_x$, and $x_9 = p_y$, the original problem can be rewritten into the following form manually:

$$\max \quad 9x + 15y - 6a - c_1 b - 10(c_x + c_y).$$
$$\boldsymbol{x} \text{ s.t.} \begin{cases} x_8+x_9-x_3-x_4=0 \\ x_1-x_8-x_5=0 \\ x_2-x_9-x_6=0 \\ x_7x_8+2x_5-2.5x_1\leqslant 0 \\ x_7x_9+2x_6-1.5x_2\leqslant 0 \\ x_7x_8+x_7x_9-3x_3-x_4=0 \\ \boldsymbol{x}_{\mathrm{m}}\leqslant \boldsymbol{x}\leqslant \boldsymbol{x}_{\mathrm{M}} \end{cases}$$

The constraints can be expressed in a linear matrix form manually as

$$\boldsymbol{A}_{\mathrm{eq}} = \begin{bmatrix} 0 & 0 & -1 & -1 & 0 & 0 & 0 & 1 & 1 \\ 1 & 0 & 0 & 0 & -1 & 0 & 0 & -1 & 0 \\ 0 & 1 & 0 & 0 & 0 & -1 & 0 & 0 & -1 \end{bmatrix}, \quad \boldsymbol{B}_{\mathrm{eq}} = \begin{bmatrix} 0 \\ 0 \\ 0 \end{bmatrix}.$$

The nonlinear constraints can be described by the MATLAB function:

```
function [c,ceq]=c5mpool(x)
c=[x(7)*x(8)+2*x(5)-2.5*x(1); x(7)*x(9)+2*x(6)-1.5*x(2)];
ceq=x(7)*x(8)+x(7)*x(9)-3*x(3)-x(4);
```

The mathematical form of the objective function is

$$f(\boldsymbol{x}) = 9x_1 + 15x_2 - 6x_3 - c_1x_4 - 10(x_5 + x_6).$$

Now, the upper and lower bounds of the decision variables can be extracted and function `fmincon()` can then be called to solve the original problem. With repeated runs of the following statements, typically the solutions obtained are the same as those in Example 5.24. But there are occasions where local optimum solutions are found.

```
>> c1=13; c2=600; P.options=optimset; P.solver='fmincon';
   P.objective=@(x)-(9*x(1)+15*x(2)-6*x(3)-c1*x(4)-10*(x(5)+x(6)));
   x0=ones(9,1); xM=500*x0; P.lb=0*x0; xM(1:2)=[c2; 200];
   P.Aeq=[0,0,-1,-1,0,0,0,1,1; 1,0,0,0,-1,0,0,-1,0;
          0,1,0,0,0,-1,0,0,-1];
   P.Beq=[0; 0; 0]; P.ub=xM; P.nonlcon=@c5mpool;
   P.x0=100*rand(9,1); [x,f0,flag]=fmincon(P)
```

5.5.4 Optimization design of heat exchange network

Heat exchange and heating systems are often encountered in practical problems. Three heaters are to be designed as shown in Figure 5.10. The temperature of the inlet air is $T_{c,\rm in} = 100°\rm F$, and the temperature of the outlet air is $T_{c,\rm out} = 500°\rm F$. The middle three heaters are to be designed so that the heating temperatures are respectively $T_{\rm in,1} = 300°\rm F$, $T_{\rm in,2} = 400°\rm F$, and $T_{\rm in,3} = 600°\rm F$. In order to save energy, three heaters are expected such that the sum of the heated areas $A_1 + A_2 + A_3$ is minimized.

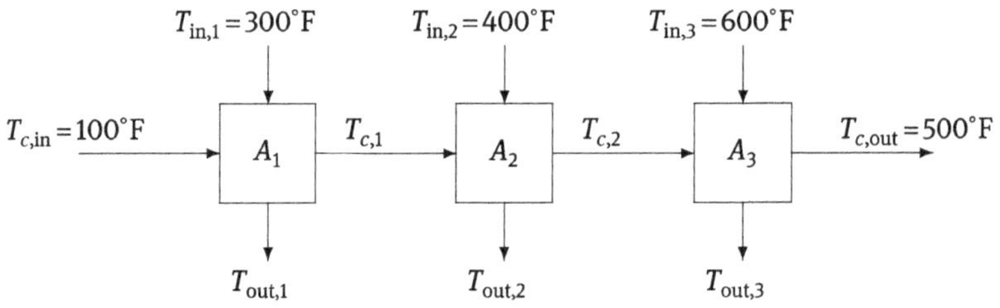

Figure 5.10: Illustration of a heating system.

An optimization model is provided in [2] as

$$\min_{\boldsymbol{A},\boldsymbol{T}_c,\boldsymbol{T}_{\rm out}\ \text{s.t.}\ \begin{cases} T_{c,1}+T_{\rm out,1}-T_{c,\rm in}-T_{\rm in,1}\leqslant 0\\ -T_{c,1}+T_{c,2}+T_{\rm out,2}-T_{\rm in,1}\leqslant 0\\ T_{\rm out,3}-T_{c,2}-T_{\rm in,3}+T_{c,\rm out}\leqslant 0\\ A_1-A_1T_{\rm out,1}+\gamma T_{c,1}/U_1-\gamma T_{c,\rm in}/U_1\leqslant 0\\ A_2T_{c,1}-A_2T_{\rm out,2}-\gamma T_{c,1}/U_2+\gamma T_{c,2}/U_2\leqslant 0\\ A_3T_{c,2}-A_3T_{\rm out,3}-\gamma T_{c,2}/U_3+\gamma T_{c,\rm out}/U_3\leqslant 0\\ 100\leqslant A_1\leqslant 10\,000,\ 1\,000\leqslant A_2,\ A_3\leqslant 10\,000\\ 10\leqslant T_{c,1},\ T_{c,2},\ T_{\rm out,1},\ T_{\rm out,2},\ T_{\rm out,3}\leqslant 1\,000\end{cases}} A_1+A_2+A_3, \tag{5.5.10}$$

where $\gamma = 10^5$, $U_1 = 120$, $U_2 = 80$, and $U_3 = 40$.

Example 5.26. Solve the optimization problem for the heating system design problem provided in Figure 5.10.

Solutions. If YALMIP Toolbox is used, some decision variables and constants are defined, then the objective function and constraints can be entered and the optimization problem be solved directly. There is no need to carry out manual conversions. Manual rewriting of the problems may lead to errors.

```
>> A=sdpvar(3,1); Tc=sdpvar(2,1); Tout=sdpvar(3,1); gam=1e5;
   U=[120,80,40]; Tcin=100; T1in=300; T2in=400;
   T3in=600; Tcout=500;
   cons=[Tc(1)+Tout(1)-Tcin-T1in<=0,
        -Tc(1)+Tc(2)+Tout(2)-T1in<=0,
         Tout(3)-Tc(2)-T3in+Tcout<=0,
         A(1)-A(1)*Tout(1)+gam*Tc(1)/U(1)-gam*Tcin/U(1)<=0,
         A(2)*Tc(1)-A(2)*Tout(2)-gam*Tc(1)/U(2)+gam*Tc(2)/U(2)<=0,
         A(3)*Tc(2)-A(3)*Tout(3)-gam*Tc(2)/U(3)+gam*Tcout/U(3)<=0,
         100<=A(1)<=10000, 1000<=A([2,3])<=10000,
         10<=Tc<=1000, 10<=Tout<=1000];
   opt=sum(A);
   optimize(cons,opt), value(A), value(Tc), value(Tout)
```

The result obtained is:

$$\boldsymbol{A} = [1\,026.9, 1\,000, 5\,485.3]^{\mathrm{T}}, \quad \boldsymbol{T}_c = [265.0597, 280.5887]^{\mathrm{T}},$$
$$\boldsymbol{T}_{\mathrm{out}} = [134.9403, 284.4710, 380.5887]^{\mathrm{T}}.$$

It is the same as presented in [23].

In [22], an optimization problem of a two-unit heat exchange network is proposed. This is a test problem in nonlinear programming. Since the engineering background of the problem is complicated, here no specific introduction is given. Discussions one presented on the mathematical problem and its solutions.

Considering a constrained optimization problem, the objective function can be repressed as

$$J = \left[\frac{20\,000}{2\sqrt{T_{11}T_{12}}/3 + (T_{11} + T_{12})/6}\right]^{0.6} + \left[\frac{20\,000}{2\sqrt{T_{21}T_{22}}/3 + (T_{21} + T_{22})/6}\right]^{0.6}. \tag{5.5.11}$$

Since the constraints of the problem are complicated, the following constraints are listed:

$$f_1^{\mathrm{I}} + f_2^{\mathrm{I}} = 10, \quad f_1^{\mathrm{I}} + f_{12}^{\mathrm{B}} - f_1^{\mathrm{E}} = 0, \quad f_2^{\mathrm{I}} + f_{21}^{\mathrm{B}} - f_2^{\mathrm{E}} = 0, \tag{5.5.12}$$
$$f_1^{\mathrm{O}} + f_{21}^{\mathrm{B}} - f_1^{\mathrm{E}} = 0, \quad f_2^{\mathrm{O}} + f_{12}^{\mathrm{B}} - f_2^{\mathrm{E}} = 0, \tag{5.5.13}$$
$$150 f_1^{\mathrm{I}} + t_2^{\mathrm{O}} f_{12}^{\mathrm{B}} - t_1^{\mathrm{I}} f_1^{\mathrm{E}} = 0, \quad 150 f_2^{\mathrm{I}} + t_1^{\mathrm{O}} f_{21}^{\mathrm{B}} - t_2^{\mathrm{I}} f_2^{\mathrm{E}} = 0, \tag{5.5.14}$$
$$f_1^{\mathrm{E}}(t_1^{\mathrm{O}} - t_1^{\mathrm{I}}) = 1\,000, \quad f_2^{\mathrm{E}}(t_2^{\mathrm{O}} - t_2^{\mathrm{I}}) = 600, \tag{5.5.15}$$

$$T_{11} = 500 - t_1^{\mathrm{O}}, \quad T_{12} = 250 - t_1^{\mathrm{I}}, \quad T_{21} = 350 - t_2^{\mathrm{O}}, \quad T_{22} = 200 - t_2^{\mathrm{I}}, \tag{5.5.16}$$

$$T_{11}, T_{12}, T_{21}, T_{22} \geqslant 10, \tag{5.5.17}$$

$$0 \leqslant f_1^{\mathrm{I}}, f_2^{\mathrm{I}}, f_{12}^{\mathrm{B}}, f_{21}^{\mathrm{B}}, f_1^{\mathrm{O}}, f_2^{\mathrm{O}} \leqslant 10, \tag{5.5.18}$$

$$2.941 \leqslant f_1^{\mathrm{E}} \leqslant 10, \quad 3.158 \leqslant f_2^{\mathrm{E}} \leqslant 10, \quad 150 \leqslant t_1^{\mathrm{I}} \leqslant 240, \tag{5.5.19}$$

$$250 \leqslant t_1^{\mathrm{O}} \leqslant 490, \quad 150 \leqslant t_2^{\mathrm{I}} \leqslant 190, \quad 210 \leqslant t_2^{\mathrm{O}} \leqslant 340. \tag{5.5.20}$$

It can be seen that there are 12 variables to be optimized in the original problem. The decision vector can be introduced, for instance, letting $x_1 = f_1^{\mathrm{I}}$, $x_2 = f_2^{\mathrm{I}}$, $x_3 = f_1^{\mathrm{O}}$, $x_4 = f_2^{\mathrm{O}}$, $x_5 = f_1^{\mathrm{E}}$, $x_6 = f_2^{\mathrm{E}}$, $x_7 = f_{12}^{\mathrm{B}}$, $x_8 = f_{21}^{\mathrm{B}}$, $x_9 = t_1^{\mathrm{I}}$, $x_{10} = t_2^{\mathrm{I}}$, $x_{11} = t_1^{\mathrm{O}}$, and $x_{12} = t_2^{\mathrm{O}}$. An example is given below to demonstrate the problem solutions.

Example 5.27. Solve the nonlinear programming problem described above.

Solutions. From definition, the four intermediate variables T_{ij} can be computed from (5.5.16) as

$$T_{11} = 500 - x_{11}, \quad T_{12} = 250 - x_9, \quad T_{21} = 350 - x_{12}, \quad T_{22} = 200 - x_{10}.$$

With (5.5.11), the objective function value can be computed. With these preparations, the following MATLAB function can be written to describe the objective function:

```
function J=c5mheat(x)
T11=500-x(11); T12=250-x(9); T21=350-x(12); T22=200-x(10);
J=(20000/(2/3)*sqrt(T11*T12)+1/6*(T11+T12))^0.6+...
  (20000/(2/3)*sqrt(T21*T22)+1/6*(T21+T22))^0.6;
```

From (5.5.12) and (5.5.13), the linear equalities can be described by

$$x_1 + x_2 = 10, \quad x_1 + x_7 - x_5 = 0, \quad x_2 + x_8 - x_6 = 0,$$
$$x_3 + x_8 - x_5 = 0, \quad x_4 + x_7 - x_6 = 0,$$

from which the matrix form $\boldsymbol{A}_{\mathrm{eq}}\boldsymbol{x} = \boldsymbol{B}_{\mathrm{eq}}$ can be established, where

$$\boldsymbol{A}_{\mathrm{eq}} = \begin{bmatrix} 1 & 1 & 0 & 0 & 0 & 0 & 0 & 0 & 0 & 0 & 0 & 0 \\ 1 & 0 & 0 & 0 & -1 & 0 & 1 & 0 & 0 & 0 & 0 & 0 \\ 0 & 1 & 0 & 0 & 0 & -1 & 0 & 1 & 0 & 0 & 0 & 0 \\ 0 & 0 & 1 & 0 & -1 & 0 & 0 & 1 & 0 & 0 & 0 & 0 \\ 0 & 0 & 0 & 1 & 0 & -1 & 1 & 0 & 0 & 0 & 0 & 0 \end{bmatrix}, \quad \boldsymbol{B}_{\mathrm{eq}} = \begin{bmatrix} 10 \\ 0 \\ 0 \\ 0 \\ 0 \end{bmatrix}.$$

From (5.5.14) and (5.5.15), the nonlinear equality constraints are written as

$$150x_1 + x_{12}x_7 - x_9x_5 = 0, \quad 150x_2 + x_{11}x_8 - x_{10}x_6 = 0,$$
$$x_5(x_{11} - x_9) - 1000 = 0, \quad x_6(x_{12} - x_{10}) - 600 = 0.$$

From (5.5.17), the nonlinear inequality constraints can be formulated as

$$-T_{11}+10 \leqslant 0, \quad -T_{12}+10 \leqslant 0, \quad -T_{21}+10 \leqslant 0, \quad -T_{22}+10 \leqslant 0.$$

There is no need to convert the inequality constraints into an expression of $\boldsymbol{x}$. With the above constraints, the following MATLAB function can be written to describe them:

```
function [c,ceq]=c5mheat1(x)
T11=500-x(11); T12=250-x(9); T21=350-x(12); T22=200-x(10);
c=-[T11; T12; T21; T22]+10;
ceq=[150*x(1)+x(12)*x(7)-x(9)*x(5); 150*x(2)+x(11)*x(8)-x(10)*x(6);
     x(5)*(x(11)-x(9))-1000; x(6)*(x(12)-x(10))-600];
```

Besides, from (5.5.18), (5.5.19), and (5.5.20), the lower and upper bounds of the decision variables can be established as

$$\boldsymbol{x}_\mathrm{m} = [0,0,0,0,2.941,3.158,0,0,150,150,250,210]^\mathrm{T},$$
$$\boldsymbol{x}_\mathrm{M} = [10,10,10,10,10,10,10,10,240,190,490,340]^\mathrm{T}.$$

Since function `fmincon()` is used, local optimum solutions may be obtained. It is recommended to use the global solver `fmincon_global()` to solve the following problem directly:

```
>> A=[]; B=[]; Beq=[10; 0; 0; 0; 0];
   Aeq=[1,1,0,0,0,0,0,0,0,0,0,0; 1,0,0,0,-1,0,1,0,0,0,0,0;
        0,1,0,0,0,-1,0,1,0,0,0,0; 0,0,1,0,-1,0,0,1,0,0,0,0;
        0,0,0,1,0,-1,1,0,0,0,0,0];
   xm=[0,0,0,0,2.941,3.158,0,0, 150,150,250,210];
   xM=[10,10,10,10,10,10,10,10,240,190,490,340]; n=12; N=10;
   x=fmincon_global(@c5mheat,0,10,n,N,A,B,Aeq,Beq,xm,xM,@c5mheat1)
```

The global optimum solution $\boldsymbol{x} = [0,10,10,0,10,10,10,0,210,150,310,210]^\mathrm{T}$ can be found, that is, $f_1^\mathrm{I} = f_2^\mathrm{O} = f_{21}^\mathrm{B} = 0$, $f_2^\mathrm{I} = f_1^\mathrm{O} = f_1^\mathrm{E} = f_2^\mathrm{E} = f_{12}^\mathrm{B} = 10$, $t_1^\mathrm{I} = 210$, $t_2^\mathrm{I} = 150$, $t_1^\mathrm{O} = 310$, and $t_2^\mathrm{O} = 210$. The results are exactly the same as given in [22].

5.5.5 Solving nonlinear equations with optimization techniques

In Chapter 2, nonlinear equation solution problems were fully covered. In Section 3.5.4, a conversion method was introduced and illustrated to show the relationship between equations and optimization problems. Here, an alternative method is used to illustrate how to convert a nonlinear equation problem into a nonlinear constrained programming problem, so as to find the equation with an optimization problem solver.

Assume that the equations $\boldsymbol{f}(\boldsymbol{x}) = \boldsymbol{0}$ are given. It is natural to introduce decision variables $\boldsymbol{s}$, whose elements are the same as the numbers in $\boldsymbol{x}$. Therefore, the following optimization model can be set up:

$$\min_{\boldsymbol{x},\boldsymbol{s} \text{ s. t. } \begin{cases} \boldsymbol{f}(\boldsymbol{x})-\boldsymbol{s}\leqslant\boldsymbol{0} \\ -\boldsymbol{f}(\boldsymbol{x})-\boldsymbol{s}\leqslant\boldsymbol{0} \end{cases}} \boldsymbol{s}. \tag{5.5.21}$$

The two decision vectors $\boldsymbol{x}$ and $\boldsymbol{s}$ can be combined to form a new decision vector $\boldsymbol{x} = [\boldsymbol{x};\boldsymbol{s}]^{\mathrm{T}}$. The equation solution problems can be converted into constrained optimization problems. For simplicity, a scalar s can be used to show the problem. An example is used below to demonstrate the solutions of nonlinear equations.

Example 5.28. Solve the following simultaneous equations with two variables studied in Example 3.32. For simplicity, the original problem is displayed here:[23]

$$\begin{cases} 4x_1^3 + 4x_1x_2 + 2x_2^2 - 42x_1 = 14, \\ 4x_2^3 + 2x_1^2 + 4x_1x_2 - 26x_2 = 22. \end{cases}$$

Solutions. In fact, for the simultaneous equations with two variables, an extra scalar decision variable s is introduced such that $x_3 = s$. The original problem can be expressed as the following constrained optimization problem:

$$\min_{\boldsymbol{x},\boldsymbol{s} \text{ s. t. } \begin{cases} 4x_1^3+4x_1x_2+2x_2^2-42x_1-14-x_3\leqslant0 \\ 4x_2^3+2x_1^2+4x_1x_2-26x_2-22-x_3\leqslant0 \\ -4x_1^3-4x_1x_2-2x_2^2+42x_1+14-x_3\leqslant0 \\ -4x_2^3-2x_1^2-4x_1x_2+26x_2+22-x_3\leqslant0 \end{cases}} x_3.$$

Therefore, the following MATLAB function can be written to describe the nonlinear constraints:

```
function [c,ceq]=c5me2o(x)
ceq=[];
c=[4*x(1)^3+4*x(1)*x(2)+2*x(2)^2-42*x(1)-14-x(3);
   4*x(2)^3+2*x(1)^2+4*x(1)*x(2)-26*x(2)-22-x(3);
   -4*x(1)^3-4*x(1)*x(2)-2*x(2)^2+42*x(1)+14-x(3);
   -4*x(2)^3-2*x(1)^2-4*x(1)*x(2)+26*x(2)+22-x(3)];
```

With the constraints, the following commands can be used to find a solution for the given nonlinear equations. If another initial search point is used, other solutions of the equations may also be found.

```
>> P.objective=@(x)x(3); P.nonlcon=@c5me2o;
   P.solver='fmincon'; P.options=optimset;
   P.x0=rand(3,1); [x,f0,flag]=fmincon(P)
```

5.6 Exercises

5.1 Solve the following optimization problem.[45] Is there any high-precision method in solving this optimization problem?

$$\max_{x,y,z \text{ s.t. } \begin{cases} 8+5z^3x-4z^8y+3x^2y-xy^2=0 \\ 1-z^9-z^3x+y+3z^5xy+7x^2y+2xy^2=0 \\ -1-5z-5z^9x-5z^8y-2z^9xy+x^2y+4xy^2=0 \end{cases}} z.$$

5.2 Solve the following optimization problem:

$$\max_{x,y \text{ s.t. } \begin{cases} y\leqslant 2x^4-8x^3+8x^2+2 \\ y\leqslant 4x^4-32x^3+88x^2-96x+36 \\ 0\leqslant x\leqslant 3,\ 0\leqslant y\leqslant 4 \end{cases}} x+y.$$

5.3 Solve the following nonconvex quadratic programming problem:

$$\max_{x,y \text{ s.t. } \begin{cases} 6x_1+3x_2+3x_3+2x_4+x_5\leqslant 6.5 \\ 10x_1+10x_3+y\leqslant 20 \\ 0\leqslant x_i\leqslant 1,\ y>0 \end{cases}} \boldsymbol{c}^{\mathrm{T}}\boldsymbol{x}+dy+\frac{1}{2}\boldsymbol{x}^{\mathrm{T}}\boldsymbol{Q}\boldsymbol{x},$$

where $\boldsymbol{c}^{\mathrm{T}}=[-10.5,-7.5,-3.5,-2.5,-1.5]$, $\boldsymbol{Q}=\boldsymbol{I}$, $d=1$.

5.4 Solve the following optimization problem:

$$\min_{\boldsymbol{x} \text{ s.t. } \begin{cases} 0.003079x_1^3x_2^3x_5-\cos^3 x_6\geqslant 0 \\ 0.1017x_3^3x_4^3-x_5^2\cos^3 x_6\geqslant 0 \\ 0.09939(1+x_5)x_1^3x_2^2-\cos^2 x_6\geqslant 0 \\ 0.1076(31.5+x_5)x_3^3x_4^2-x_5^2\cos^2 x_6\geqslant 0 \\ x_3x_4(x_5+31.5)-x_5[2(x_1+5)\cos x_6+x_1x_2x_5]\geqslant 0 \\ 0.2\leqslant x_1\leqslant 0.5, 14\leqslant x_2\leqslant 22, 0.35\leqslant x_3\leqslant 0.6 \\ 16\leqslant x_4\leqslant 22, 5.8\leqslant x_5\leqslant 6.5, 0.14\leqslant x_6\leqslant 0.2618 \end{cases}} \frac{1}{2\cos x_6}\left[x_1x_2(1+x_5)+x_3x_4\left(1+\frac{31.5}{x_5}\right)\right].$$

5.5 Solve the following nonlinear programming problem:[22]

$$\min_{\boldsymbol{x} \text{ s.t. } \begin{cases} -0.0022053x_3x_5+0.0056858x_2x_5+0.0006262x_1x_4-6.665593\leqslant 0 \\ 0.0022053x_3x_5-0.0056858x_2x_5-0.0006262x_1x_4-85.334407\leqslant 0 \\ 0.0071317x_2x_5+0.0021813x_3^2+0.0029955x_1x_2-29.48751\leqslant 0 \\ -0.0071317x_2x_5-0.0021813x_3^2-0.0029955x_1x_2+9.48751\leqslant 0 \\ 0.0047026x_3x_5+0.0019085x_3x_4+0.0012547x_1x_3-15.699039\leqslant 0 \\ -0.0047026x_3x_5-0.0019085x_3x_4-0.0012547x_1x_3+10.699039\leqslant 0 \\ 78\leqslant x_1\leqslant 102,\ 33\leqslant x_2\leqslant 45,\ 27\leqslant x_3,x_4,x_5\leqslant 45 \end{cases}} 37.293239x_1+0.8356891x_1x_5+5.3578547x_3^2-40\,792.141.$$

5.6 Solve the following optimization problems:[22]

(1)
$$\min_{\boldsymbol{x} \text{ s.t. } \begin{cases} x_2-3x_1-3u_1=0 \\ x_1+2u_1\leqslant 4 \\ x_2+2u_2\leqslant 4 \\ x_1\leqslant 3,\ u_2\leqslant 1 \\ x_1,x_2,u_1,u_2\geqslant 0 \end{cases}} x_1^{0.6}+x_2^{0.6}-6x_1-4u_1+3u_2;$$

(2)
$$\min_{\boldsymbol{x} \text{ s.t. } \begin{cases} x_2-3x_1-3u_1=0 \\ x_3-2x_2-2u_2=0 \\ 4u_1-u_3\leqslant 0 \\ x_1+2u_1\leqslant 4 \\ x_2+u_2\leqslant 4 \\ x_3+u_3\leqslant 6 \\ x_1\leqslant 3,\ u_2\leqslant 2,\ x_3\leqslant 4 \\ x_1,x_2,x_3,u_1,u_2,u_3\geqslant 0 \end{cases}} x_1^{0.6}+x_2^{0.6}+x_3^{0.4}+2u_1+5u_2-4x_3-u_3.$$

5.7 Solve the following quadratic programming problem:[22]

$$\min_{\boldsymbol{x} \text{ s.t. } \begin{cases} -1+0.0025(x_4+x_6)\leqslant 0 \\ -1+0.0025(-x_4+x_5+x_7)\leqslant 0 \\ -1+0.01(-x_5+x_8)\leqslant 0 \\ 100x_1-x_1x_6+833.33252x_4-83\,333.333\leqslant 0 \\ x_2x_4-x_2x_7-1\,250x_4+1\,250x_5\leqslant 0 \\ x_3x_5-x_3x_8-2\,500x_5+1\,250\,000\leqslant 0 \\ 100\leqslant x_1\leqslant 10\,000,\ 1\,000\leqslant x_2,x_3\leqslant 10\,000 \\ 10\leqslant x_4,x_5,x_6,x_7,x_8\leqslant 1\,000 \end{cases}} x_1+x_2+x_3.$$

5.8 Solve the following optimization problem:

$$\min_{\boldsymbol{x} \text{ s.t. } \begin{cases} 0.0193x_3-x_1\leqslant 0 \\ 0.00954x_3-x_2\leqslant 0 \\ 750\times 1\,728-\pi x_3^2x_4-4\pi x_3^3/3\leqslant 0 \\ x_4-240\leqslant 0 \\ 0.0625\leqslant x_1,x_2\leqslant 6.1875, 10\leqslant x_3,x_4\leqslant 200 \end{cases}} 0.6224x_1x_2x_3x_4+1.7781x_2x_3^2+3.1661x_1^2x_4+19.84x_1^2x_3.$$

5.9 Solve the following optimization problem:[26]

$$\min_{\boldsymbol{q},k \text{ s.t. } \begin{cases} g(\boldsymbol{q})\leqslant 0 \\ 800-800k\leqslant q_1\leqslant 800+800k \\ 4-2k\leqslant q_2\leqslant 4+2k \\ 6-3k\leqslant q_3\leqslant 6+3k \end{cases}} k,$$

where $g(\boldsymbol{q}) = 10q_2^2q_3^3+10q_2^3q_3^2+200q_2^2q_3^2+100q_2^3q_3+q_1q_2q_3^2+q_1q_2^2q_3+1\,000q_2q_3^3+8q_1q_3^2+1\,000q_2^2q_3+8q_1q_2^2+6q_1q_2q_3-q_1^2+60q_1q_3+60q_1q_2-200q_1$.

5.10 Solve Exercises 4.12 and 4.13 in Chapter 4, and find the global optimum solutions of the nonconvex quadratic problems.

5.11 Solve the following optimization problem:

$$\min \quad -2x_1 + x_2 - x_3,$$
$$\boldsymbol{x} \text{ s. t. } \begin{cases} x_1+x_2+x_3 \leqslant 4 \\ 3x_2+x_3 \leqslant 6 \\ x_1 \leqslant 2,\ x_3 \leqslant 3 \\ x_1,x_2,x_3 \geqslant 0 \\ \boldsymbol{x}^{\mathrm{T}}\boldsymbol{B}^{\mathrm{T}}\boldsymbol{B}\boldsymbol{x} - 2\boldsymbol{r}^{\mathrm{T}}\boldsymbol{B}\boldsymbol{x} + \|\boldsymbol{r}\|^2 - 0.25\|\boldsymbol{b}-\boldsymbol{v}\|^2 \geqslant 0 \end{cases}$$

where

$$\boldsymbol{B} = \begin{bmatrix} 0 & 0 & 1 \\ 0 & -1 & 0 \\ -2 & 1 & -1 \end{bmatrix}, \quad \boldsymbol{b} = \begin{bmatrix} 3 \\ 0 \\ -4 \end{bmatrix}, \quad \boldsymbol{v} = \begin{bmatrix} 0 \\ -1 \\ -6 \end{bmatrix}, \quad \boldsymbol{r} = \begin{bmatrix} 1.5 \\ -0.5 \\ -5 \end{bmatrix}.$$

5.12 Solve the following programming problems and try to find the global optimum solutions:[6]

$$\max \quad 5x_1 + \mathrm{e}^{-2x_2} - \mathrm{e}^{-x_2} + x_1x_3 + 4x_3 + 6x_4 + \frac{5x_5}{x_5+1} + \frac{6x_6}{x_6+1}.$$
$$\boldsymbol{x} \text{ s. t. } \begin{cases} x_1+x_2+x_3+x_4+x_5+x_5 \leqslant 10 \\ x_1+x_3+x_4 \leqslant 5 \\ x_1-x_2^2+x_3+x_5+x_6^2 \leqslant 5 \\ x_2+2x_4+x_5+0.8x_8=5 \\ x_3^2+x_5^2+x_6^2=5 \end{cases}$$

5.13 Solve the following bilevel linear programming problem:[23]

$$\min \quad -2x_1 + x_2 + 0.5y_1.$$
$$\boldsymbol{x},\boldsymbol{y} \text{ s. t. } \begin{cases} \boldsymbol{x} \geqslant 0,\ \boldsymbol{y} \geqslant 0 \\ \min \quad -4y_1 + y_2 + x_1 + x_2 \\ \boldsymbol{y} \text{ s. t. } \begin{cases} -2x_1+y_1-y_2 \leqslant -2.5 \\ x_1-3x_2+y_2 \leqslant 2 \\ y_1,y_2 \geqslant 0 \end{cases} \end{cases}$$

5.14 Solve the following bilevel quadratic programming problem:[3]

$$\min \quad x^2 + (y-10)^2.$$
$$x,y \text{ s. t. } \begin{cases} -x+y \leqslant 0 \\ 0 \leqslant x \leqslant 15 \\ \min \quad (x+2y-30)^2 \\ y \text{ s. t. } \begin{cases} x+y \leqslant 20 \\ 0 \leqslant y \leqslant 20 \end{cases} \end{cases}$$

5.15 Solve the following semiinfinite programming problem:[41]

$$\min \quad \sum_{i=1}^{4} x_i^2 - x_i.$$
$$\boldsymbol{x} \text{ s. t. } \begin{cases} x_4 \sin(30t_1 \sin x_1 + 30t_2 \cos x_2)/5 \\ \quad + x_3 \sin(t_1t_2/10)/10 + t_3x_1 + t_4x_2 + t_5x_3 + t_6x_4 - 4 \leqslant 0 \\ -1 \leqslant t_1,t_2,\ldots,t_6 \leqslant 1 \end{cases}$$

5.16 Solve the following nonlinear programming problem:

$$\min_{\boldsymbol{x} \text{ s.t. } \begin{cases} -x_1^2-x_2^2-x_3^2-x_4^2-x_1+x_2-x_3+x_4+8\geqslant 0 \\ -x_1^2-2x_2^2-x_3^2-2x_4^2+x_1+x_4+10\geqslant 0 \\ -2x_1^2-x_2^2-x_3^2-2x_4^2+x_2+x_4+5\geqslant 0 \end{cases}} x_1^2+x_2^2+2x_3^2+x_4-5x_1-5x_2-21x_3+7x_4.$$

5.17 Solve the following nonlinear programming problem:

$$\min_{\boldsymbol{x} \text{ s.t. } \begin{cases} -2x_1^2-3x_2^4-x_3-4x_4^2-5x_5+127\geqslant 0 \\ 7x_1-3x_2-10x_3^2-x_4+x_5+282\geqslant 0 \\ 23x_1-x_2^2-6x_6^2+8x_7+196\geqslant 0 \\ -4x_1^2-x_2^2+3x_1x_2-2x_3^2-5x_6+11x_7\geqslant 0 \end{cases}} \begin{array}{l} (x_1-10)^2+5(x_2-12)^2+x_3^4+3(x_4-11)^2 \\ \quad +10x_5^2+7x_6^2+x_7^2-4x_6x_7-10x_6-8x_7. \end{array}$$

5.18 Consider the application problem in Example 5.24, solve it for the other two cases in [23]: (1) $c_1=16$, $c_2=100$, and (2) $c_1=16$, $c_2=600$.

5.19 Solve the following optimization problem,[22] where the objective function is

$$\min \left(\frac{Q_1}{U_1A_1}\right)^{0.6}+\left(\frac{Q_2}{U_2A_2}\right)^{0.6}+\left(\frac{Q_3}{U_3A_3}\right)^{0.6}.$$

The constraints are

$$\begin{gathered} Q_1=C(T_1-T_\mathrm{m}),\quad Q_2=C(T_2-T_1),\quad Q_3=C(T_\mathrm{M}-T_2), \\ Q_1=C(t_1'-t_1),\quad Q_2=C(t_2'-t_2),\quad Q_3=C(t_3'-t_3), \\ A_1=\frac{(t_1-T_\mathrm{m})-(t_1'-T_1)}{\ln((t_1-T_\mathrm{m})/(t_1'-T_1))}, \\ A_2=\frac{(t_2-T_1)-(t_2'-T_2)}{\ln((t_2-T_1)/(t_2'-T_2))}, \\ A_3=\frac{(t_3-T_2)-(t_3'-T_\mathrm{M})}{\ln((t_3-T_2)/(t_3'-T_\mathrm{M}))}, \\ T_\mathrm{m}\leqslant T_1,T_2\leqslant T_\mathrm{M},\quad t_1\leqslant t_1',\quad t_2\leqslant t_2',\quad t_3\leqslant t_3', \end{gathered}$$

and the constants $T_\mathrm{m}=100$, $T_\mathrm{M}=500$, $C=10^5$, $t_1'=300$, $t_2'=400$, $t_3'=600$, $U_1=120$, $U_2=80$, and $U_3=40$.

Note that the decision variables can be selected as T_1, T_2, t_1, t_2, t_3, Q_1, Q_2, and Q_3.

5.20 Solve again the problem in Example 5.26 with MATLAB Optimization Toolbox functions. Compare the results and see whether better solutions can be found.

5.21 Convert the equation solution problems in Exercises 2.12 and 2.14 into constrained optimization problems, and compare the efficiency and accuracy.

5.22 Normally, nonconvex problems have no analytical solutions. Therefore, there is no way to find the global optimum solutions for them. In many examples in this chapter, the term “global optimum solution” was used. It satisfies the constraints and yields the smallest possible objective function value. A better term “best known solution” should be used. Try to get the solutions again and see whether better results can be found.

6 Mixed integer programming

In the programming problems discussed so far, the decision variables can be selected as any real numbers. In some actual optimization problems, for instance, the decision variables x_i may be the number of people, or the number of items in the transportation problem. The actual decision variables should be restricted to integers. In this case, the programming problems may be changed into integer or mixed-integer programming problems.

In Section 6.1, the fundamental concepts of integer and mixed-integer programming are proposed. Then, the computational complexity of the integer programming problems is discussed. In Section 6.2, enumeration methods in solving integer programming problems are introduced; these methods are suitable to handle linear and nonlinear integer programs, as well as discrete and 0–1 programming problems. Attempts are made on mixed-integer programming problems. In Section 6.3, examples are given to demonstrate the solutions of various mixed-integer linear, discrete and mixed-integer nonlinear programming problems. Linear matrix inequality-based linear mixed-integer programming problems are also demonstrated. In Section 6.4, a special kind of mixed-integer programming problems, namely mixed 0–1 programming problems, are discussed. In Section 6.5, some applications of mixed-integer programming problems are presented, including the applications of sudoku, assignment, traveling salesman and knapsack problems.

6.1 Introduction to integer programming

Integer programming problems are the commonly encountered problems in the field of optimization and operations research. In this section, the fundamental concepts of integer and mixed-integer programming are proposed. The graphical method is used to show the solution of a simple integer linear programming problem. The computational complexity of integer programming problems is also discussed.

6.1.1 Integer and mixed-integer programming problems

Definition 6.1. If all the decision variables in an optimization problem are assumed to be integers, the problem is referred to as an integer programming problem. If some of the decision variables are assumed to be integers, while there is no such restriction on others, the problem is referred to as a mixed-integer programming problem.

Here a simple example is given. The graphical method can be used to show the solution to an integer programming problem with two decision variables. The concept of feasible solutions is also demonstrated.

https://doi.org/10.1515/9783110667011-006

Example 6.1. Consider again the optimization problem in Example 4.1:

$$\min_{x,y \text{ s.t. } \begin{cases} -x+2y\leqslant 4 \\ 3x+2y\leqslant 12 \\ x\geqslant 0,\ y\geqslant 0 \end{cases}} -x-2y.$$

If the decision variables x and y are expected to be integers, solve the problem again with the graphical method.

Solutions. Two straight lines, $-x + 2y = 4$ and $3x + 2y = 12$, can be drawn in the x–y plane. The feasible region can be bounded with the other two lines, $x = 0$ and $y = 0$, in the shaded area, as shown in Figure 6.1. Integer programming feasible solutions are the integer points in the shaded area. All the feasible solutions are labeled in the figure.

```
>> syms x; fplot([2+x/2, 6-3*x/2],[-1,5])
```

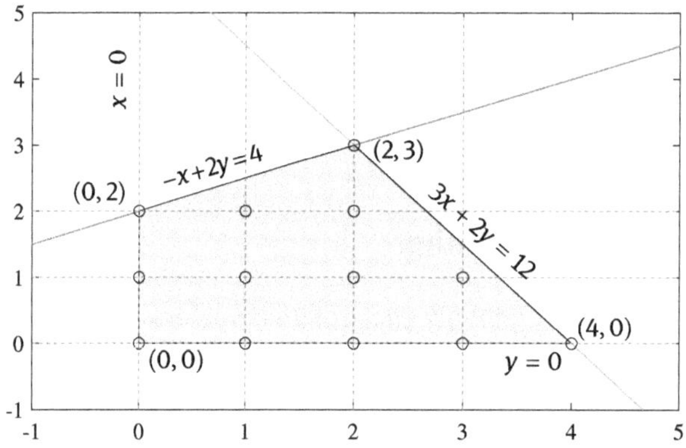

Figure 6.1: Feasible region demonstration.

Since a linear integer programming problem is convex, the optimum solution is located at one of the four corners, or vertexes. For this specific problem, point (2, 3) is the global optimum solution, the objective function value is −8.

6.1.2 Computational complexity of integer programming problems

In computer science fields, computational complexity in terms of time complexity is used to assess the merit or the simplicity of the algorithms. For instance, the multiplication of two $n\times n$ matrices can be carried out with a triple loop, and the computational complexity is represented as an $n^3 + n^2$ problem. If the computational complexity can be expressed as a polynomial function of n, the problem is referred to as a P problem. If the complexity cannot be expressed as deterministic polynomial of n, the problem is referred to as an NP (non-deterministic polynomial) problem. If the complexity cannot be reduced, the problem is known as an NP-hard problem.

It can be seen that in the integer programming problems, if each of the n decision variables has two options, there are 2^n different combinations. If each variable has 50 options, the number of combination may reach 50^n. If the enumeration method is used, the computational complexity cannot be expressed as a polynomial of n. An integer programming problem is then an NP-complete problem. Search algorithms can sometimes be used to reduce the computational complexity of the problems.

6.2 Enumeration methods for integer programming

Integer programming problems have various solutions methods. Some NP-complete problems can be converted to P problems with specific methods. Since the number of options for the decision variable is finite, the enumeration method can be used to solve problems. In this section, the concept of the enumeration method is introduced first, then various examples are demonstrated by the enumeration method.

6.2.1 An introduction to the enumeration method

Definition 6.2. The so-called enumeration method is to used assess all the possible combinations of the decision variables. The feasible combinations can be found, and the objective function values can be sorted. A feasible solution with the least objective function value can be found. The solution is referred to as a global optimum solution. The enumeration method is also known as a brute-force method.

If the intervals of the decision variables are known, theoretically speaking, the enumeration method can be used to list all the possible combinations of the decision variables. Testing each combination whether the constraints are satisfied or not, the feasible combinations are found. From the feasible combinations, the object function values are computed and sorted. From the theoretical point of view, the enumeration method can be used to list all the possible combinations of the decision variables. After comparing the objective function values at the feasible solutions, the global optimum problem can be solved. This method seems to be simple and straightforward. When there are too many the decision variables, or the they have too many combinations, the computational load is too high to be implemented on any computer. The related mathematical problem is an NP-hard problem. The enumeration method cannot be used to solve the problems universally. We can only use it to solve small-scale problems.

Example 6.2. Solve the problem in Example 6.1 with the enumeration method.

Solutions. The following commands can be used to generate mesh grids in matrices $\boldsymbol{x}$ and $\boldsymbol{y}$. The constraints can be tested, and the subscripts `ii` satisfying the constraints form the feasible solutions are shown in Table 6.1.

```
>> [x,y]=meshgrid(0:5);                    % all the combinations
   ii=find(-x+2*y<=4 & 3*x+2*y<=12)  % find the feasible solutions
   x1=x(ii); y1=y(ii); T=[x1'; y1'; -x1'-2*y1'] % list them
```

With function `meshgrid()`, the mesh grid obtained yields the following matrices:

$$\boldsymbol{x}=\begin{bmatrix}0&1&2&3&4&5\\0&1&2&3&4&5\\0&1&2&3&4&5\\0&1&2&3&4&5\\0&1&2&3&4&5\\0&1&2&3&4&5\end{bmatrix},\quad \boldsymbol{y}=\begin{bmatrix}0&0&0&0&0&0\\1&1&1&1&1&1\\2&2&2&2&2&2\\3&3&3&3&3&3\\4&4&4&4&4&4\\5&5&5&5&5&5\end{bmatrix}.$$

Table 6.1: Information of all the feasible solutions.

ii	1	2	3	7	8	9	13	14	15	16	19	20	25
x_i	0	0	0	1	1	1	2	2	2	2	3	3	4
y_i	0	1	2	0	1	2	0	1	2	3	0	1	0
f_i	0	−2	−4	−1	−3	−5	−2	−4	−6	−8	−3	−5	−4

The subscripts `ii` satisfying all the constraints are expressed in a one-dimensional vector. For instance, $\boldsymbol{x}$ is a 6×6 matrix, when `ii` is 16, it represents the 4th row and 3rd column element. That is, $x=2$, $y=3$. It can be seen from the table that the value of the objective function is -8. In the listed feasible solutions, the value is the smallest. The point $x=2$ and $y=3$ is the global optimum solution of the problem, which is the same as that in Figure 6.1.

If the amount of data in the table are too large, it may not be possible to find the smallest value by observations. Sorting function `sort()` can be used to list the objective function values from the smallest to the largest. The one ranked in the first place is the global minimum solution. In the function call of `sort()`, apart from the objective function, the sorting order is also provided. The value corresponding to `i0(1)` is the one with the smallest value. Therefore, the corresponding x and y can be extracted. The result obtained is $\boldsymbol{v}=[2,3,-8]$, indicating that $x=2$, $y=3$ is the global solution, with the objective function value of -8.

```
>> z=-x1-2*y1; [z0 i0]=sort(z); v=[x1(i0(1)),y(i0(1)),z0(1)]
```

Example 6.3. Consider the problem in Example 4.6. For convenience of comparison, the problem is shown here again:

$$\min_{x \text{ s.t.} \begin{cases} 2x_2+x_3+4x_4+2x_5 \leqslant 54 \\ 3x_1+4x_2+5x_3-x_4-x_5 \leqslant 62 \\ x_1,x_2 \geqslant 0, x_3 \geqslant 3.32, x_4 \geqslant 0.678, x_5 \geqslant 2.57 \end{cases}} -2x_1 - x_2 - 4x_3 - 3x_4 - x_5.$$

If the decision variables x_i are all integers, the problem is classified as an integer programming problem. Solve the problem.

Solutions. For small-scale problems, the enumeration method can be adopted. Assuming that the upper bounds $\boldsymbol{x}_{\mathrm{M}}$ are all 25, function `ndgrid()` can be used to generate all the combinations for the five decision variables. When generating the mesh grid of x_3, the vector used is `4:N`, since the lower bound of x_3 is 3.32. The starting point is 4. Therefore, the vector starting from 4 is used. Similar explanations for vectors x_4 and x_5 are used.

```
>> N=25;
   [x1,x2,x3,x4,x5]=ndgrid(0:N,0:N,4:N,1:N,3:N); % all possible ones
   i=find((2*x2+x3+4*x4+2*x5<=54)&...
          (3*x1+4*x2+5*x3-x4-x5<=62));                % feasible solutions
   x1=x1(i); x2=x2(i); x3=x3(i); x4=x4(i); x5=x5(i); % extract them
   f=-2*x1-x2-4*x3-3*x4-x5; [fmin,ii]=sort(f);        % sorting
   i0=ii(1); x=[x1(i0),x2(i0),x3(i0),x4(i0),x5(i0)]  % optimum one
```

Of course, all the feasible solutions can be extracted. The objective function values at these points are evaluated and sorted. The one ranked first is the global minimum solution. The global optimum solution obtained is $\boldsymbol{x} = [19, 0, 4, 10, 5]^{\mathrm{T}}$.

There are two things to note:

(1) The interval tried now is $x_1 \in [0, 25]$. The interval cannot be arbitrarily extended. If one wants to extended it from 25 to 30, the computational load may be too heavy, since the memory demanded may be too large. For the 5 variables, the memory required is $31^5 \times 5 \times 8/2^{20} = 1092.1$ MB of space. For such problems, enumeration methods are not suitable.

(2) Apart from the global optimum solutions, there may be other combinations where suboptimal solutions are obtained. The top 10 in the sorted objective functions are $\boldsymbol{x}_1 = [-89, -88, -88, -88, -88, -88, -88, -88, -87, -87]$.

```
>> L=15; fx=fmin(1:L)' % extract the top 15 feasible solutions
   in=ii(1:L); x=[x1(in),x2(in),x3(in),x4(in),x5(in),fmin(1:15)]
```

It can be seen that the smallest objective function value is −89. Besides, there are other points with function values of −88. Apart from the global minimum solutions, the suboptimal solutions are also listed in Table 6.2.

Table 6.2: Optimal and suboptimal solutions.

x_1	x_2	x_3	x_4	x_5	f	solutions	x_1	x_2	x_3	x_4	x_5	f	solutions
19	0	4	10	5	−89	optimal	18	0	4	11	3	−88	suboptimal
17	0	5	10	4	−88	suboptimal	15	0	6	10	4	−88	suboptimal
12	0	8	9	5	−88	suboptimal	19	0	4	9	7	−88	suboptimal
16	0	6	8	8	−88	suboptimal	20	0	4	7	11	−88	suboptimal
15	0	6	10	3	−87	suboptimal	13	0	7	10	3	−87	suboptimal
11	0	8	10	3	−87	suboptimal	10	0	9	9	4	−87	suboptimal
8	0	10	9	4	−87	suboptimal	5	0	12	8	5	−87	suboptimal
18	0	4	10	5	−87	suboptimal							

Example 6.4. Solve the following integer programming problem:

$$\min \quad x_1^2 + x_2^2 + 2x_3^2 + x_4^2 - 5x_1 - 5x_2 - 21x_3 + 7x_4.$$
$$\boldsymbol{x} \text{ s. t. } \begin{cases} -x_1^2-x_2^2-x_3^2-x_4^2-x_1+x_2-x_3+x_4+8\geqslant 0 \\ -x_1^2-2x_2^2-x_3^2-2x_4^2+x_1+x_4+10\geqslant 0 \\ -2x_1^2-x_2^2-x_3^2-2x_4^2+x_2+x_4+5\geqslant 0 \end{cases}$$

Solutions. Selecting the decision variable bounds of interest as $-N \sim N$, and selecting $N = 30$, the enumeration method can be tried to find the global optimum solution at $\boldsymbol{x} = [0,1,2,0]^{\mathrm{T}}$, and the corresponding objective function value is −38. Apart from the optimum solution, some suboptimal solutions can be found, such as $[0,0,2,0]$, $[0,1,2,1]$, $[0,1,1,-1]$, and $[1,2,1,0]$, the corresponding objective function values are −38, −34, −30, −29, and −29.

```
>> N=30; [x1 x2 x3 x4]=ndgrid(-N:N); % all possible combinations
   ii=find(-x1.^2-x2.^2-x3.^2-x4.^2-x1+x2-x3+x4+8>=0 & ...
           -x1.^2-2*x2.^2-x3.^2-2*x4.^2+x1+x4+10>=0 & ...
           -2*x1.^2-x2.^2-x3.^2-2*x4.^2+x2+x4+5>=0);   % constraints
   x1=x1(ii); x2=x2(ii); x3=x3(ii); x4=x4(ii);           % feasible
   ff=x1.^2+x2.^2+2*x3.^2+x4.^2-5*x1-5*x2-21*x3+7*x4; % sorting
   [fm,ii]=sort(ff); k=ii(1:5); X=[x1(k),x2(k),x3(k),x4(k)]
   fm(1:5) % extract top 5
```

Note that the enumeration method can only be used to solve problems where the combinations of the decision variables can be enumerated, such as in integer programming problems. It cannot be used in solving mixed-integer problems or ordinary optimization problems, since the variables are continuous. It is not possible to list all possible combinations for continuous problems. The solutions then can only be found with searching methods.

Compared with the searching methods discussed earlier, the computational complexity is not significantly increased in the enumeration method when there are nonlinearities.

6.2.2 Discrete programming

If in an optimization problem the decision variables are only allowed to take some discrete values, and the discrete values may not be integers, a discrete programming problem appears. Since the decision variables can only be some finite discrete values, the enumeration method may be tried. Finding all the combinations satisfying the constraints, all such combinations can be assessed, from which the global optimum solutions may be found. Here an example is given to demonstrate the solution method for discrete programming problems.

Example 6.5. Solve the following discrete programming problem:[10]

$$\min_{\boldsymbol{x} \text{ s.t. } \begin{cases} x_1^2-6x_1+x_2-11\leqslant 0 \\ -x_1x_2+3x_2+e^{x_1-3}-1\leqslant 0 \end{cases}} 2x_1^2+x_2^2-16x_1-10x_2,$$

where x_1 is a multiple of 0.25, x_2 is a multiple of 0.1, and $x_2 \geqslant 3$.

Solutions. The enumeration method may not only be used to handle integer programming problems, it is suitable to handle discrete optimization problems. Assuming that the searching interval of x_1 is $(-20, 20)$, while that for x_2 is $[3, 20]$, the following commands can be used and the global minimum solution is found at $(4, 5)$. Besides, some other suboptimal points can be found at $(4, 5.1)$, $(4, 4.9)$, $(4, 4.8)$, $(4, 5.2)$, and so on, the corresponding objective function values are slightly larger than that for $(4, 5)$. In Table 6.3, those ranked in the top 16 are listed, and shown as optimal and suboptimal solutions.

```
>> [x1 x2]=meshgrid(-20:0.25:20,3:0.1:20); % all possible combinations
   ii=find(x1.^2-6*x1+x2-11<=0 & ...
            -x1.*x2+3*x2+exp(x1-3)-1<=0);       % find feasible solutions
   x1=x1(ii); x2=x2(ii);
   ff=2*x1.^2+x2.^2-16*x1-10*x2; [fm,ij]=sort(ff); % sorting
   k=ij(1:16); X=[x1(k) x2(k) fm(1:16)] % extract top 16 solutions
```

Table 6.3: Optimal and suboptimal solutions.

x_1	x_2	f	x_1	x_2	f	x_1	x_2	f	x_1	x_2	f
4	5	−57	4	5.1	−56.99	4	4.9	−56.99	4	4.8	−56.96
4	5.2	−56.96	4	5.3	−56.91	4	4.7	−56.91	3.75	5	−56.875
4.25	5	−56.875	3.75	5.1	−56.865	4.25	5.1	−56.865	3.75	4.9	−56.865
4.25	4.9	−56.865	4	4.6	−56.84	4	5.4	−56.84	3.75	4.8	−56.835

6.2.3 0–1 programming

In some particular application fields, the decision variables may only be set to 0 and 1. For instance, when a decision is made, there are only two options, "do it" and "do not do it". The former can be expressed as 1 while the other is 0. Therefore, the concept of 0–1 programming needs to be introduced. In this section, the definitions of 0–1 programming and mixed 0–1 programming problems are proposed. Examples are then given to demonstrate the application of the enumeration method in solving 0–1 programming problems.

Definition 6.3. If the decision variables in an optimization problem can only be selected as 0 and 1, the problem is referred to as 0–1 programming problem, or as binary programming problem. If some of the decision variables are not required to be 0 and 1, the problem is known as a mixed 0–1 programming problem.

Examples given below show the solution of 0–1 programming problems with the enumeration method.

Example 6.6. Solve the following 0–1 programming problem:

$$\min_{\boldsymbol{x}\ \text{s.t.}\ \begin{cases} 3x_1-x_1x_4x_5-x_2x_3x_5+x_3x_4x_5+2\leqslant 4 \\ 2x_1-4x_1x_2x_3-7x_1x_4x_5-8x_2x_3x_5-x_3x_4x_5+14\leqslant 11 \\ -6x_1-8x_1x_2x_3+6x_1x_4x_5-8x_2x_3x_5+6x_3x_4x_5+13\leqslant 18 \\ x_1,x_2,x_3,x_4,x_5\in\{0,1\} \end{cases}} \quad 3x_1+5x_1x_2x_3+8x_1x_4x_5+8x_2x_3x_5-4x_3x_4x_5.$$

Solutions. If x_i can only be selected as 0 and 1, function `ndgrid()` can also be used to generate all combinations for a small-scale problem like this. There are altogether $2^5 = 32$ different combinations. It can be seen that there are 6 combinations satisfying all three constraints. The corresponding objective functions values are evaluated and sorted, such that the global optimum solution $\boldsymbol{x} = [0,1,1,1,1]^\mathrm{T}$ can be found, with $f_0 = 4$. The total elapsed time is 0.042 seconds. Besides, some other suboptimal solutions are also obtained, as shown in Table 6.4.

```
>> [x1 x2 x3 x4 x5]=ndgrid(0:1); tic
   ii=find(3*x1-x1.*x4.*x5-x2.*x3.*x5+x3.*x4.*x5+2<=4 &...
            2*x1-4*x1.*x2.*x3-7*x1.*x4.*x5- ...
                  8*x2.*x3.*x5-x3.*x4.*x5+14<=11 &...
            -6*x1-8*x1.*x2.*x3+6*x1.*x4.*x5- ...
                  8*x2.*x3.*x5+6*x3.*x4.*x5+13<=18);
   x1=x1(ii); x2=x2(ii); x3=x3(ii); x4=x4(ii); x5=x5(ii);
   f=3*x1+5*x1.*x2.*x3+8*x1.*x4.*x5+8*x2.*x3.*x5-4*x3.*x4.*x5;
   [f1,i1]=sort(f); i0=i1(1); f0=f1(i0), size(f)
   x=[x1(i0),x2(i0),x3(i0),x4(i0),x5(i0)], toc
   xx=[x1(i1) x2(i1) x3(i1) x4(i1) x5(i1) f1]
```

Table 6.4: Feasible solution sorting of the Example 6.6.

x_1	x_2	x_3	x_4	x_5	f	explanation	x_1	x_2	x_3	x_4	x_5	f	explanation
0	0	1	0	0	4	optimum	0	0	0	0	0	8	suboptimal
0	1	0	0	0	11	feasible	1	1	0	0	0	11	feasible
1	0	0	0	0	16	feasible	1	0	1	0	0	20	feasible

Example 6.7. Solve the 0–1 optimization problem with the enumeration method:

$$\max \quad 99x_1 + 90x_2 + 58x_3 + 40x_4 + 79x_5 + 92x_6 + 102x_7 + 74x_8 + 67x_9 + 80x_{10}.$$
$$x \text{ s.t. } \begin{cases} 30x_1+8x_2+6x_3+5x_4+20x_5+12x_6+25x_7+24x_8+32x_9+29x_{10} \leqslant 100 \\ x_i \in \{0,1\},\ i=1,2,\ldots,10 \end{cases}$$

Solutions. The problem here is a linear programming problem. The constraints and objective function are linear. Apart from the previously discussed method, the coefficients in the constraints are expressed as a column vector $\boldsymbol{v}$. The number of combinations satisfying the constraints is 575. The coefficients in the objective function are expressed in another column vector $\boldsymbol{w}$. Vectorized computation can be applied to evaluate all the objective function values at the feasible solutions. Then, sorting can be performed, such that the one ranked first is the global optimum solution. The solution is $\boldsymbol{x}_0 = [0,1,1,1,1,1,1,1,0,0]^{\mathrm{T}}$, and the value of the objective function is $y_0 = 535$. The top 10 feasible combinations are shown in Table 6.5.

```
>> [y1 y2 y3 y4 y5 y6 y7 y8 y9 y10]=ndgrid(0:1); n=length(y1(:))
   X=[y1(:) y2(:) y3(:) y4(:) y5(:) y6(:) y7(:) ...
      y8(:) y9(:) y10(:)];
   v=[30; 8; 6; 5; 20; 12; 25; 24; 32; 29]; ii=find(X*v<=100);
   X1=X(ii,:); w=[99; 90; 58; 40; 79; 92; 102; 74; 67; 80];
   y=X1*w; [y1,i1]=sort(y,'descend'); length(ii)
   i0=i1(1:10); y0=y(i0), Tab=[X1(i0,:) y0]
```

Table 6.5: Top 10 feasible solutions in Example 6.6.

x_1	x_2	x_3	x_4	x_5	x_6	x_7	x_8	x_9	x_{10}	f	x_1	x_2	x_3	x_4	x_5	x_6	x_7	x_8	x_9	x_{10}	f
0	1	1	1	1	1	1	1	0	0	535	1	1	0	1	1	1	1	0	0	0	502
0	1	1	0	1	1	1	0	0	1	501	0	1	1	0	1	1	1	1	0	0	495
1	1	1	0	1	1	0	1	0	0	492	0	1	0	1	1	1	1	0	0	1	483
1	1	1	1	0	1	1	0	0	0	481	0	1	0	1	1	1	1	1	0	0	477
1	1	0	1	1	1	0	1	0	0	474	0	1	1	0	1	1	0	1	0	1	473

For 0–1 programming problems, if the enumeration method is to be used, since each decision variable has two options, the problem with n decision variables has

a total of 2^n possible combinations. If n is not too large, the enumeration method can be considered. It may ensure the global optimum solution. For instance, if there are 10 decision variables, the number of combination is 1 024, while for $n = 20$, the number of combinations is 1 048 576, both smaller than the number combinations used in Example 6.3. The enumeration method can be used to solve such programming problems.

6.2.4 Trials on mixed-integer programming problems

If some decision variables are assumed to be integers, while others have no such restrictions, the problem is changed into a mixed-integer programming problem. In general, for mixed-integer programming problems, the enumeration method cannot be used, since it is not possible to list the continuous variables. In this section, two examples are tried, to find solutions with the enumeration method, and then demonstrate the problems with the enumeration methods.

Example 6.8. Consider again the integer programming problem in Example 6.3. If the 1st, 4th, and 5th decision variables are required to be integers, solve the mixed-integer linear programming problem.

Solutions. Since a finite number of decision variables are required to be integers, the original problem cannot be solved with the enumeration method. Since the 2nd and 3rd decision variables are continuous, we cannot enumerate them. We can still try to use an integer programming-based method to solve the original problem. For instance, we can find the integer combinations of the 1st, 4th, and 5th decision variables, and solve a linear programming problem with a loop structure. This is a time demanding problem.

Letting $p_1 = x_2$, $p_2 = x_3$, the original problem can be modified as

$$\min_{\boldsymbol{p} \text{ s.t. } \begin{cases} 2p_1+p_2\leqslant 54-4x_4-2x_5 \\ 4p_1+5p_2\leqslant 62-3x_1+x_4+x_5 \\ p_1\geqslant 0,\ p_2\geqslant 3.32,\ x_1\geqslant 0,\ x_4\geqslant 0.678,\ x_5\geqslant 2.57 \end{cases}} \quad -p_1 - 4p_2 - 2x_1 - 3x_4 - x_5,$$

where x_1, x_4, and x_5 can be regarded as given constants, which are provided in the outer loops. For standard linear programming problems, the objective function with constants cannot be handled directly, they can be deleted first, and then added back after solution. With such an idea, the following commands can be issued. There are altogether 14 950 combinations. For each of them, function `linprog()` is called once to compute the optimum solution, then $-2x_1 - 3x_4 - x_5$ can be added back to the objective function.

```
>> N=25; [x10,x40,x50]=meshgrid(0:N,1:N,3:N);
   x10=x10(:); x40=x40(:); x50=x50(:); size(x10)
```

```
f=[-1,-4]; f0=[]; A=[2,1; 4,5]; xx=[]; tic
for i=1:length(x10)
   x1=x10(i); x4=x40(i); x5=x50(i);
   B=[54-4*x4-2*x5; 62-3*x1+x4+x5];
   [p,fx,flag]=linprog(f,A,B,[],[],[0; 3.32]);
   if flag==0
      xx=[xx; x1,p(1),p(2),x4,x5]; f0=[f0; fx-2*x1-3*x4-x5];
end, end, toc
```

This piece of code finds 2460 feasible solutions, and the elapsed time is 243.4 seconds. If the feasible solutions are sorted, the global optimum solution can be found as $\boldsymbol{x}_0 = [19, 0, 3.8, 11, 3]^{\mathrm{T}}$, and the value of the objective function is −89.2.

```
>> [f1 ii]=sort(f0); i0=ii(1); f1(1), x0=xx(i0,:)
```

Example 6.9. Solve the following mixed 0–1 programming problem:[32]

$$\min \quad 5y_1 + 6y_2 + 8y_3 + 10x_1 - 7x_3 - 18\ln(x_2+1) - 19.2\ln(x_1 - x_2 + 1) + 10.$$
$$\boldsymbol{x},\boldsymbol{y} \text{ s.t. } \begin{cases} 0.8\ln(x_2+1)+0.96\ln(x_1-x_2+1)-0.8x_3 \geqslant 0 \\ \ln(x_2+1)+1.2\ln(x_1-x_2+1)-x_3-2y_3 \geqslant -2 \\ x_2-x_1 \leqslant 0 \\ x_2-2y_1 \leqslant 0 \\ x_1-x_2-2y_2 \leqslant 0 \\ y_1+y_2 \leqslant 1 \\ 0 \leqslant \boldsymbol{x} \leqslant [2,2,1]^{\mathrm{T}},\ \boldsymbol{y} \in \{0,1\} \end{cases}$$

Solutions. Since y_1, y_2, and y_3 can only be selected as 0 or 1, their combinations can be generated, and for each combination, the nonlinear programming problem regarding the decision variable $\boldsymbol{x}$ can be solved once. The linear constraints in this problem can be written in matrix form as

$$\boldsymbol{A} = \begin{bmatrix} -1 & 1 & 0 \\ 0 & 1 & 0 \\ 1 & -1 & 0 \end{bmatrix}, \quad \boldsymbol{B} = \begin{bmatrix} 0 \\ 2y_1 \\ 2y_2 \end{bmatrix},$$

the nonlinear inequality constraints can be expressed as below, and there are no nonlinear equality constraints:

$$\begin{cases} -0.8\ln(x_2+1) - 0.96\ln(x_1 - x_2 + 1) + 0.8x_3 \leqslant 0, \\ -\ln(x_2+1) - 1.2\ln(x_1 - x_2 + 1) + x_3 + 2y_3 - 2 \leqslant 0. \end{cases}$$

The constraint $y_1 + y_2 \leqslant 1$ can be used as the testing condition in the loop, while the $\boldsymbol{y}$ vector can be used as the additional parameter of the inner optimization problem. The following nonlinear constraints can be expressed in a MATLAB file:

```
function [c,ceq]=c6mbp1(x,y)
ceq=[]; % no equality constraints; return an empty matrix
```

```
c=[-0.8*log(x(2)+1)-0.96*log(x(1)-x(2)+1)+0.8*x(3);
   -log(x(2)+1)-1.2*log(x(1)-x(2)+1)+x(3)+2*y(3)-2];
```

Now all the combinations of y_1~y_3 can be found, then, with command `find()`, all the possible combinations satisfying $y_1 + y_2 \leqslant 1$ can be listed. For each of them, the solver `fmincon()` can be called once, so that all the solutions of the original problem can be found. Some of the solutions are local ones. Sorting the objective function values, the global optimum solution can be found as $\boldsymbol{x}_0 = [1.3010, 0, 1, 0, 1, 0]^\mathrm{T}$, with the smallest objective function value of 6.0098. The elapsed time is 0.29 seconds. The solution and other local ones are shown in Table 6.6. The solution recommended in [32] is $\boldsymbol{x} = [1.301, 0, 1, 1, 0, 1]^\mathrm{T}$, with an objective function of 13.0098. Obviously, this solution is merely a local one, far poorer than the global optimum solution.

Table 6.6: Feasible solution combinations in Example 6.9.

x_1	x_2	x_3	y_1	y_2	y_3	f	x_1	x_2	x_3	y_1	y_2	y_3	f
1.301	0	1	0	1	0	6.0098	1.5	1.5	0.91629	1	0	0	7.0927
0	0	0	0	0	0	10	1.301	0	1	1	0	1	13.0098
1.301	0	1	0	1	1	14.01	0	0	0	0	0	1	18

For this particular example, the number of `fmincon()` calls is small, only 6, one may not notice any difference in the solution process.

```
>> [y1 y2 y3]=meshgrid(0:1); xx=[]; yy=[]; ff=optimset;
   ii=find(y1+y2<=1); y1=y1(ii); y2=y2(ii); y3=y3(ii); tic
   A=[-1,1,0; 0,1,0; 1,-1,0];
   Aeq=[]; Beq=[]; xm=[0;0;0]; xM=[2;2;1];
   for i=1:length(y1)
      B=[0; 2*y1(i); 2*y2(i)]; y=[y1(i); y2(i); y3(i)];
      f=@(x,y)5*y(1)+6*y(2)+8*y(3)+10*x(1)-7*x(3)-...
               18*log(x(2)+1)-19.2*log(x(1)-x(2)+1)+10;
      x0=rand(3,1);
      [x1,f0,flag]=fmincon(f,x0,A,B,Aeq,Beq,xm,xM,@c6mbp1,ff,y);
      if flag==1, xx=[xx; x1',y']; yy=[yy; f0]; end
   end
   [y0 i1]=sort(yy); i0=i1(1); f0=y0(1), x0=xx(i0,:), toc
```

It can be seen that although some of the mixed programming problems can be solved in this way, the cost of solving is rather heavy. Besides, for complicated nonlinear mixed-integer programming problems, searching for solutions may be even more time consuming. Better algorithms and tools should be introduced. In the remaining parts in this chapter, mixed-integer programming solvers will be demonstrated.

6.3 Solutions of mixed-integer programming problems

It has been mentioned earlier that the enumeration method cannot be used in solving medium- or large-scale problems. An attempt to combine the enumeration method and nonlinear programming solver is of very low efficiency, and it is not a good choice. Better and more efficient integer and mixed-integer programming problem solvers are expected. In this section, representation of a mixed-integer linear programming problem is introduced first, then the commonly used branch-and-bound method is introduced. With MATLAB solvers, mixed-integer nonlinear programming problems can be directly solved.

6.3.1 Mixed-integer linear programming

Mixed-integer linear programming problems are a class of simple mixed-integer programming problems. In this section, the mathematical form of the mixed-integer linear programming problem is presented. Then, a MATLAB-based solver will be used to solve example problems.

Definition 6.4. The mathematical form of a mixed-integer linear programming problem is

$$\min_{\boldsymbol{x} \text{ s.t.} \begin{cases} \boldsymbol{A}\boldsymbol{x}\leqslant\boldsymbol{B} \\ \boldsymbol{A}_{\text{eq}}\boldsymbol{x}=\boldsymbol{B}_{\text{eq}} \\ \boldsymbol{x}_{\text{m}}\leqslant\boldsymbol{x}\leqslant\boldsymbol{x}_{\text{M}} \\ \hat{\boldsymbol{x}} \text{ are integers} \end{cases}} \boldsymbol{f}^{\text{T}}\boldsymbol{x}, \tag{6.3.1}$$

where $\hat{\boldsymbol{x}}$ is a subset of $\boldsymbol{x}$. If $\hat{\boldsymbol{x}}$ is the whole $\boldsymbol{x}$, the original problem is an integer linear programming problem.

For mixed-integer linear programming problems, the following major tools can be adopted. One is solving with the new `intlinprog()` function, where the problem can be expressed in a traditional way, with a structured variable, or with a problem-based description pattern. YALMIP Toolbox is another useful tool to handle mixed-integer linear programming problems. The tool for nonlinear programming problems, `new_bnb20()` function, can also be used, and will be demonstrated later.

A function `intlinprog()` is provided in the recent versions of Optimization Toolbox in MATLAB. The syntaxes are

$[\boldsymbol{x}, f_{\rm m}, \texttt{key}, \texttt{out}]=\texttt{intlinprog}(\texttt{problem})$

$[\boldsymbol{x}, f_{\rm m}, \texttt{key}, \texttt{out}]=\texttt{intlinprog}(f, \texttt{intcon}, \boldsymbol{A}, \boldsymbol{B})$

$[\boldsymbol{x}, f_{\rm m}, \texttt{key}, \texttt{out}]=\texttt{intlinprog}(f, \texttt{intcon}, \boldsymbol{A}, \boldsymbol{B}, \boldsymbol{A}_{\rm eq}, \boldsymbol{b}_{\rm eq})$

$[\boldsymbol{x}, f_{\rm m}, \texttt{key}, \texttt{out}]=\texttt{intlinprog}(f, \texttt{intcon}, \boldsymbol{A}, \boldsymbol{B}, \boldsymbol{A}_{\rm eq}, \boldsymbol{b}_{\rm eq}, \boldsymbol{x}_{\rm m}, \boldsymbol{x}_{\rm M})$

$[\boldsymbol{x}, f_{\rm m}, \texttt{key}, \texttt{out}]=\texttt{intlinprog}(f, \texttt{intcon}, \boldsymbol{A}, \boldsymbol{B}, \boldsymbol{A}_{\rm eq}, \boldsymbol{b}_{\rm eq}, \boldsymbol{x}_{\rm m}, \boldsymbol{x}_{\rm M}, \texttt{opt})$

It can be seen in the syntaxes that, compared with the solver `linprog()`, an extra argument `intcon` is provided. This is the serial number vector, indicating which of the decision variables are supposed to be integers. In the structured variable `problem`, the difference with function `linprog()` is that an extra member `intcon` should be used, and the `solver` member should be set to `'intlinprog'`. The following commands can be used:

```
problem.solver='intlinprog'
problem.options=optimoptions('intlinprog')
```

There are limitations in the solver `intlinprog()`, since the results obtained for the supposed integer variables are usually not exact integers. The function call $\boldsymbol{x}$`(intcon)=round(`$\boldsymbol{x}$`(intcon))` can be used to fine-tune the solutions, such that integer decision variables can be found.

Example 6.10. Solve again the integer linear programming problem in Example 6.3. Besides, if the 1st, 4th, and 5th decision variables are supposed to be integers, solve the mixed-integer linear programming problem.

Solutions. In fact, there are three ways to describe the original problem; one is to use structured variable to describe the original problem. The following commands can be issued to solve the problem:

```
>> clear P; P.solver='intlinprog';           % structured description
   P.options=optimoptions('intlinprog');
   P.lb=[0; 0; 3.32; 0.678; 2.57]; P.f=[-2 -1 -4 -3 -1];
   P.Aineq=[0 2 1 4 2; 3 4 5 -1 -1]; P.Bineq=[54; 62];
   P.intcon=1:5; [x,f,a,b]=intlinprog(P)   % solve linear program
   x=round(x)                               % fine-tune the integers
```

It can be seen that the result obtained is $\boldsymbol{x} = [19, 0, 4, 10, 5]$, which is the same as obtained in Example 6.3. In fact, the solution here is more reliable, since the artificial upper bound $N \leqslant 25$ used in the enumeration method is removed, so that large-scale searching is possible with the solver `intlinprog()`.

If the 1st, 4th, and 5th decision variables are expected to be integers, the vector `intcon` can be set to $[1, 4, 5]$. Therefore, the following commands can be used to solve this mixed-integer linear programming problem. The solution obtained is $\boldsymbol{X} =$

$[19, 0, 3.8, 11, 3]^T$, the same as in Example 6.8. The elapsed time is 0.082 seconds, much smaller than that in Example 6.8, which was 243.4 seconds. The efficiency is much higher than of the enumeration method-based solvers.

```
>> P.intcon=[1 4 5];
   tic, [x,f,flag,b]=intlinprog(P), toc %mixed-integer program
   x(P.intcon)=round(x(P.intcon))        %fine-tune the solution
```

If structured variable is not used to describe the original problem, the traditional way can be used to describe it, and with the following commands, the solution can also be obtained. The solution quality and efficiency are almost the same.

```
>> f=[-2 -1 -4 -3 -1]; intcon=[1,4,5];
   A=[0 2 1 4 2; 3 4 5 -1 -1]; B=[54; 62];
   Aeq=[]; Beq=[]; xm=[0; 0; 3.32; 0.678; 2.57];
   [x,f0,flag,b]=intlinprog(f,intcon,A,B,Aeq,Beq,xm)
   x(intcon)=round(x(intcon))   %fine-tune the results
```

Example 6.11. Solve again the integer linear programming problem in Example 6.3 with the problem-based description and solution method.

Solutions. This linear programming problem can be described by the problem-based method. Then, the problem can be solved directly. The result here is exactly the same as that obtained using the enumeration method. The elapsed time is about 0.147 seconds, so the efficiency is higher than that of other methods.

```
>> P=optimproblem; tic
   x=optimvar('x',5,1,'LowerBound',0,'Type','integer');
   P.Objective=-2*x(1)-x(2)-4*x(3)-3*x(4)-x(5);
   P.Constraints.c1=2*x(2)+x(3)+4*x(4)+2*x(5)<=54;
   P.Constraints.c2=3*x(1)+4*x(2)+5*x(3)-x(4)-x(5)<=62;
   P.Constraints.c3=x(3)>=4; P.Constraints.c4=x(4)>=1;
   P.Constraints.c5=x(5)>=3;
   sol=solve(P); x=round(sol.x), toc
```

With such a problem-based description method, mixed-integer linear programming problems can also be solved. However, the descriptions of mixed-integer problems are complicated, since one has to split the decision variable into integers and non-integers, then build up the model manually. It is not recommended to solve mixed-integer programming methods with problem-based descriptions.

Example 6.12. Consider the transportation problem in Example 4.14. If the decision variables are integers, solve the problem again.

Solutions. In Example 4.14, function `transport_linprog()` was written, and a switch was provided. The fourth input argument can be used to solve integer linear programming problems.

```
>> C=[10,5,6,7; 8,2,7,6; 9,3,4,8]; b=[1500 2500 3000 3500];
   a=[2500 2500 5000]; x=transport_linprog(-C,a,b,1) %maximization
   f=sum(C(:).*X(:))    %maximum profit
```

6.3.2 Integer programming with YALMIP Toolbox

YALMIP Toolbox can also be used in solving mixed-integer linear programming problems. The specific point is to declare integer decision variables with command `intvar()`. The integer linear programming problem model can be constructed. Finally, the solution can be found. However, the method seems to be complicated. It can also be used to solve mixed-integer programming problems, while the method is even more complicated. Examples are used to illustrate solutions of the problems.

Example 6.13. Use YALMIP Toolbox to solve again the integer and mixed-integer programming problems in Example 6.3.

Solutions. The integer programming problem can be modeled easily. The integer decision variable can be described with function `intvar()`, then the following commands can be used to solve the integer programming problem:

```
>> A=[0 2 1 4 2; 3 4 5 -1 -1]; B=[54; 62];
   xm=[0; 0; 3.32; 0.678; 2.57]; x=intvar(5,1);
   const=[A*x<=B, x>=xm]; obj=-[2 1 4 3 1]*x;
   tic, sol=optimize(const,obj); toc
   x=round(double(x)), value(obj) %solve and extract the solutions
```

The solution is $\boldsymbol{x} = [19, 0, 4, 10, 5]^{\mathrm{T}}$, with the objective function value being −89, the same as in Example 6.3.

To solve the mixed-integer programming problem, we need to define the decision variable as two integer decision variables and two real variables. Then, the decision vector can be composed. The following commands are employed to define the problem, and the solution found is $\boldsymbol{x} = [19, 0, 3.8, 11, 3]^{\mathrm{T}}$, which is still the same as that in Example 6.3. The elapsed time is 0.73 second, so the efficiency is lower than of the solver `intlinprog()`.

```
>> x1=intvar(3,1); x2=sdpvar(2,1); x=[x1(1); x2; x1(2:3)];
   const=[A*x<=B, x>=xm]; obj=-[2 1 4 3 1]*x;
   tic, sol=optimize(const,obj); toc, x=double(x) %extract solution
```

6.3.3 Mixed-integer nonlinear programming

The enumeration methods can only be used to solve small-scale problems. The mixed-integer programming problems are also complicated to solve. The solver `intlinprog()` is powerful, but it is only applicable to mixed-integer linear programming problems. It cannot be used to handle nonlinear ones. New solution algorithms and tools are needed in solving nonlinear problems.

In practical applications, branch-and-bound algorithms are usually used. A mixed-integer nonlinear programming solver was written by Koert Kuipers at Groningen University in the Netherlands. The function name was `BNB20()`. The function can be downloaded from MathWorks website.[30]

Since the function has not been updated for over 20 years, the support to the recent versions of MATLAB is not satisfactory, since an anonymous function and structured variables are not supported. Also the input arguments and returned arguments are not fully supported in the old versions. Modifications are needed in the solver provided in the book, and the name of the new solver is `new_bnb20()`, with the syntaxes

$[\boldsymbol{x},f,$`flag,errmsg]=new_bnb20(problem)`

$[\boldsymbol{x},f,$`flag]=new_bnb20(fun,`$\boldsymbol{x}_0$`,intcon,`$\boldsymbol{A},\boldsymbol{B},\boldsymbol{A}_{\rm eq},\boldsymbol{B}_{\rm eq},\boldsymbol{x}_{\rm m},\boldsymbol{x}_{\rm M}$`)`

$[\boldsymbol{x},f,$`flag]=new_bnb20(fun,`$\boldsymbol{x}_0$`,intcon,`$\boldsymbol{A},\boldsymbol{B},\boldsymbol{A}_{\rm eq},\boldsymbol{B}_{\rm eq},\boldsymbol{x}_{\rm m},\boldsymbol{x}_{\rm M}$`,CFun)`

where most of the syntaxes are the same as of other Optimization Toolbox functions. The solver `fmincon()` in the Optimization Toolbox is used, possibly with additional parameters. In these revised syntaxes, it is kept as close as possible to the MATLAB solvers. Vector $\boldsymbol{x}$ and f are the decision variable and objective function, respectively. If `flag` is 1, the solution is successful, −1 for unsuccessful. When `flag` is −2, there are problems in the calling syntax. The specific error information is returned from the `errmsg` string. The vector `intcon` is the same as that defined earlier. Note that in the calling process, vectors $\boldsymbol{x}_{\rm m}$ and $\boldsymbol{x}_{\rm M}$ must be provided, and they should be deterministic values, not `Inf`'s.

If the original problem is described by the structured variable `problem`, the members are the same as the those in `fmincon()` function. Besides, an extra member of `intcon` is allowed, and the setting is the same as described earlier. In other syntaxes, the input arguments are the same as for the `fmincon()` function.

Example 6.14. Solve the linear programming problem in Example 6.3 with the `new_bnb20()` solver.

Solutions. With the modified `new_bnb20()` function, an anonymous function is allowed to describe the objective function. The upper bounds cannot be set to infinity any more. Large upper bounds should be selected, such as 20 000. Now, if the lower bounds are selected as fractional numbers, the new solver is allowed to take ceiling rounding action, so that fractional bounds can also be used. The following commands can be employed, and the solution is the same as that obtained in Example 6.3. The elapsed time is 2.02 seconds. It can be seen that the solver is not efficient enough for linear problems.

```
>> f=@(x)-[2 1 4 3 1]*x; xm=[0,0,3.32,0.678,2.57]';
   A=[0 2 1 4 2; 3 4 5 -1 -1]; Aeq=[]; Beq=[];B=[54; 62];
   xM=20000*ones(5,1); intcon=1:5; x0=ceil(xm);  %bounds
   tic, [x,f0,flag]=new_bnb20(f,x0,intcon,A,B,Aeq,Beq,xm,xM), toc
```

If a structured variable is used to describe the system, the following commands can be used, and the result is the same:

```
>> clear P; P.objective=f; P.lb=xm; P.x0=x0; P.ub=xM;
   P.Aineq=A; P.Bineq=B; P.intcon=intcon; %structured variable
   tic, [x,f0,flag]=new_bnb20(P), tic      %direct solution
```

If one still needs x_1, x_4, and x_5 as integers, while the other two decision variables are real, the `intcon` variable should be modified accordingly, to `intcon=[1,4,5]`. The following commands can be used to solve the problem, the solution found is $\boldsymbol{X} = [19, 0, 3.8, 11, 3]^{\mathrm{T}}$, and the elapsed time is 2.83 seconds. It can be seen that the efficiency in handling mixed-integer linear programming problems is not as good as that of the dedicated solvers:

```
>> intcon=[1,4,5];
   tic, [x,f0,flag]=new_bnb20(f,x0,intcon,A,B,Aeq,Beq,xm,xM), toc
```

If a structured variable is used to describe the problem, the following statements can be issued, and the solution obtained is exactly the same:

```
>> P.intcon=[1,4,5]; [x,fm,flag]=new_bnb20(P) %mixed-integer
```

Example 6.15. Solve again the integer programming problem in Example 6.4.

Solutions. For this integer programming problem, the constraints should be written in the following MATLAB function. Note that all the inequalities are $\geqslant 0$ ones, so both sides must be applied by -1 to convert them into standard form, then

```
function [c,ceq]=c6mnl1(x)
ceq=[];    %empty matrix returned
c=-[-x(1)^2-x(2)^2-x(3)^2-x(4)^2-x(1)+x(2)-x(3)+x(4)+8;
    -x(1)^2-2*x(2)^2-x(3)^2-2*x(4)^2+x(1)+x(4)+10;
    -2*x(1)^2-x(2)^2-x(3)^2-2*x(4)^2+x(2)+x(4)+5];
```

From the constraints, if a larger search area is considered, the following commands should be used in solving the problem. After a long waiting time, the optimum solution found is $\boldsymbol{x} = [0,1,2,0]^{\mathrm{T}}$, and the elapsed time is 120 seconds, much longer than for the enumeration method. If the upper and lower bounds are set to ±1 000, it may be too time consuming, and the solution may not be found.

```
>> clear P; P.nonlcon=@c6mnl1;
   P.objective=f; P.lb=-100*ones(4,1); P.ub=-P.lb;
   f=@(x)x(1)^2+x(2)^2+2*x(3)^2+x(4)^2-5*x(1)-5*x(2)-21*x(3)+7*x(4);
   P.intcon=1:4; P.x0=10*rand(4,1); P.solver='fmincon';
   tic, [x,f0,flag]=new_bnb20(P), toc
```

Example 6.16. Solve the integer nonlinear programming problem:[32]

$$\min_{\boldsymbol{x} \text{ s.t. } \begin{cases} x_1-2x_2+12+x_3\geqslant 0 \\ -x_1^2+3x_2-8-x_3\geqslant 0 \\ x_1\geqslant 0, x_2\geqslant 0, x_3\geqslant 0 \end{cases}} \quad x_1^3+x_2^2-4x_1+4+x_3^4.$$

Solutions. Since there are nonlinear constraints, the following MATLAB function can be written to describe them:

```
function [c,ce]=c6exnl2(x)
ce=[]; % describe the nonlinear inequality constraints
c=[-x(1)+2*x(2)-12-x(3); x(1)^2-3*x(2)+8+x(3)];
```

The following commands can be used to solve this integer programming problem and the solution found is $\boldsymbol{x} = [1,3,0]^{\mathrm{T}}$:

```
>> clear P; P.intcon=[1;2;3]; P.nonlcon=@c6mnl2;
   P.objective=@(x)x(1)^3+x(2)^2-4*x(1)+4+x(3)^4; % structured
   P.lb=[0;0;0]; P.ub=100*[1;1;1]; P.x0=P.ub;
   [x,fm,flag]=new_bnb20(P)      % direct solution
```

Since the original problem is small-scale, the enumeration method can be used to find the global optimum solution. The result is the same as those obtained earlier. Besides other feasible solutions can be found as shown in Table 6.7. They cannot be obtained with other searching methods.

```
>> N=200; [x1 x2 x3]=meshgrid(0:N);
   ii=find(x1-2*x2+12+x3>=0 & -x1.^2+3*x2-8-x3>=0);
   x1=x1(ii); x2=x2(ii); x3=x3(ii);
   ff=x1.^3+x2.^2-4*x1+4+x3.^4; [fm,ij]=sort(ff);
   k=ij(1:12); [x1(k) x2(k) x3(k) fm(1:12)]  % enumeration
```

Table 6.7: Optimal solution and other feasible solutions in Example 6.16.

x_1	x_2	x_3	f	x_1	x_2	x_3	f	x_1	x_2	x_3	f	x_1	x_2	x_3	f
1	3	0	10	0	3	0	13	0	3	1	14	1	4	0	17
1	4	1	18	0	4	0	20	2	4	0	20	0	4	1	21
1	5	0	26	1	5	1	27	0	5	0	29	2	5	0	29

6.3.4 A class of discrete programming problems

Similar to the above mentioned solutions, some discrete programming problems can also be solved. Through appropriate conversion, some of the discrete programming problems can be transformed into integer programming problems. It should be noted that not all discrete programming problems can be solved in this way. In this section, examples are used to show the solutions of discrete programming problems.

Example 6.17. Solve the discrete optimization problem in Example 6.5.[10]

Solutions. MATLAB cannot be used to solve discrete programming problems directly. In this example, since the increment is given, that is, x_1 has increment 0.25, while x_2 has increment 0.1, it is natural to introduce two new decision variables $y_1 = 4x_1$, $y_2 = 10x_2$, that is, by variable substitution, $x_1 = y_1/4$ and $x_2 = y_2/10$. The original problem can be converted into an integer programming problem of decision variables y_i:

$$\min_{\boldsymbol{y}\ \text{s.t.}\ \begin{cases} y_1^2/16-6y_1/4+y_2/10-11\leqslant 0 \\ -y_1y_2/40+3y_2/10+e^{y_1/4-3}-1\leqslant 0 \\ y_2\geqslant 30 \end{cases}} \quad 2y_1^2/16 + y_2^2/100 - 4y_1 - y_2.$$

A MATLAB function can be written to describe the nonlinear constraints:

```
function [c,ceq]=c6mdisp(y)
ceq=[]; %nonlinear constraints
c=[y(1)^2/16-6*y(1)/4+y(2)/10-11;
   -y(1)*y(2)/40+3*y(2)/10+exp(y(1)/4-3)-1];
```

With the solver `new_bnb20()`, the following commands can be issued to solve the optimization problem directly. Assume that the upper and lower bounds of y_1 are set to ±200, the upper bound of y_2 is 200, and the lower bound is 30 (that is, $3 \leqslant x_2$). The solution is $\boldsymbol{x} = [4, 5]^{\mathrm{T}}$, which is different from (4, 4.75) recommended in [32]. The objective function obtained here is slightly smaller than the one given, and all the constraints are satisfied. Therefore, this solution is better than that in the reference. The elapsed time is only 0.025 seconds.

```
>> clear P; P.intcon=[1;2]; P.x0=[12;30];
   P.objective=@(y)2*y(1)^2/16+y(2)^2/100-4*y(1)-y(2); %structured
```

```
P.nonlcon=@c6mdisp; P.lb=[-200;30]; P.ub=200*[1; 1];
tic, [y,ym,flag]=new_bnb20(P); toc
x=[y(1)/4,y(2)/10] %solution and variable substitution
```

Note that if the discrete variables are not multiples of a certain number, the problem cannot be mapped into an integer programming problem, and the solver here cannot be used. Solution by enumeration might be the only way to solve such problems.

6.3.5 Solutions of ordinary discrete programming problems

In the discrete programming problems described so far, discrete decision variables were given in multiples of a number. If they are not multiples of a single number, the method in the previous section cannot be employed to convert the problem into a mixed-integer programming problem. If the number of discrete decision variables is not too large, and the discrete options are not too many, loops can be used to manipulate the samples, so as to find the global optimum solutions. An example is given to demonstrate solution of complicated discrete programming problems.

Example 6.18. Consider the test problem in Example 5.16. If decision variables x_1 and x_2 are integers; x_3 and x_5 may only take values from $\boldsymbol{v}_1 = [2.4, 2.6, 2.8, 3.1]$; while x_4 and x_6 may only select values from $\boldsymbol{v}_2 = [45, 50, 55, 60]$. Solve the discrete programming problem.

Solutions. From the given new constraints, it is known that x_1 and x_2 are regular integers, which can be handled easily. Variables x_3 and x_5 are not multiples and so are difficult to process. The variables x_4 and x_6 are multiples of 5, and can be handled in a mixed-integer programming problem, too.

Since the solver cannot be used with discrete variables which are not multiples, they cannot be used in the form of subscripts of the discrete vector. Such variables should be removed from a decision vector. Since x_4 and x_6 are multiples of 5, they can be divided by 5, to convert into a mixed-integer programming problem. The range of the decision variables is defined as the interval [9, 12]. Therefore, the problem can be transformed into a problem with 8 decision variables. The other two can be defined as global variables.

With such an idea, the variable `intcon` is set to `[1:4]`.

With the mathematical model, the objective function and the constraints in Example 5.16 can be reused. Internal manipulations inside the function can be made so that the $\boldsymbol{x}$ vector can be reconstructed. The manipulation is not difficult, and a MATLAB function can be used to describe the objective function:

```
function f=c6mglo1(x)
global v1 v2; x=x(:).';
```

```
l=100; x=[x(1:2),v1,x(3)*5,v2,x(4)*5,x(5:end)];
f=l*(x(1)*x(2)+x(3)*x(4)+x(5)*x(6)+x(7)*x(8)+x(9)*x(10));
```

For convenience, the $\boldsymbol{x}$ vector is converted into a row vector, and the above method can be used to reconstruct it, such that the objective function can be found. The discrete variables v_1 and v_2, that is, x_3 and x_5 in the mathematical model, can be passed in the function as global variables.

Similarly, the following function can be written to express the constraints. The same method is used to reconstruct vector $\boldsymbol{x}$, and function `c5mglo1()` is reused to compute the constraints.

```
function [c,ceq]=c6mglo2(x)
global v1 v2; x=x(:).';
x=[x(1:2),v1,x(3)*5,v2,x(4)*5,x(5:end)]; [c,ceq]=c5mglo1(x);
```

With these preparations, the following commands can be used to solve the original problem. Loop structures can be used for the discrete variables. For simplicity $v_1 > v_2$ is implied, and the combinations constructed otherwise can be expelled. The solutions can finally be found.

```
>> global v1 v2; v=[2.4,2.6,2.8,3.1];
   clear P; P.objective=@c6mglo1; P.nonlcon=@c6mglo2;
   P.lb=[1,30,9,9,1,30,1,30]'; P.ub=[5,65,12,12,5,65,5,65]';
   P.x0=P.lb; P.intcon=1:4;
   P.solver='fmincon'; P.options=optimset; t0=cputime, vv=[];
   for v1=v, for v2=v, if v1>v2, [v1,v2]
      [x,fv,flag,err]=new_bnb20(P); x=x(:).';
      vv=[vv; v1,v2,x(1:2),v1,x(3)*5,v2,x(4)*5,x(5:8),fv];
   end, end, end, cputime-t0, vv
```

Six different combinations are made in the loops, where two of them lead to convergent solutions. The results are listed in Table 6.8. It can be seen that the first one is the global optimum solution, and the elapsed time is 6.19 seconds.

Note that the solution method here is successful for a discrete programming problem. It must be noted that the method is only applicable if the numbers of discrete vari-

Table 6.8: Optimum solutions in a discrete programming problem.

x_1	x_2	x_3	x_4	x_5	x_6	x_7	x_8	x_9	x_{10}	f
3	60	3.1	55	2.6	50	2.2046	44.0911	1.7498	34.9951	63 893.4358
3	60	3.1	55	2.8	50	2.2046	44.0911	1.7498	34.9951	64 893.4358

ables and their combinations are not too large. Otherwise, other methods should be considered. In Chapter 9, a genetic algorithm-based intelligent optimization method is demonstrated for such problems.

6.4 Mixed 0–1 programming problems

As described earlier, 0–1 programming is a special case of integer programming problems, where the decision variables have only two options, 1 and 0. Therefore, the solutions of 0–1 programming problems look simpler, since each variable x_i may be 0 or 1, and we need to find the smallest objective function from the feasible solutions. In fact, with the increase of the problem size, the computational load may increase significantly. For instance, if the number of decision variables is n, the number of combinations is 2^n. When n is large, it is not possible to solve the problem in this way. Alternative methods should be tried.

There is no dedicated solver for 0–1 programming problems provided in MATLAB. In the old version there was a function called `bintprog()` to solve 0–1 linear programming problems, but the function vanished in the new versions. The 0–1 programming problem is a special case of integer programming problems. The upper and lower bounds of the decision variables are set to 1 and 0, respectively, so the problem can be converted into an ordinary integer programming problem. Therefore, the solvers `intlinprog()` can be used to solve 0–1 linear programming problems, while the solver `new_bnb20()` can be used to solve nonlinear 0–1 programming problems.

6.4.1 0–1 linear programming problems

Mixed 0–1 linear programming problems can be solved directly, with the solver `intlinprog()`. The upper and lower bounds of the decision variables can be set to 1 and 0 directly. Besides, if the problem-based method is used, the decision variables can be declared directly as binary ones. Function `solve()` can be used directly to solve the mixed 0–1 programming problem. Here, an example is used to show the solution of mixed 0–1 linear programming problems.

Example 6.19. Solve the 0–1 linear programming problem in Example 6.7 with a search method.

Solutions. It can be seen from the mathematical model that the original problem has only one linear inequality constraint, and no equality constraints. The matrix form of the constraint is $\boldsymbol{Ax} \leqslant \boldsymbol{B}$, where

$$\boldsymbol{A} = [30, 8, 6, 5, 20, 12, 25, 24, 32, 29], \quad \boldsymbol{B} = 100.$$

The following commands can be used to solve the optimization problem. The solution found is $x = [0,1,1,1,1,1,1,1,0,0]^{\mathrm{T}}$, and the minimized objective function value is $f_0 = -535$, the same as shown in Example 6.7. The elapsed time is 0.017 seconds, similar to that for the enumeration method.

```
>> A=[30,8,6,5,20,12,25,24,32,29]; B=100;
   f=-[99,90,58,40,79,92,102,74,67,80];
   Aeq=[]; Beq=[]; xM=ones(10,1); xm=0*xM; intcon=1:10;
   tic, [x,f0,flag]=intlinprog(f,intcon,A,B,Aeq,Beq,xm,xM), toc
```

Example 6.20. Solve the 0–1 linear programming problem with 28 decision variables

$$\max_{\boldsymbol{x} \text{ s.t. } \boldsymbol{A}\boldsymbol{x} \leqslant \begin{bmatrix} 600 \\ 600 \end{bmatrix}} \boldsymbol{f}^{\mathrm{T}}\boldsymbol{x},$$

where

$$\boldsymbol{A} = \begin{bmatrix} 45 & 0 & 85 & 150 & 65 & 95 & 30 & 0 & 170 & 0 & 40 & 25 & 20 & 0 \\ 30 & 20 & 125 & 5 & 80 & 25 & 35 & 73 & 12 & 15 & 15 & 40 & 5 & 10 \end{bmatrix.$$
$$\begin{matrix} 0 & 25 & 0 & 0 & 25 & 0 & 165 & 0 & 85 & 0 & 0 & 0 & 0 & 100 \\ 10 & 12 & 10 & 9 & 0 & 20 & 60 & 40 & 50 & 36 & 49 & 40 & 19 & 150 \end{matrix}\right],$$

$$\begin{aligned} \boldsymbol{f}^{\mathrm{T}} = [&1\,898, 440, 22\,507, 270, 14\,148, 3\,100, 4\,650, 30\,800, 615, 4\,975, \\ &1\,160, 4\,225, 510, 11\,880, 479, 440, 490, 330, 110, 560, 24\,355, \\ &2\,885, 11\,748, 4\,550, 750, 3\,720, 1\,950, 10\,500]. \end{aligned}$$

Solutions. It is obvious that for this problem, if the enumeration method is used, 28 arrays of 28-dimensions are needed. The storage required is as high as $8 \times 28 \times 2^{28}/2^{30} = 56$ GB. Therefore, the enumeration method cannot be used to solve this problem.

(1) For the given 0–1 linear programming problem, since a maximization problem is to be solved, it is natural to multiply the coefficient vector $\boldsymbol{f}^{\mathrm{T}}$ by -1, and the original problem can be converted into a minimization one. The following commands can be used, and then `intlinprog()` solver can be used to solve the problem. To solve the 0–1 program, one should set `intcon` to `1:28`, indicating that all the decision variables are integers. Further, the lower and upper bounds are set to 0 and 1, respectively. The following commands can be used to solve the 0–1 programming problem directly:

```
>> A=[45,0,85,150,65,95,30,0,170,0,40,25,20,0,0,25,...
          0,0,25,0,165,0,85,0,0,0,0,100;
      30,20,125,5,80,25,35,73,12,15,15,40,5,10,10,12,...
          10,9,0,20,60,40,50,36,49,40,19,150];
   B=[600; 600];
   f=-[1898,440,22507,270,14148,3100,4650,30800,615,4975,1160,...
```

```
    4225,510,11880,479,440,490,330,110,560,24355,2885,11748,...
    4550,750,3720,1950,10500];
  intcon=1:28; xM=ones(28,1); xm=zeros(28,1); Aeq=[]; Beq=[];
  tic, [x,f0,flag]=intlinprog(f,intcon,A,B,Aeq,Beq,xm,xM), toc
  x0=round(x), f0=-f*x0, norm(x-x0)  % fine tuning the solutions
```

The optimum solution found is

$$\boldsymbol{x} = [0,0,1,0,1,1,1,1,0,1,0,1,1,1,0,0,0,0,1,0,1,0,1,1,0,1,0,0]^{\mathrm{T}},$$

with the maximum objective function value of $f_0 = 141\,278$. The elapsed time is 0.167 seconds, and the norm of the error is 3.1165×10^{-15}. It can be seen that the solution is quite accurate.

(2) Problem-based method can be used to describe and solve the problem. The results obtained are exactly the same as those obtained above, and the elapsed time is 0.12 seconds, similar to the previous method.

```
>> P=optimproblem; tic
   x=optimvar('x',28,1,'Type','integer','Lower',0,'Upper',1);
   P.Objective=f*x; P.Constraints.c=A*x<=B;
   sol=solve(P); x0=round(sol.x), toc
```

(3) YALMIP Toolbox can be used to solve the same problem, and the result obtained is the same. The elapsed time is 5.02 seconds.

```
>> x=binvar(28,1); const= A*x<=B;
   tic, optimize(const,f*x); round(value(x)), toc
```

Example 6.21. Solve the following mixed 0–1 linear programming problem:

$$\min_{\boldsymbol{x},\boldsymbol{y}\ \text{s.t.}\ \begin{cases} x_1-2x_2+x_3-2x_4-x_5+y_1+2y_3-y_4+2y_5-y_6\leqslant 1.01\\ x_1-2x_2-2x_3-2x_4+x_5+y_1+y_2-y_3-2y_4-2y_5-y_6\leqslant -5.09\\ -x_1-x_2-2x_3-2x_4+x_5-x_6+2y_2+2y_3-2y_4-2y_5-y_6\leqslant -3.63\\ -2x_2+x_3+x_5-2x_6-y_2-y_3+2y_3+y_4+2y_5-y_6\leqslant 2.08\\ -x_2+x_3-2x_4-2x_5+x_6-2y_1-y_2-y_3-2y_4\leqslant -5.28\\ -3x_1+3x_3-2x_4-2x_5+3x_6+3y_1-3y_2+2y_4-y_5-2y_6=3.88\\ -3x_1+3x_2-2x_3+3x_4+x_5+x_6-3y_1+2y_2+3y_3+y_4-2y_5-2y_6=-2.54\\ 0\leqslant x_i\leqslant 1,\ y_i\in\{0,1\},\ i=1,2,\dots,6 \end{cases}} \quad x_1+x_3-x_4+2x_6+y_1-y_2+2y_3+y_4-2y_5-y_6.$$

Solutions. In natural considerations, the $\boldsymbol{x}$ vector should be extended to contain all 12 decision variables. If a problem-based description is used, vector $\boldsymbol{x}$ can be defined as a continuous vector, and $\boldsymbol{y}$ can be defined as a 0–1 vector. Then, the objective function and linear constraints can be expressed naturally. The `solve()` function can be used to solve the problem directly. It should be noted that the above defined $\boldsymbol{c}_1$ is the expression describing inequality constraints, while $\boldsymbol{c}_2$ is a vector for equality constraints. They cannot be combined into one vector, otherwise there will be error messages.

```
>> P=optimproblem; tic
   x=optimvar('x',6,1,'LowerBound',0,'Upperbound',1);
   y=optimvar('y',6,1,'Type','integer','Lower',0,'Upper',1);
   P.Objective=x(1)+x(3)-x(4)+2*x(6)+y(1)-y(2)+...
               2*y(3)+y(4)-2*y(5)-y(6);
   c1=[x(1)-2*x(2)+x(3)-2*x(4)-x(5)+y(1)+2*y(3)-...
            y(4)+2*y(5)-y(6)<=1.01;
      x(1)-2*x(2)-2*x(3)-2*x(4)+x(5)+y(1)+y(2)-y(3)-2*y(4)-...
           2*y(5)-y(6)<=-5.09;
      -x(1)-x(2)-2*x(3)-2*x(4)+x(5)-x(6)+2*y(2)+2*y(3)-...
           2*y(4)-2*y(5)-y(6)<=-3.63;
      -2*x(2)+x(3)+x(5)-2*x(6)-y(2)-y(3)+2*y(3)+y(4)+...
           2*y(5)-y(6)<=2.08;
      -x(2)+x(3)-2*x(4)-2*x(5)+x(6)-2*y(1)-y(2)-y(3)-...
           2*y(4)<=-5.28];
   c2=[-3*x(1)+3*x(3)-2*x(4)-2*x(5)+3*x(6)+3*y(1)-3*y(2)+2*y(4)-...
           y(5)-2*y(6)==3.88;
      -3*x(1)+3*x(2)-2*x(3)+3*x(4)+x(5)+x(6)-3*y(1)+2*y(2)+...
           3*y(3)+y(4)-2*y(5)-2*y(6)==-2.54];
   P.Constraints.c1=c1; P.Constraints.c2=c2;
   sol=solve(P); x=sol.x, y=round(sol.y), toc
   f0=x(1)+x(3)-x(4)+2*x(6)+y(1)-y(2)+2*y(3)+y(4)-2*y(5)-y(6)
```

It can be seen from the above result that $\boldsymbol{x} = [0,1,1,0.7727,0,0.1418]^{\mathrm{T}}$, $\boldsymbol{y} = [1,0,0,1,1,1]^{\mathrm{T}}$, with the objective function value of $f_0 = -0.4891$. The elapsed time is 0.51 seconds.

Example 6.22. Solve the mixed 0–1 programming problem in Example 6.21, with the conventional method.

Solutions. If th conventional method is used, one may set $x_{6+i} = y_i$, $i = 1,2,\ldots,6$. Therefore, the inequality constraints are

$$\boldsymbol{A}_{\mathrm{eq}} = \begin{bmatrix} 1 & -2 & 1 & -2 & -1 & 0 & 1 & 0 & 2 & -1 & 2 & -1 \\ 1 & -2 & -2 & -2 & 1 & 0 & 1 & 1 & -1 & -2 & -2 & -1 \\ -1 & -1 & -2 & -2 & 1 & -1 & 0 & 2 & 2 & -2 & -2 & -1 \\ 0 & -2 & 1 & 0 & 1 & -2 & 0 & -1 & 2 & 1 & 2 & -1 \\ 0 & -1 & 1 & -2 & -2 & 1 & -2 & -1 & -1 & -2 & 0 & 0 \end{bmatrix}, \quad \boldsymbol{B} = \begin{bmatrix} 1.01 \\ -5.09 \\ -3.63 \\ 2.08 \\ -5.28 \end{bmatrix}.$$

The equality constraints in matrix form can be found as

$$A_{eq} = \begin{bmatrix} -3 & 0 & 3 & -2 & -2 & 3 & 3 & -3 & 0 & 2 & -1 & -2 \\ -3 & 3 & -2 & 3 & 1 & 1 & -3 & 2 & 3 & 1 & -2 & -2 \end{bmatrix}, \quad B_{eq} = \begin{bmatrix} 3.88 \\ -2.54 \end{bmatrix}.$$

Using function intlinprog() to solve the problem, the result is the same.

6.4.2 0–1 nonlinear programming problems

Mixed 0–1 nonlinear programming problems can be handled with the above mentioned new_bnb20() function. The lower and upper bounds, x_m and x_M, can be set respectively to vectors of zeros and ones. Then, the solver can be used to solve the original problem directly. Examples are given below to demonstrate the solutions of mixed 0–1 nonlinear programming problems.

Example 6.23. Solve again the 0–1 linear programming problem in Example 6.20.

Solutions. The following commands can be used to input the problem into MATLAB workspace and solve it. The result is the same as that obtained earlier in Example 6.20. The elapsed time is 5.31 seconds, indicating that the efficiency is not as good as for the dedicated intinprog() solver.

```
>> A=[45,0,85,150,65,95,30,0,170,0,40,25,20,0,0,25,...
        0,0,25,0,165,0,85,0,0,0,0,100;
     30,20,125,5,80,25,35,73,12,15,15,40,5,10,10,12,...
        10,9,0,20,60,40,50,36,49,40,19,150];
   B=[600; 600];
   f=-[1898,440,22507,270,14148,3100,4650,30800,615,4975,1160,...
     4225,510,11880,479,440,490,330,110,560,24355,2885,11748,...
     4550,750,3720,1950,10500];
   F=@(x)f*x(:); x0=rand(28,1);
   intcon=1:28; xM=ones(28,1); xm=zeros(28,1); Aeq=[]; Beq=[];
   tic, [x,f0,flag]=new_bnb20(F,x0,intcon,A,B,Aeq,Beq,xm,xM), toc
```

Example 6.24. Solve the 0–1 programming problem in Example 6.6.

Solutions. From the first three inequality constraints, it is not hard to write the following MATLAB function:

```
function [c,ceq]=c6mbin2(x)
ceq=[]; % no equality constraints, returns an empty matrix
c=[3*x(1)-x(1)*x(4)*x(5)-x(2)*x(3)*x(5)+x(3)*x(4)*x(5)+2-4;
   2*x(1)-4*x(1)*x(2)*x(3)-7*x(1)*x(4)*x(5)-...
```

```
      8*x(2)*x(3)*x(5)-x(3)*x(4)*x(5)+14-11;
   -6*x(1)-8*x(1)*x(2)*x(3)+6*x(1)*x(4)*x(5)-...
      8*x(2)*x(3)*x(5)+6*x(3)*x(4)*x(5)+13-18];
```

From the given constraints and the objective function, the following statements can be used to solve the 0–1 programming problem directly. The optimum solution is $\boldsymbol{x} = [0, 1, 1, 1, 1]^{\mathrm{T}}$, with the objective function value of $f_0 = 4$. The elapsed time is 0.63 seconds. It can be seen that for this small-scale problem, with only 32 combinations, the efficiency of the enumeration method is the highest, and the global optimum solution can be ensured. There is no need to use searching methods.

```
>> xM=ones(5,1); xm=0*xM; intcon=1:5; %integer programming
   f=@(x)3*x(1)+5*x(1)*x(2)*x(3)+8*x(1)*x(4)*x(5)+...
         8*x(2)*x(3)*x(5)-4*x(3)*x(4)*x(5);
   A=[]; B=[]; Aeq=[]; Beq=[]; x0=rand(5,1); tic
   [x,f0,flag]=new_bnb20(f,x0,intcon,A,B,Aeq,Beq,xm,xM,@c6mbin2)
   toc
```

Example 6.25. Solve the mixed 0–1 nonlinear integer program in Example 6.9.

Solutions. Since there are both linear and nonlinear constraints, `intlinprog()` function cannot be used to solve the problem. A nonlinear solver should be used for the problem. Similar to the cases discussed earlier, two decision vectors $\boldsymbol{x}$ and $\boldsymbol{y}$ are involved, while in MATLAB solvers, only one decision vector can be handled. A new decision vector $\boldsymbol{x}$ should be created, with the first three components retained, and the next three defined as $x_4 = y_1$, $x_5 = y_2$, and $x_6 = y_3$. The optimization problem can be manually modified as

$$\min \quad 5x_4 + 6x_5 + 8x_6 + 10x_1 - 7x_3 - 18\ln(x_2+1) - 19.2\ln(x_1 - x_2 + 1) + 10.$$
$$\boldsymbol{x}\ \text{s.t.}\ \begin{cases} -0.8\ln(x_2+1) - 0.96\ln(x_1-x_2+1) + 0.8x_3 \leqslant 0 \\ -\ln(x_2+1) - 1.2\ln(x_1-x_2+1) + x_3 + 2x_6 - 2 \leqslant 0 \\ x_2 - x_1 \leqslant 0 \\ x_2 - 2x_4 \leqslant 0 \\ x_1 - x_2 - 2x_5 \leqslant 0 \\ x_4 + x_5 \leqslant 1 \\ 0 \leqslant \boldsymbol{x} \leqslant [2,2,1,1,1,1]^{\mathrm{T}} \end{cases}$$

The following MATLAB function can be written to describe the nonlinear inequality constraints:

```
function [c,ceq]=c6mmibp(x)
ceq=[]; %nonlinear inequality constraints
c=[-0.8*log(x(2)+1)-0.96*log(x(1)-x(2)+1)+0.8*x(3);
   -log(x(2)+1)-1.2*log(x(1)-x(2)+1)+x(3)+2*x(6)-2];
```

The matrices in the linear inequality constraints are

$$A = \begin{bmatrix} -1 & 1 & 0 & 0 & 0 & 0 \\ 0 & 1 & 0 & -2 & 0 & 0 \\ 1 & -1 & 0 & 0 & -2 & 0 \\ 0 & 0 & 0 & 1 & 1 & 0 \end{bmatrix}, \quad B = \begin{bmatrix} 0 \\ 0 \\ 0 \\ 1 \end{bmatrix}.$$

Therefore, a structured variable can be used to describe the mixed-integer nonlinear programming problem. Then, the solver can be called to solve the problem. The solution found is $x = [1.301, 0, 1, 0, 1, 0]^T$, with the objective function value of 6.098, the same as for the global optimum solution found in Example 6.9. The total elapsed time is 0.21 seconds, and the efficiency is slightly higher than that in Example 6.9.

```
>> clear P; P.intcon=4:6; P.x0=[0 0 0 0 0 0]'; %structured variable
   P.objective=@(x)5*x(4)+6*x(5)+8*x(6)+10*x(1)-7*x(3) ...
               -18*log(x(2)+1)-19.2*log(x(1)-x(2)+1)+10;
   P.ub=[2 2 1 1 1 1]'; P.lb=[0 0 0 0 0 0]'; P.Bineq=[0;0;0;1];
   P.Aineq=[-1 1 0 0 0 0; 0 1 0 -2 0 0; 1 -1 0 0 -2 0;
            0 0 0 1 1 0];
   P.nonlcon=@c6mmibp; tic, [x,fm,flag]=new_bnb20(P), toc %solution
```

6.5 Mixed-integer programming applications

There are various applications of integer and mixed-integer programming problems in science and engineering. In this section, several examples are explored, including the problem of how one can convert sudoku into a 0–1 linear programming problem, as well as how to solve typical knapsack, assignment, and traveling salesman problems.

6.5.1 Optimal material usage

Many problems in daily life, scientific research, and engineering practice can be mathematically modeled with linear programming problems. Here a practical example is given of how to use the least materials to satisfy the production demands. The modeling and solution with MATLAB are discussed.

Example 6.26. Assume that a manufacturer needs 1 000 pipes of 0.98-meter length and 2 000 pipes of 0.78-meter length. It has only 5-meter long pipes. At least how many 5-meter long pipes are needed to complete the request?

Solutions. Before establishing the mathematical model, we have to study how many plans can be made to cut the 5-meter long pipes, according to the requirement of 0.98-meter and 0.78-meter long products. There are altogether 6 different plans, as shown in Table 6.9.

Table 6.9: The six intersection plans.

method	0.98-meter	0.78-meter	waste	method	0.98-meter	0.78-meter	waste
1	5	0	0.1	2	4	1	0.3
3	3	2	0.5	4	2	3	0.7
5	1	5	0.12	6	0	6	0.32

If a decision variable x_i is selected as the number of 5-meter pipes cut according to the ith plan, it is obvious that the total number of 5-meter pipes needed is $x_1 + x_2 + \cdots + x_6$. The goal is to minimize this objective function. Apart from the objective function, the constraints should be written according to Table 6.9. If x_i values are known, the number of 0.98-meter long pipes should satisfy the constraint $5x_1 + 4x_2 + 3x_3 + 2x_4 + 1x_5 + 0x_6 \geqslant 1\,000$, while the number of 0.78-meter long pipes satisfies $0x_1 + 1x_2 + 2x_3 + 3x_4 + 5x_5 + 6x_6 \geqslant 2\,000$. Here $\geqslant$ inequality is used, since there might be extra pipes. The original problem can be represented directly as an integer linear programming problem:

$$\min \quad x_1 + x_2 + x_3 + x_4 + x_5 + x_6.$$
$$\boldsymbol{x} \text{ s. t. } \begin{cases} 5x_1+4x_2+3x_3+2x_4+x_5 \geqslant 1\,000 \\ x_2+2x_3+3x_4+5x_5+6x_6 \geqslant 2\,000 \\ x_i \geqslant 0,\ i=1,2,\ldots,6 \end{cases}$$

Then, with the problem-based method, the whole optimization problem can be described. Next, the function `solve()` is used to solve the linear programming problem. The solution obtained is $\boldsymbol{x}_0 = [120, 0, 0, 0, 400, 0]^{\mathrm{T}}$. The interpretation of the result is that, if the first plan is taken, 120 pipes are needed, and in the 5th plan, 400 pipes are used. In fact, from the given plans, it can be seen that the first and fifth have the least waste. Therefore, if in the first plan, the needed 0.98-meter products are cut, the fifth plan can be used to cut 0.78-meter products. Other plans need not be considered.

```
>> P=optimproblem;
   x=optimvar('x',6,1,'Lower',0); P.Objective=sum(x);
   P.Constraints.c1=5*x(1)+4*x(2)+3*x(3)+2*x(4)+x(5)>=1000;
   P.Constraints.c2=x(2)+2*x(3)+3*x(4)+5*x(5)+6*x(6)>=2000;
   sols=solve(P); x0=sols.x, x0=round(x0);
```

6.5.2 Assignment problem

The assignment problem is a special 0–1 linear programming problem. In practice, if there are n tasks to be assigned to n persons, out of m persons to choose, we assume that $m \geqslant n$ and that the m persons have different efficiency in completing different tasks. How can the tasks be assigned to the persons, such that the efficiency is the highest (or the cost is the smallest)? This is the assignment problem to be solved. More precisely, the three assumptions are made as follows:[27]

(1) The number of tasks is smaller than or equal to the numbers of workers;
(2) Each worker may take at most one task;
(3) Each task must be taken by one worker.

Definition 6.5. Assume that the cost for the ith task completed by the jth worker is c_{ij}, and it is known. To make sure the total cost is minimized, the mathematical model of the assignment problem is established as

$$\min \sum_{i=1}^{n}\sum_{j=1}^{m} c_{ij}x_{ij}. \tag{6.5.1}$$

$$\boldsymbol{X} \text{ s.t. } \begin{cases} \sum_{j=1}^{m} x_{ij}=1,\ i=1,2,\ldots,n \\ \sum_{i=1}^{n} x_{ij}\leqslant 1,\ j=1,2,\ldots,m \\ x_{ij}\in\{0,1\},\ i=1,2,\ldots,n,\ j=1,2,\ldots,m \end{cases}$$

It can be seen that assignment is a special case of transportation problem, $s_i = 1$, $d_i = 1$, and the decision variable x_{ij} is 0 or 1; $x_{ij} = 1$ means assigning the ith task to the jth worker. This is a typical 0–1 linear programming problem. Next, an example is given, and the solution with MATLAB is demonstrated.

Example 6.27. The recorded time for a 50 meter swim of five swimmers is provided in Table 6.10 for different styles. The unit is seconds. How to find four swimmers from them to form a 4×50 meters medley relay team, such that the expected completion time is minimized?[51]

Table 6.10: The records of five swimmers.

styles	swimmer 1	swimmer 2	swimmer 3	swimmer 4	swimmer 5
backstroke	37.7	32.9	33.8	37	35.4
breaststroke	43.4	33.1	42.2	34.7	41.8
butterfly stroke	33.3	28.5	38.9	30.4	33.6
crawling	29.2	26.4	29.6	28.5	31.1

Solutions. The cost matrix can be input into MATLAB workspace, and the following statements can be used to solve the assignment problem directly:

```
>> C=[37.7,32.9,33.8,37.0,35.4; 43.4,33.1,42.2,34.7,41.8;
      33.3,28.5,38.9,30.4,33.6; 29.2,26.4,29.6,28.5,31.1];
   x=optimvar('x',4,5,'Type','integer','Lower',0,'Upper',1);
   P=optimproblem; P.Objective=sum(sum(C.*x));
```

```
P.Constraints.c1=sum(x,2)==1; %assign one player to each task
P.Constraints.c2=sum(x,1)<=1; %at most one task for each player
sol=solve(P); X=round(sol.x), f0=sum(sum(C.*X))
```

The assignment matrix and objective function value can be found as

$$\boldsymbol{X}=\begin{bmatrix}0&0&1&0&0\\0&0&0&1&0\\0&1&0&0&0\\1&0&0&0&0\end{bmatrix},\quad f_0=126.2.$$

The interpretation of the result is as follows: assign the first (backstroke) task to swimmer 3, the second task (breaststroke) to swimmer 4, the third task (butterfly stroke) to swimmer 2, and the fourth task (front crawl) to swimmer 1. There is no task to be assigned to swimmer 5. The relay team thus assigned is the fastest, and the time may reach $f_0 = 126.2$ seconds.

6.5.3 Traveling salesman problem

Traveling salesman problem (TSP) is an often studied problem in operations research. A simple description of the TSP is that there are n cities and the distance between any two cities is given and represented as d_{ij}. If a salesman is in a certain city, he wants to visit all the other cities, and then return to his city. The restriction is that all the cities must be visited, and only once. How can he select his route such that the total distance traveled is the shortest? Of course, this is an NP-hard problem. If n is not too large, the enumeration method can be tried, while if n is large, a search method should be used to find the optimum solution, or at least a feasible solution.

Definition 6.6. If the distances between any two cites are d_{ij}, a matrix can be created. The decision $x_{ij} = 1$ means that the salesman travels from city i to city j. The shortest distance route is the solution of the following linear programming problem:

$$\min_{\boldsymbol{X}\ \text{s.t.}\ \begin{cases}\sum\limits_{j=1}^{n}x_{ij}=1,\ i=1,2,\dots,n\\ \sum\limits_{i=1}^{n}x_{ij}=1,\ j=1,2,\dots,n\\ x_{ij}\in\{0,1\},\ i=1,2,\dots,n,\ j=1,2,\dots,n\\ \text{the solution is a closed-path of } n \text{ nodes}\end{cases}} \sum_{i=1}^{n}\sum_{j=1}^{n} d_{ij}x_{ij}. \tag{6.5.2}$$

It seems that the TSP problem is similar to an assignment problem described in Definition 6.5. Please note that there is an extra constraint. The following example is used to show the meaning of such a constraint and its handling methods.

Example 6.28. If there are 11 cities, located at $(4,49)$, $(9,30)$, $(21,56)$, $(26,26)$, $(47,19)$, $(57,38)$, $(62,11)$, $(70,30)$, $(76,59)$, $(76,4)$, and $(96,4)$, solve the TSP.

Solutions. The coordinates of the cities can be entered into MATLAB first. The distance between any two cities can be evaluated using the Euclidian distance formula

$$d_{ij} = \sqrt{(x_i - x_j)^2 + (y_i - y_j)^2}, \quad i = 1,2,\dots,n,\ j = 1,2,\dots,n,$$

where $n = 11$. Of course, the diagonal elements are 0, and the theoretical values should be set to ∞. It is not appropriate to set them to `Inf` in describing the objective function, they should be set to large values, for instance, $d_{ii} = 10\,000$. Therefore, the following commands can be used to construct the Euclidian distance matrix $\boldsymbol{D}$:

```
>> x0=[4,9,21,26,47,57,62,70,76,76,96];
   y0=[49,30,56,26,19,38,11,30,59,4,32]; n=11;
   for i=1:n, for j=1:n   % compute the Euclidian distance
      D(i,j)=sqrt((x0(i)-x0(j))^2+(y0(i)-y0(j))^2);
   end, end, D=D+10000*eye(n);
```

If the last constraint in Definition 6.6 is not considered temporarily, the number of routes is n, and the following commands can be used to solve the linear programming problem directly:

```
>> P=optimproblem;  % create an optimization model
   x=optimvar('x',n,n,'Type','integer','Lower',0,'Upper',1);
   P.Objective=sum(D(:).*x(:)); P.Constraints.c1=sum(x(:))==n;
   P.Constraints.x=sum(x,1)==1; P.Constraints.y=sum(x,2)==1;
   sol=solve(P); x1=round(sol.x), sparse(x1)
```

It can be seen from a sparse matrix description that there are 11 routes found. They are (3,1), (4,2), (1,3), (2,4), (6,5), (8,6), (10,7), (5,8), (11,9), (7,10), and (9,11). The routes are shown graphically in Figure 6.2. It can be seen that the solution is not the same as expected. For instance, there is a closed-path between cities 1 and 3. There are also local closed-paths. It is not a single connected path to all the cities. The local closed-paths are referred to subtours in the TSP problems.

If all the subtours are removed and the objective function is minimized, the solution to the TSP problem is found. How can we eliminate the subtours? If the constraint “the solution is a closed-path of n nodes” is used to eliminate the subtours, many extra constraints must be created. It is not easy to implement them. Therefore, some of the extra constraints should be added. If a subtour is found, an extra constraint can be added to avoid it. If constraints to all the subtours are handled, the TSP can be solved again. An iteration process is used to implement the above process, until the total number of subtours is reduced to 1. The TSP problem is then solved.

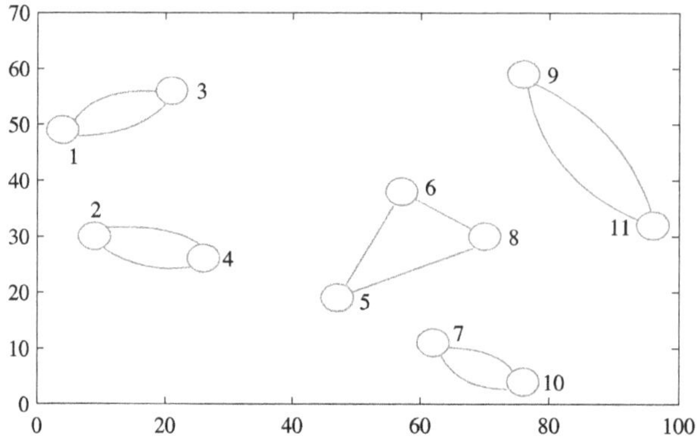

Figure 6.2: The solutions with subtours.

Subtour detection can be made by checking a closed-path whose end point equals to the starting point. If a subtour with m nodes is found, an extra constraint is set by letting the number of connections between the m paths equal to $m - 1$. An existing subtour can be eliminated in this way. However, this manipulation method may introduce new subtours. An iterative process should be applied such that the subtours can be eliminated gradually.

If a solution matrix $\boldsymbol{x}$ is found, the following MATLAB function can be written to search for subtours. The ith subtour is returned in `stours{i}`. The number of subtours can be extracted with command `length(stours)`.

```
function stours=find_subtours(x)
[n,~]=size(x); u=1; visited=zeros(n,1);
for k=1:n, s0(k)=k; d0(k)=find(x(k,:)==1); end
for i=1:n, if visited(i)==1, continue; end
   s1=s0(i); v=s1; d=d0(i); mat=[s1,d]; visited(i)=1;
   while d~=s1, d1=d; visited(d)=1; d=d0(d); mat=[mat; d1,d]; end
   stoursu=mat; u=u+1;
end
```

With the subtour detecting function, the following general-purpose function in MATLAB can be written to solve the TSP:

```
function stours=tsp_solve(x0,y0)
x0=x0(:); y0=y0(:); n=length(x0);
r=0.025*max(max([x0 y0])-min([x0 y0]));  % circle radius
for i=1:n, for j=1:n
   D(i,j)=sqrt((x0(i)-x0(j))^2+(y0(i)-y0(j))^2); % Euclidian distance
end, end
```

```
D=D+100000*eye(n); t=linspace(0,2*pi,100); P=optimproblem;
x=optimvar('x',n,n,'Type','integer','Lower',0,'Upper',1);
P.Objective=sum(sum(D.*x));
P.Constraints.x=sum(x,1)==1; P.Constraints.y=sum(x,2)==1;
sol=solve(P); x1=round(sol.x); sparse(x1); cstr=[];
while (1)              % use loop to eliminate subtours
   stours=find_subtours(x1); length(stours) % find subtours
   if length(stours)==1, break; end % if only one subtour, terminate loop
   for i=1:length(stours)            % process each subtour
      u=stours{i}(:,1); v=stours{i}(:,2); s=0; n0=length(u);
      for j=1:n0, s=s+x(u(j),v(j)); end
      cstr=[cstr; s<=n0-1];          % add a constraint
   end
   P.Constraints.cc=cstr;            % solve again
   sol=solve(P); x1=round(sol.x); sparse(x1);
end
for i=1:length(x0), line(x0(i)+r*cos(t),y0(i)+r*sin(t)), end
stours=find_subtours(x1); u=stours{1}(:,1); v=stours{1}(:,2);
line([x0(u) x0(v(end))],[y0(u) y0(v(end))]); % draw the optimal routes
```

In the function, the coordinate vectors of the cities are expressed by $\boldsymbol{x}_0$ and $\boldsymbol{y}_0$. The returned `stours` is a cell array, storing the start and end nodes of the final route, as demonstrated later.

Example 6.29. Eliminate the subtours in Example 6.28 and solve the TSP.

Solutions. The coordinates of the nodes can be entered first, and the general-purpose solver can be used to solve the problem. The TSP solution can be found as shown in Figure 6.3. Since this is a symmetric problem, either direction yields the same result.

```
>> x0=[4,9,21,26,47,57,62,70,76,76,96];
   y0=[49,30,56,26,19,38,11,30,59,4,32];
   tours=tsp_solve(x0,y0), T=[tours{1}]'
```

The final subtour matrix can be displayed as follows, where the first row is the start node and the second row is the end node of the route:

$$\boldsymbol{T}=\begin{bmatrix}1&2&4&5&7&10&11&9&8&6&3\\2&4&5&7&10&11&9&8&6&3&1\end{bmatrix}.$$

From the matrix it is found that the route is

$$1\to2\to4\to5\to7\to10\to11\to9\to8\to6\to3\to1.$$

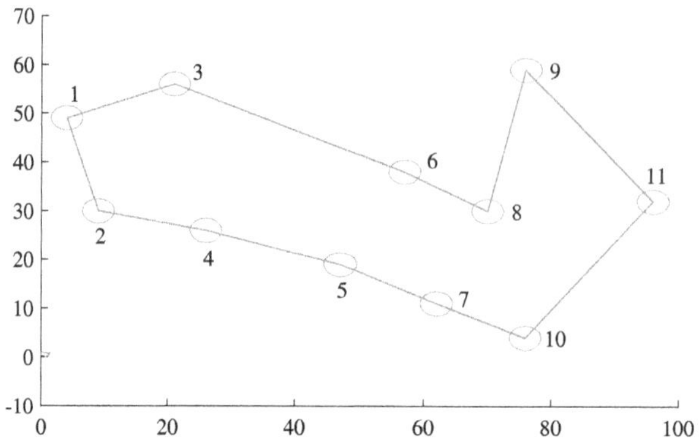

Figure 6.3: Solution illustration of the TSP.

It should be noted that this is only a simple-to-implement TSP solver. It is not suitable for handling large-scale TSPs. In [15], the Mona Lisa portrait with 100 000 cites was solved via TSP. Interested readers may refer to other references for more details.

6.5.4 Knapsack problems

Knapsack problems is an important application topic in the fields of operations research and combinatorial optimization. They have various applications, since many of practical problems can be converted into the framework of knapsack problems. In this section, an introduction of knapsack problems is provided, then examples are used to demonstrate the modeling and solutions of knapsack problems.

Definition 6.7. Assume that there are n kinds of items to select. The weight of the ith kind of item is w_i, the price is p_i, and the number of items selected is x_i. The knapsack problem is that, if the total weight selected does not exceed the allowed one, W, how to select items such that the total price is maximized. In standard knapsack problems, each kind of item cannot be selected more than once. The mathematical model of knapsack problem is

$$\max_{\boldsymbol{X}\text{ s.t. }\begin{cases}\sum_{i=1}^{n} w_i x_i \leqslant W\\ x_i\in\{0,1\},\ i=1,2,\dots,n\end{cases}} \sum_{i=1}^{n} p_i x_i. \tag{6.5.3}$$

Definition 6.8. If the constraints $x_i \in \{0, 1\}$ are eliminated from Definition 6.7, that is, each kind of item can be selected more than 1, the original problem is referred to as unbounded knapsack problem. If there is an upper bound for each kind of item, the problem is called a bounded knapsack problem.

Example 6.30. Assume that there are 10 kinds of item, whose weights and prices are given in Table 6.11. If each kind of item can be selected at most once, and the total weight should not exceed 40 kg, solve the 0–1 knapsack problem such that the total price is maximized.

Table 6.11: Weights and prices of 10 kinds of items.

item number	1	2	3	4	5	6	7	8	9	10
numbers	x_1	x_2	x_3	x_4	x_5	x_6	x_7	x_8	x_9	x_{10}
weights w_i (kg)	19	15	4.6	14.6	10	4.4	1.2	11.4	17.8	19.6
prices p_i (dollars)	89	59	19	43	100	72	44	16	7	64

Solutions. Assume that the ith kind of item is selected x_i times, where $x_i = \{0, 1\}$, then the following commands can be used to solve the 0–1 knapsack problem. The solution obtained is $\boldsymbol{x}_0 = [1, 0, 1, 0, 1, 1, 1, 0, 0, 0]^{\mathrm{T}}$, with the maximum price of \$324, and total weight of 39.2 kg. It can be seen from the result that items 1, 3, 5, 6, and 7 are selected.

```
>> w=[19,15,4.6,14.6,10,4.4,1.2,11.4,17.8,19.6]; W=40;
   p=[89,59,19,43,100,72,44,16,7,64]; n=10;
   P=optimproblem('ObjectiveSense','max');
   x=optimvar('x',n,1,'Type','integer','Lower',0,'Upper',1);
   P.Objective=p*x; P.Constraints.c=w*x<=W;
   sol=solve(P); x0=round(sol.x), p*x0, w*x0
```

Example 6.31. It can be seen that in Table 6.11, kinds 6 and 7 are relatively light-weight and expensive. Therefore, if they are allowed to be selected arbitrarily, only these two, or one, kinds are selected. Now solve again the problem, if kinds 6 and 7 are restricted to be select at most twice.

Solutions. In the new modeling process, the upper bounds of the items are not set to 1. The constraints $x_{6,7} \leqslant 2$ should be proposed. Therefore, the following commands can be used to set up an integer linear programming problem. The solution found is $\boldsymbol{x}_0 = [0, 0, 0, 0, 3, 2, 1, 0, 0, 0]^{\mathrm{T}}$, with the total price of \$488, and total weight of 40 kg. There are 5 items of kind 3, 2 items of kind 6, and 1 item for kind 7.

```
>> P=optimproblem('ObjectiveSense','max'); n=10; W=40;
   x=optimvar('x',n,1,'Type','integer','Lower',0); % remove binary
   P.Objective=p*x; P.Constraints.c=w*x<=W;
   P.Constraints.upp=[x(6)<=2, x(7)<=2];
   sol=solve(P); x0=round(sol.x), p*x0, w*x0
```

6.5.5 Sudoku problems

A number game called sudoku was invented by a retired American architect Howard Garns (1905–1989), with an original name of "Number Place".[38] The name "sudoku" was the Japanese name, when the game was first made popular in Japan. Normally, sudoku needs to place numbers inside a 9×9 mesh grid. The initial form of a sudoku puzzle is shown in Figure 6.4(a). The existing numbers are referred to as clues. This puzzle is number 6 in [38]. The player should make inferences based on the rules and clues in the puzzle, and complete filling numbers in the entire 9×9 mesh grid.

			1				3	
	9	4	3			7		
1		6				8	2	
			5					
6	2	8				5	1	9
					6			
	4	1				2		5
		9			2	4	8	
	8				5			

(a) puzzle 6

8	5	7	1	2	4	9	3	6
2	9	4	3	6	8	7	5	1
1	3	6	9	5	7	8	2	4
9	7	3	5	8	1	6	4	2
6	2	8	4	7	3	5	1	9
4	1	5	2	9	6	3	7	8
7	4	1	8	3	9	2	6	5
5	6	9	7	1	2	4	8	3
3	8	2	6	4	5	1	9	7

(b) sudoku solution

Figure 6.4: Initial form and solution of the 6th sudoku puzzle.

The rules in sudoku are relatively simple. It requires that in each row and column, unique numbers 1 to 9 are placed. Besides in each bold mesh grid, referred to as boxes here, the cells in the 3×3 matrix should have unique numbers 1 to 9. The sudoku problem is to use the existing clues to fill in all the other numbers in the 9×9 matrix. A simple example of the inference process is explained here. Look at the first three rows, in rows 1 and 3, there exists digit "1", and they are located in the left and middle boxes. Therefore, there must be a "1" in row 2, located in the upper-right box; now look at the middle row in the box. The $(2,7)$ grid is held by number 7, therefore, 1 can only be placed in one of the $(2,8)$ and $(2,9)$ grids; it can be seen from column 3 that since $(5,8)$ is 1, digit 1 cannot be filled in the 8th column, at $(2,8)$, therefore, there is only one choice for number 1. It must be filled in at $(2,9)$. Through the inference process, a new clue is generated. Besides, since rows 1 and 2 have 3, and they are located respectively in the right and middle boxes, row 3 must have number 3 and it should be located in the left box. Since $(3,1)$ and $(3,3)$ are occupied, $(3,2)$ grid is the only place to fill in 3.

Another clue is then added. Using the existing clues, all the blank grids can be filled with appropriate numbers so that the sudoku puzzle can be solved.

For a well designed sudoku puzzle, there should be a unique solution, while ordinary sudoku puzzles may have many solutions. For instance, if only (1, 1) is known and set to 1, while all the other grids are left blank, there should be many ways in completing the sudoku problem.

In [37], an algorithm is introduced such that the 9×9 sudoku problem can be converted into a 0–1 programming problem. The basic idea is to compose a $9 \times 9 \times 9$ three-dimensional decision array $\boldsymbol{x}$, whose elements can only be 0 and 1. Besides, the following constraints are introduced:

(1) To ensure each mesh grid (i, j) contains $1, 2, \ldots, 9$, and the numbers are not repeated,

$$\sum_{k=1}^{9} x_{i,j,k} = 1, \quad i = 1, 2, \ldots, 9,\ j = 1, 2, \ldots, 9. \tag{6.5.4}$$

(2) To ensure each row and each column contains $1, 2, \ldots, 9$, and the numbers are not repeated,

$$\sum_{i=1}^{9} x_{i,j,k} = 1, \quad \sum_{j=1}^{9} x_{i,j,k} = 1, \quad k = 1, 2, \ldots, 9. \tag{6.5.5}$$

(3) To ensure each box contains $1, 2, \ldots, 9$, and the numbers are not repeated,

$$\sum_{i=1}^{3}\sum_{j=1}^{3} x_{i+u,j+v,k} = 1, \quad u, v = \{0, 3, 6\}. \tag{6.5.6}$$

Apart from these constraints, in order to ensure the existing clues do not participate in the optimization process, their upper and lower bounds are all set to 1. Therefore, a feasible solution can be found by solving 0–1 linear programming problems. When the solution is found, the 3rd dimension elements in $\boldsymbol{x}$ can further be manipulated. That is, by weighting the 3rd layer elements, which means multiplying the elements in the kth layer by k, and summing up the elements according to the sum for the 3rd dimension, the solution of the sudoku problem can be found.

There are two ways in representing initial sudoku problems, one is by a matrix, the other is by an indexed matrix with three i, j, s numbers to represent a matrix element. The three numbers indicate $b_{i,j} = s$. If there are m clues, an $m \times 3$ matrix $\boldsymbol{B}$ can be used to represent an initial sudoku puzzle.

In [37], the code for a 9×9 sudoku solver via 0–1 programming block is presented. Extension and modification are made to the code, and a general-purpose MATLAB solver is written to handle $n^2 \times n^2$ sudoku problems, where $n = 2, 3, \ldots$ Some commands in the source code provided in [37] are simplified and normalized, without affecting the original code. The revised MATLAB function is

```
function S=solve_sudoku(B,n)
if nargin==0, B=[]; n=3; end
if size(B)~=3, [i,j,s]=find(B); B=[i,j,s]; end %to indexed matrix
if nargin==1, n=sqrt(length(sparse(B(:,1),B(:,2),B(:,3)))); end
P=optimproblem; %no objective function since a feasible one is needed
x=optimvar('x',n^2,n^2,n^2,'Type','integer','Lower',0,'Upper',1);
P.Constraints.x=sum(x,1)==1; %only one element is 1 in each direction
P.Constraints.y=sum(x,2)==1; P.Constraints.z=sum(x,3)==1;
D=[]; for u=0:n:n^2-1, for v=0:n:n^2-1  %constraints in each box
   a=x(u+1:u+n,v+1:v+n,:);
   s=sum(sum(a,1),2)==ones(1,1,n^2); D=[D; s(:)];
end, end
P.Constraints.D=D;
for u=1:size(B,1), x.Lower(B(u,1),B(u,2),B(u,3))=1; end
sol=solve(P); x=round(sol.x); %set the weights in layer k, and sum up
for k=2:n^2, x(:,:,k)=k*x(:,:,k); end, S=sum(x,3); %completed
```

In [37], a display function of a 9×9 sudoku problem is provided. The function is extended to display an $n^2 \times n^2$ sudoku. The listings of the new function `show_sudoku()` is not given here. Interested users may read the source code in the toolbox provided with the book.

Example 6.32. Assuming that puzzle 6 in [38] is considered, solve and display the sudoku puzzle.

Solutions. From the sudoku shown in Figure 6.4(a), the initial matrix can be entered into MATLAB workspace and displayed. With the initial sudoku problem, the solver `solve_sudoku()` can be called directly to solve it. Function `show_sudoku()` can be called to display the results, as shown in Figure 6.4(b).

```
>> X=[0,0,0,1,0,0,0,3,0; 0,9,4,3,0,0,7,0,0; 1,0,6,0,0,0,8,2,0;
      0,0,0,5,0,0,0,0,0; 6,2,8,0,0,0,5,1,9; 0,0,0,0,0,6,0,0,0;
      0,4,1,0,0,0,2,0,5; 0,0,9,0,0,2,4,8,0; 0,8,0,0,0,5,0,0,0];
   show_sudoku(X), S=solve_sudoku(X), show_sudoku(S)
```

Example 6.33. Assume that an initial 16×16 matrix is given in Figure 6.5(a). Find a feasible solution for the sudoku puzzle.

Solutions. It is obvious that this sudoku puzzle may have multiple solutions. The following commands can be used to generate the initial matrix in Figure 6.5(a). Displaying it and then with function `solve_sudoku()`, a feasible solution can be found as shown in Figure 6.5(b).

```
>> M=magic(4); X=blkdiag(M,M,M,M); show_sudoku(X)
   S=solve_sudoku(X), show_sudoku(S)
```

16	2	3	13												
5	11	10	8												
9	7	6	12												
4	14	15	1												
				16	2	3	13								
				5	11	10	8								
				9	7	6	12								
				4	14	15	1								
								16	2	3	13				
								5	11	10	8				
								9	7	6	12				
								4	14	15	1				
												16	2	3	13
												5	11	10	8
												9	7	6	12
												4	14	15	1

(a) a sudoku puzzle

16	2	3	13	6	10	14	4	1	15	11	7	12	9	8	5
5	11	10	8	12	16	2	7	14	6	9	4	13	3	1	15
9	7	6	12	1	5	11	15	3	8	13	16	2	10	4	14
4	14	15	1	8	3	13	9	10	12	2	5	7	6	11	16
15	8	12	4	16	2	3	13	11	9	7	10	14	1	5	6
13	6	9	14	5	11	10	8	2	1	16	3	15	12	7	4
11	3	1	5	9	7	6	12	15	13	4	14	8	16	2	10
7	10	16	2	4	14	15	1	12	5	8	6	11	13	9	3
12	15	8	7	10	4	9	6	16	2	3	13	1	5	14	11
14	1	13	3	15	12	7	2	5	11	10	8	6	4	16	9
10	4	11	16	14	8	1	5	9	7	6	12	3	15	13	2
2	9	5	6	3	13	16	11	4	14	15	1	10	8	12	7
6	5	14	15	7	9	12	10	8	4	1	11	16	2	3	13
1	12	7	9	13	15	4	16	6	3	14	2	5	11	10	8
3	16	4	11	2	1	8	14	13	10	5	15	9	7	6	12
8	13	2	10	11	6	5	3	7	16	12	9	4	14	15	1

(b) a feasible solution

Figure 6.5: A 16×16 sudoku problem and solution.

6.6 Exercises

6.1 Use the enumeration method to solve the following integer programming problem:[36]

$$\min \quad -3x_1 - 3x_1^2 + 8x_2 - 7x_2^2 - 5x_3 - 8x_3^2 + 2x_4 + 4x_4^2 - 4x_5 - 7x_5^2.$$
$$\boldsymbol{x} \text{ s.t. } \begin{cases} 7x_1+7x_1^2+4x_2+4x_2^2-8x_3-7x_4+2x_4^2-5x_5+2x_5^2 \leqslant -6 \\ 8x_1-5x_1^2+4x_2-7x_2^2-4x_3+8x_3^2+7x_4-6x_4^2-2x_5-7x_5^2 \leqslant -2 \\ -x_1-3x_1^2-2x_2+x_2^2-2x_3+8x_3^2-5x_4-3x_4^2+5x_5-7x_5^3 \leqslant 9 \\ -5 \leqslant x_i \leqslant 5,\ i=1,2,3,4,5 \end{cases}$$

6.2 Solve the following integer programming problems:

(1) $$\max \quad 592x_1 + 381x_2 + 273x_3 + 55x_4 + 48x_5 + 37x_6 + 23x_7;$$
$$\boldsymbol{x} \text{ s.t. } \begin{cases} \boldsymbol{x} \geqslant \boldsymbol{0} \\ 3\,534x_1+2\,356x_2+1\,767x_3+589x_4+528x_5+451x_6+304x_7 \leqslant 119\,567 \end{cases}$$

(2) $$\max \quad 120x_1 + 66x_2 + 72x_3 + 58x_4 + 132x_5 + 104x_6.$$
$$\boldsymbol{x} \text{ s.t. } \begin{cases} x_1+x_2+x_3=30 \\ x_4+x_5+x_6=18 \\ x_1+x_4=10 \\ x_2+x_5 \leqslant 18 \\ x_3+x_6 \geqslant 30 \\ x_{1,\dots,6} \geqslant 0 \end{cases}$$

6.3 Solve the following linear integer programming problem:

$$\max \quad x_1+x_2+x_3+x_4+x_5+x_6+x_7.$$

$$\boldsymbol{x}\ \text{s. t.}\ \begin{cases} x_1+x_4+x_5+x_6+x_7\geqslant 20\\ x_4+x_5+x_6=18\\ x_1+x_4=10\\ x_2+x_5\leqslant 18\\ x_3+x_6\geqslant 30\\ x_1,\dots,x_6\geqslant 0\end{cases}$$

6.4 Solve the following nonlinear integer programming problems,[32] and validate the results with the enumeration method:

(1) $$\min \quad \left(\frac{1}{6.931}-\frac{x_2x_3}{x_1x_4}\right)^2;$$

$\boldsymbol{x}$ s. t. $12\leqslant x_i\leqslant 32$

(2) $$\min \quad (x_1-10)^2+5(x_2-12)^2+x_3^4+3(x_4-11)^2+10x_5^6+7x_6^2+x_7^4-10x_6-8x_7.$$

$$\boldsymbol{x}\ \text{s. t.}\ \begin{cases} -2x_1^2-3x_2^4-x_3-4x_4^2-5x_5+127\geqslant 0\\ 7x_1-3x_2-10x_3^2-x_4+x_5+282\geqslant 0\\ 23x_1-x_2^2-6x_6^2+8x_7+196\geqslant 0\\ -4x_1^2-x_2^2+3x_1x_2-2x_3^2-5x_6+11x_7\geqslant 0\end{cases}$$

6.5 Solve the following 0–1 linear integer programming problems, and validate the results with the enumerate method:

(1) $$\min \quad 5x_1+7x_2+10x_3+3x_4+x_5;$$

$$\boldsymbol{x}\ \text{s. t.}\ \begin{cases} x_1-x_2+5x_3+x_4-4x_5\geqslant 2\\ -2x_1+6x_2-3x_3-2x_4+2x_5\geqslant 0\\ -2x_2+2x_3-x_4-x_5\leqslant 1\end{cases}$$

(2) $$\min \quad -3x_1-4x_2-5x_3+4x_4+4x_5+2x_6.$$

$$\boldsymbol{x}\ \text{s. t.}\ \begin{cases} x_1-x_6\leqslant 0\\ x_1-x_5\leqslant 0\\ x_2-x_4\leqslant 0\\ x_2-x_5\leqslant 0\\ x_3-x_4\leqslant 0\\ x_1+x_2+x_3\leqslant 2\end{cases}$$

6.6 Solve the following integer programming problem,[23] and validate the result with the enumerate method:

$$\min \quad 7y_1+10y_2.$$

$$\boldsymbol{y}\ \text{s. t.}\ \begin{cases} y_1^{1.2}y_2^{1.7}-7y_1-9y_2\leqslant 24\\ -y_1-2y_2\leqslant 5\\ -3y_1+y_2\leqslant 1\\ 4y_1-3y_2\leqslant 11\\ y_1,y_2\in\{1,2,3,4,5\}\end{cases}$$

6.7 Apart from Example 6.9, two other similar mixed 0–1 programming problems in [32], with recommended solutions are supplied. Solve the two problems and see whether better solutions can be found:

(1) $$\min \quad 5y_1+8y_2+6y_3+10y_4+6y_5-10x_1-15x_2-15x_3+15x_4 + 5x_5-20x_6+e^{x_1}+e^{x_2/1.2}-60\ln(x_4+x_5+1)+140.$$

$$\boldsymbol{x}\ \text{s.t.}\ \begin{cases} -\ln(x_4+x_5+1)\leqslant 0\\ e^{x_1}-10y_1\leqslant 1,\ e^{x_2/1.2}-10y_2\leqslant 1\\ 1.25x_3-10y_3\leqslant 0,\ x_4+x_5-10y_4\leqslant 0\\ -3x_3-2x_6-10y_5\leqslant 0,\ x_3-x_6\leqslant 0\\ -x_1-x_2-2x_3+x_4+2x_6\leqslant 0\\ -x_1-x_2-0.75x_3+x_4+2x_6\leqslant 0\\ 2x_3-x_4-2x_6\leqslant 0,\ -0.5x_4+x_5\leqslant 0,\ 0.2x_4+x_5\leqslant 0\\ y1+y2=1,\ y_4+y_5\leqslant 1\\ \mathbf{0}\leqslant\boldsymbol{x}\leqslant[2,2,2,\infty,\infty,3]^{\mathrm{T}}\\ y_i=\{0,1\},\ i=1,2,3,4,5 \end{cases}$$

Reference solution $\boldsymbol{x}=[0,2,1.078,0.652,0.326,1.078]^{\mathrm{T}}, \boldsymbol{y}=[0,1,1,1,0]^{\mathrm{T}}, f_0 = 73.035$;

(2) $$\min \quad 5y_1+8y_2+6y_3+10y_4+6y_5+7y_6+4y_7+5y_9-10x_1-15x_2-15x_3 + 80x_4+25x_5+35x_6-40x_7+15x_8-35x_9+e^{x_1}+e^{x_2/1.2} - 65\ln(x_4+x_5+1)-90\ln(x_5+1)-80\ln(x_6+1)+120.$$

$$\boldsymbol{x}\ \text{s.t.}\ \begin{cases} -1.5\ln(x_5+1)-\ln(x_6+1)-x_8\leqslant 0\\ -\ln(x_4+x_5+1)\leqslant 0\\ -x_1-x_2+x_3+2x_4+0.8x_5+0.8x_6-0.5x_7-x_8-2x_9\leqslant 0\\ -x_1-x_2+2x_4+0.8x_5+0.8x_6-2x_7-x_8-2x_9\leqslant 0,\ -x_3+0.4x_4\leqslant 0\\ -2x_4-0.8x_5-0.8x_6+2x_7+x_8+2x_9\leqslant 0,\ -x_4+x_7+x_9\leqslant 0\\ -0.4x_5-0.4x_6-1.2x_8\leqslant 0,\ 0.16x_5+0.16x_6-1.2x_8\leqslant 0,\ x_3-0.8x_4\leqslant 0\\ e^{x_1}-10y_1\leqslant 1,\ e^{x_2/1.2}-10y_2\leqslant 1\\ x_7-110y_3\leqslant 0,\ 0.8x_5+0.8x_6-10y_4\leqslant 0\\ 2x_4-2x_7-2x_9-10y_5\leqslant 0,\ x_5-10y_6\leqslant 0\\ x_6-10y_7\leqslant 0,\ x_3+x_4-10y_8\leqslant 0\\ y1+y2=1,\ y_4+y_5\leqslant 1,\ -y_4+y_6+y_7\leqslant 0,\ y_3-y_9\leqslant 0\\ \mathbf{0}\leqslant\boldsymbol{x}\leqslant[2,2,1,2,2,2,2,1,3]^{\mathrm{T}}\\ y_i=\{0,1\},\ i=1,2,3,4,5,6,7,8 \end{cases}$$

Reference solution is

$$\boldsymbol{x}=[0,2,0.468,0.585,2,0,0,0.267,0.585]^{\mathrm{T}},\quad \boldsymbol{y}=[0,1,0,1,0,1,0,1]^{\mathrm{T}}.$$

6.8 Solve the following integer programming problem:[23]

$$\min \quad 3y-5x.$$

$$\boldsymbol{y}\ \text{s.t.}\ \begin{cases} 2y^2-2y^{0.5}-2x^{0.5}y^2+11y+8x\leqslant 39\\ -y+x\leqslant 3\\ -3y_1+y_2\leqslant 1\\ 2y+3x\leqslant 24\\ 1\leqslant x\leqslant 10,\ y\in\{1,2,3,4,5,6\} \end{cases}$$

6.9 Solve the following mixed 0–1 programming problems:[23]

(1)
$$\min \quad 2x_1 + 3x_2 + 1.5y_1 + 2y_2 - 0.5y_3;$$
$$\boldsymbol{x} \text{ s. t. } \begin{cases} x_1^2+y_1=1.25 \\ x_2^{1.5}+1.5y_2=3 \\ x_1+y_1\leqslant 1.6 \\ 1.333x_2+y_2\leqslant 18 \\ x_3+x_6\leqslant 3 \\ -y_1-y_2+y_3\leqslant 0 \\ \boldsymbol{x}\geqslant \boldsymbol{0},\ \boldsymbol{y}\in\{0,1\} \end{cases}$$

(2)
$$\min \quad -0.7y + 5(x_1-0.5)^2 + 0.8;$$
$$\boldsymbol{x} \text{ s. t. } \begin{cases} e^{x_1-0.2}-x_2\leqslant 0 \\ x_2^{1.5}+1.5y_2=3 \\ x_2+1.1y\leqslant 1 \\ x_1-1.2y\leqslant 0.2 \\ 0.2\leqslant x_1\leqslant 1 \\ -2.22554\leqslant x_2,\ y\in\{0,1\} \end{cases}$$

(3)
$$\min \quad -x_1x_2x_3.$$
$$\boldsymbol{x} \text{ s. t. } \begin{cases} x_1+0.1^{y_1}0.2^{y_2}0.15^{y_3}=1 \\ x_2+0.05^{y_4}0.2^{y_5}0.15^{y_6}=1 \\ x_3+0.02^{y_7}0.06^{y_8}=1 \\ -y_1-y_2-y_3\leqslant -1 \\ -y_4-y_5-y_6\leqslant -1 \\ -y_7-y_8\leqslant -1 \\ 3y_1+y_2+2y_3+3y_4+2y_5+y_6+3y_7+2y_8\leqslant 10 \\ 0\leqslant \boldsymbol{x}\leqslant 1,\ \boldsymbol{y}\in\{0,1\} \end{cases}$$

6.10 Generate a set of random positions of 80 cites, then solve the traveling salesman problem.

6.11 Complete the statements needed in Example 6.22, and see whether global optimum solutions can be found with regular solvers.

6.12 A text file `c6mknap.txt` storing the data of 100 products is provided. The first column shows the prices and the second contains weights. If the maximum allowed weight is 50 kg, and each type of product is restricted to be selected only once, how can you select the items such that the price of total product is maximized? If there is no restrictions on the number of products, what is the new optimum solution?

6.13 Solve the sudoku problems in Figure 6.6, (a) and (b).

6.14 In a 25×25 sudoku problem, each of the 5×5 diagonal block is set to a `magic(5)` matrix. Solve the sudoku problem to find a feasible solution, and measure the execution time.

2		7		9	1			4
							1	2
6					2	5	9	
8		5		2	3	4		
9	7						2	6
		1	7	6		9		8
	8	6	2					3
7	3							
5			6	3		1		9

(a) puzzle 1

7		1				4		
5								
3			9	6				
			3	8				5
4	7							6
					9	8		2
	5			1	8			
	2	4				5		
					3		9	

(b) puzzle 2

Figure 6.6: Two 9 × 9 sudoku problems.

7 Multiobjective programming

So far, as we stressed, the objective function $f(\boldsymbol{x})$ has been a scalar one. If this assumption is extended to the case where it is a vector function, we will have to deal with an multiobjective programming problem. In this chapter, the modeling and solutions of multiobjective programming problems are presented.

In Section 7.1, a simple example is used to show the modeling and application of multiobjective programming problems. The necessity of multiobjective programming is also discussed. Besides, the graphical method is used to solve simple multiobjective programming problems. In Section 7.2, several methods are presented to convert multiobjective programming problems into those of a single objective. Then, the solvers presented in the earlier chapters can be used to tackle such optimization problems. Since the solutions of multiobjective programming are not unique, in Section 7.3, the concepts of Pareto frontier and dominating solutions are introduced, and the extraction of Pareto frontiers is introduced. In Section 7.4, a special class of multiobjective programming problems – minimax problems – is introduced.

7.1 Introduction to multiobjective programming

Multiobjective programming problem is a kind of mathematical programming problem when the objective function is no longer a scalar function. In many real applications, this type of programming problem may be encountered. In this section, a practical example of multiobjective programming problem is introduced, and then, the general mathematical form is given. Attempts are made to find graphical solutions of multiobjective programming problems.

7.1.1 Background introduction

An example is given below to show a problem which may be encountered in everyday life, and to demonstrate that the multiobjective programming problems do exist in real systems. It may introduce a new viewpoint for using optimization techniques.

Example 7.1. Assume that in a store, three types of candies, A_1, A_2 and A_3, are available, and the prices are \$4/kg, \$2.8/kg, and \$2.4/kg, respectively. A unit is having a group meeting and candies are needed. It is required that the total expenses not exceed \$20, and the total weight of candies not be less than 6 kg. Also it is required that the sum of weights of candies A_1 and A_2 be no less than 3 kg. How can we make a good purchase plan?[24]

Solutions. The first thing to do is to decide about the objective functions. In this example, the total target is to "spend less, buy more". Therefore, two objective functions

https://doi.org/10.1515/9783110667011-007

can be defined: one means spending the least money, while the other signifies that the total weight of the candies bought is the largest. The other conditions can be regarded as the constraints. Of course, these two objective functions are, to some extent, in conflict with each other. We need to consider how we can represent the problem mathematically.

Assume that the purchased amounts of candies A_1, A_2, and A_3 are respectively x_1, x_2, and x_3 kilograms. The two objective functions can be mathematically expressed as

$$\begin{cases} \min f_1(\boldsymbol{x}) = 4x_1 + 2.8x_2 + 2.4x_3 & \text{spend the least money,} \\ \max f_2(\boldsymbol{x}) = x_1 + x_2 + x_3 & \text{buy the heaviest candies.} \end{cases}$$

It is obvious that the objective function now is no longer scalar. Considering the constraints, the following can be established:

$$4x_1 + 2.8x_2 + 2.4x_3 \leqslant 20, \quad x_1 + x_2 + x_3 \geqslant 6, \quad x_1 + x_2 \geqslant 3, \quad x_1, x_2, x_3 \geqslant 0.$$

If minimization is used as the standard form, and the constraints are also unified, the constrained multiobjective programming problem can be expressed as

$$\min_{\boldsymbol{x} \text{ s.t.} \begin{cases} 4x_1+2.8x_2+2.4x_3\leqslant 20 \\ -x_1-x_2-x_3\leqslant -6 \\ -x_1-x_2\leqslant -3 \\ x_1,x_2,x_3\geqslant 0 \end{cases}} \begin{bmatrix} 4x_1 + 2.8x_2 + 2.4x_3 \\ -(x_1 + x_2 + x_3) \end{bmatrix}.$$

7.1.2 Mathematical model of multiobjective programming

It can be seen from the previous example that if the objective function is no longer a scalar, the standard form of multiobjective programming can be introduced. Physical interpretations of multiobjective programming problems are discussed.

Definition 7.1. The mathematical form of a multiobjective programming problem is

$$J = \min_{\boldsymbol{x} \text{ s.t. } \boldsymbol{G}(\boldsymbol{x})\leqslant \boldsymbol{0}} \boldsymbol{F}(\boldsymbol{x}), \tag{7.1.1}$$

where $\boldsymbol{F}(\boldsymbol{x}) = [f_1(\boldsymbol{x}), f_2(\boldsymbol{x}), \dots, f_p(\boldsymbol{x})]^{\mathrm{T}}$.

Generally speaking, these objective functions may be conflicting. If one is biased towards one of the objective functions, the others may be neglected. For instance, the desired "spend less money and buy more candies" is actually a pair of conflicting objective functions. Different decision makers may have different viewpoints and interest. Therefore, a compromise must be made, so that objectives acceptable to different parties can be selected. The solution thus obtained is referred to as a compromise solution.

7.1.3 Graphical solution of multiobjective programming problems

Graphical methods were introduced earlier for solving optimization problems with one or two decision variables. In this section, the graphical method is used to solve example problems with two objective functions.

Example 7.2. Solve the following unconstrained multiobjective programming problem with the graphical method:

$$\min_{x \text{ s.t. } 0\leqslant x\leqslant\pi/2} \begin{bmatrix} (x_1+2x_2)\sin(x_1+x_2)\mathrm{e}^{-x_1^2-x_2^2}+5x_2 \\ \mathrm{e}^{-x_2^2-4x_2}\cos(4x_1+x_2) \end{bmatrix}.$$

Solutions. The surfaces of the two objective functions are overlapped, and it may be difficult to distinguish the two surfaces. In order to distinguish them, the value of the second objective function is increased by 3. The two surfaces can be obtained as shown in Figure 7.1.

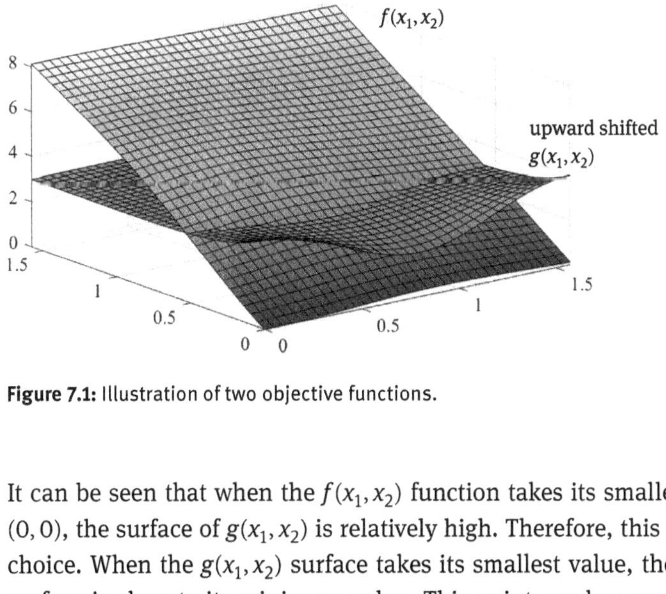

Figure 7.1: Illustration of two objective functions.

It can be seen that when the $f(x_1,x_2)$ function takes its smallest value, at the point $(0,0)$, the surface of $g(x_1,x_2)$ is relatively high. Therefore, this solution is not a good choice. When the $g(x_1,x_2)$ surface takes its smallest value, the value of the $f(x_1,x_2)$ surface is close to its minimum value. This point can be regarded as an alternative optimum solution. Since the two objective functions are sometimes in conflict, a compromise should be made to select a suitable optimum solution.

```
>> f=@(x1,x2)(x1+2*x2).*sin(x1+x2).*exp(-x1.^2-x2.^2)+5*x2;
   g=@(x1,x2)exp(-x2.^2-4*x2).*cos(4*x1+x2)+3;
   fsurf(f,[0,pi/2]), hold on; fsurf(g,[0,pi/2]), hold off
```

The commonly accepted understanding in the problem is that the acceptable minimum solution is around $x_2 = 0$. The section functions at $x_2 = 0$ can be drawn with the following statements, as shown in Figure 7.2. It can be seen that when $x_1 = 0.7854$, $F(x_1)$ has the minimum value, meanwhile $G(x_1)$ approaches its local highest point. If x_1 is shifted to the left or to the right, it seems the cost in $G(x_1)$ gets too large when the value of $F(x_1)$ is slightly reduced. Therefore, one may consider shifting x_1 to the left slightly, so that a compromise solution can be found. The curve of $G(x) + F(x)$ can also be drawn. The overall minimum solution is located near the $x_1 = 0.7708$ point.

```
>> x2=0; F=@(x1)f(x1,x2); G=@(x1)g(x1,x2);
   fplot(F,[0,pi/2]), hold on, fplot(G,[0,pi/2])
   F1=@(x1)F(x1)+G(x1); fplot(F1,[0,pi/2]); hold off
```

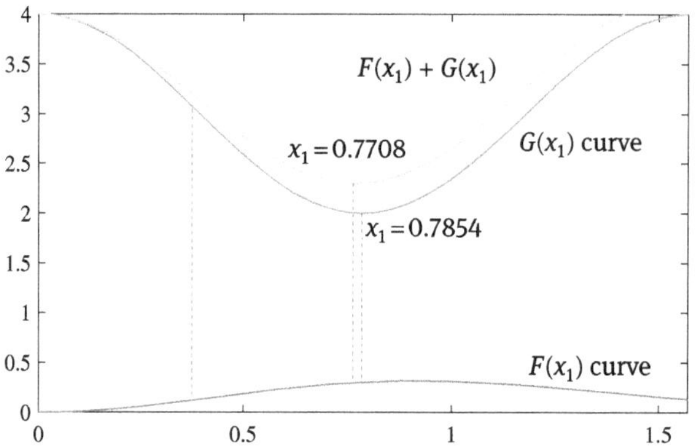

Figure 7.2: The two objective functions at $x_2 = 0$.

In fact, it can be seen that the optimality of the solution is rather subjective. The description of a multiobjective programming phenomenon itself is a subjective matter. A subjective decision from the decision maker is needed.

Example 7.3. Consider the multiobjective programming problem with three decision variables in Figure 7.1. Solve the problem with the graphical method.

Solutions. It has been mentioned earlier that the graphical method is suitable for the optimization problems with two decision variables. The problem here has three decision variables. Normally, the graphical method is not suitable for problems like this. Further studying the objective functions and constraints, it can be seen that the price of A_1 candy is rather high, and there is no direct demand on its quantity: there is only one related constraint, $x_1 + x_2 \geqslant 3$, mentioned. It is a reasonable choice that we set x_1 to 0. Besides, since the price of x_2 is higher than that of x_3, the least amount of x_2 can be bought satisfying the constraints. Therefore, $x_2 = 3$. The problem with three decision variables is reduced to that with only one decision variable x_3. The following statements can be used to draw the illustrations of the two objective functions, as shown in Figure 7.3.

```
>> x1=0; x2=3;
   f=@(x3)4*x1+2.8*x2+2.4*x3; g=@(x3)-(x1+x2+x3);
   x3=1:0.01:8; F=f(x3); G=g(x3);
   ii=find(4*x1+2.8*x2+2.4*x3>20 | -x1-x2-x3>-6 | -x1-x2>-3);
   F(ii)=NaN; G(ii)=NaN; plot(x3,F,x3,G), figure, plot(F,G)
```

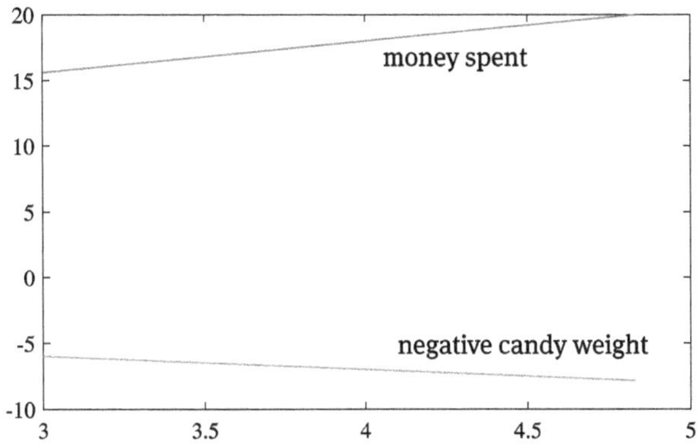

Figure 7.3: Illustrations of candy selections.

It can be seen that the two objective functions are in conflict. Any point on the curves is a reasonable option, which totally depends on the considerations of the decision maker.

For instance, the maximum amount of money is spent ($20) to buy the heaviest candies, while buying the minimum amount of candies (6 kg) means spending the least money. Any intermediate point between these choices can be considered as a compromise solution. From the boss' viewpoint, the second option should be selected, while from the attendee's viewpoint, the first one should be selected. From a neutral viewpoint, a compromise solution should be adopted. The graphical presentation of the two objective function curves is shown in Figure 7.4. This is the concept of the so-called Pareto frontier, which is later addressed.

7.2 Multiobjective programming conversions and solutions

It can be seen from the multiobjective programming problem that the solutions are not unique, and the solutions may be too subjective. In order to get objective solutions,

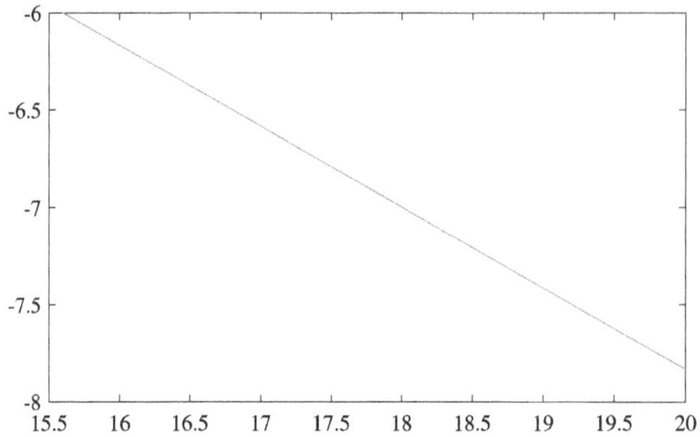

Figure 7.4: The relationship between two objective functions.

an alternative method can be used to convert a multiobjective programming problem into a problem with a single objective. For instance, the weighted sum or some other combination of objective functions can be constructed. The solvers discussed in earlier chapters can be used to find the optimum solutions, or even find the global optimum solutions.

In this section, several commonly used conversion methods are explored.

7.2.1 Least-squares solutions of multiobjective programming problems

Assuming that the objective function vector is $\boldsymbol{F}(\boldsymbol{x}) = [f_1(\boldsymbol{x}), f_2(\boldsymbol{x}), \dots, f_k(\boldsymbol{x})]^{\mathrm{T}}$, the sum of squared objective function components can be selected as the scalar objective function, so that the above multiobjective programming problem can be converted into a scalar objective function problem. The previously discussed methods can be used to solve the new problem.

Definition 7.2. The least squares converted form of a multiobjective programming problem is described as

$$\min_{\boldsymbol{x} \text{ s.t. } \boldsymbol{x}_{\mathrm{m}} \leqslant \boldsymbol{x} \leqslant \boldsymbol{x}_{\mathrm{M}}} f_1^2(\boldsymbol{x}) + f_2^2(\boldsymbol{x}) + \cdots + f_k^2(\boldsymbol{x}) \tag{7.2.1}$$

Function `lsqnonlin()` is provided in Optimization Toolbox in MATLAB. It can be used to find the least-squares solution of multiobjective programming problems directly. The syntaxes of the function are

$[\boldsymbol{x}, f_{\mathrm{n}}, \boldsymbol{f}_{\mathrm{opt}}, \texttt{flag}, \texttt{out}] = \texttt{lsqnonlin(problem)}$

$[\boldsymbol{x}, f_{\mathrm{n}}, \boldsymbol{f}_{\mathrm{opt}}, \texttt{flag}, \texttt{out}] = \texttt{lsqnonlin}(\texttt{F}, \boldsymbol{x}_0)$

$[\boldsymbol{x}, f_{\mathrm{n}}, \boldsymbol{f}_{\mathrm{opt}}, \texttt{flag}, \texttt{out}] = \texttt{lsqnonlin}(\texttt{F}, \boldsymbol{x}_0, \boldsymbol{x}_{\mathrm{m}}, \boldsymbol{x}_{\mathrm{M}})$

where F is the objective function or an anonymous function handle describing the objective function vector. The argument $\boldsymbol{x}_0$ is the initial search point. When the optimization process is completed, the solution is returned in the vector $\boldsymbol{x}$. The objective function vector at the solution can be returned in the $\boldsymbol{f}_{\rm opt}$ vector, while the norm is returned in $f_{\rm n}$. Like with other solvers, the control option OPT is sometimes very important.

The multiobjective programming least-squares problem can be expressed as a structured variable. The necessary members of the problem variable are options, x0, Objective, and solver. The member solver must be set to 'lsqnonlin'. Besides, the optional members are lb and ub, they are the lower and upper bounds of the decision variable.

Example 7.4. Solve the unconstrained nonlinear multiobjective programming problem in Example 7.2 with nonlinear least-squares method.

Solutions. The following anonymous function can be used to express the vector of two objective functions. The following commands can be used to solve the original problem directly. The result obtained is $\boldsymbol{x} = [1.5708, 0.0662]^{\rm T}$. In fact, it is seen by observing the objective functions shown in Figures 7.1 and 7.2 that the optimum solution obtained does not look reasonable. In other words, it may be indicated that the least-squares solution may not be a good choice when solving multiobjective programming problems.

```
>> f=@(x)[(x(1)+2*x(2))*sin(x(1)+x(2))*exp(-x(1)^2-x(2)^2)+5*x(2);
          exp(-x(2)^2-4*x(2))*cos(4*x(1)+x(2))]; %objective function
   xm=[0; 0]; xM=[pi/2; pi/2]; x0=xM;
   x=lsqnonlin(f,x0,xm,xM)    %direct solutions
```

In fact, if we use the function fmincon() introduced earlier, the objective function can be redefined. With the following commands, the optimum solution of the problem is obtained as $\boldsymbol{x} = [1.1832, 0]^{\rm T}$. Of course, the constrained problem may also be solved as a single objective function problem.

```
>> G=@(x)f(x)'*f(x); A=[]; B=[]; Aeq=[]; Beq=[]; x0=xm;
   x=fmincon(G,x0,A,B,Aeq,Beq,xm,xM) %equivalent solutions
```

It should be pointed out that the nonlinear least-squares conversion method is suitable for curve and data fitting problems, rather than for ordinary multiobjective programming problems. Other conversion methods will be explored later.

7.2.2 Linear weighting conversions

It is obvious that the above multiobjective programming problem cannot be solved with the single objective function optimization solvers discussed earlier. Before the solution process, the original multiobjective programming problem should be somehow converted into an optimization problem with a single objective function. The simplest way is to introduce a weight for each objective function and add them together, such that the multiobjective function can be converted into a scalar objective function:

$$f(\boldsymbol{x}) = w_1 f_1(\boldsymbol{x}) + w_2 f_2(\boldsymbol{x}) + \cdots + w_p f_p(\boldsymbol{x}), \tag{7.2.2}$$

where $w_1 + w_2 + \cdots + w_p = 1$, and $0 \leqslant w_1, w_2, \ldots, w_p \leqslant 1$.

Example 7.5. For different weights, solve the problem in Example 7.1.

Solutions. The original problem can be modified into the following linear programming problem:

$$\min_{\boldsymbol{x} \text{ s.t. } \begin{cases} 4x_1+2.8x_2+2.4x_3 \leqslant 20 \\ -x_1-x_2-x_3 \leqslant -6 \\ -x_1-x_2 \leqslant -3 \\ x_1, x_2, x_3 \geqslant 0 \end{cases}} (w_1[4, 2.8, 2.4] - w_2[1, 1, 1])\boldsymbol{x}.$$

With different weighting coefficients, the optimum purchase plans with a loop structure can be obtained as shown Table 7.1. It can be seen that if the weights are selected differently, the two extreme plans can be found. For smaller values of w_1, the solution is concentrated on the \$20 plan, with $\boldsymbol{x} = [0, 3, 4.83]$. For larger values of w_1, the concentration is on the \$15.6 plan, and the solution is $\boldsymbol{x} = [0, 3, 3]$. No matter which weight is selected, $x_1 \equiv 0$, since there is no direct constraint on x_1. Therefore, its value is made as small as possible.

```
>> f1=[4,2.8,2.4]; f2=[-1,-1,-1]; Aeq=[]; Beq=[];
   A=[4 2.8 2.4; -1 -1 -1; -1 -1 0]; B=[20;-6;-3];
   xm=[0;0;0]; ww1=[0:0.1:1]; C=[];
   for w1=ww1, w2=1-w1; % try different weights with loop
      x=linprog(w1*f1+w2*f2,A,B,Aeq,Beq,xm);
      C=[C; w1 w2 x' f1*x -f2*x]
   end
```

Example 7.6. Solve the problem in Example 7.2 with different weights.

Solutions. The weights can be selected as $w_1 = 0, 0.1, \ldots, 1$, and the following statements can convert the problem into that with a single objective function. Function

Table 7.1: Optimization plans under different weightings.

w_1	w_2	x_1	x_2	x_3	costs	weights	w_1	w_2	x_1	x_2	x_3	costs	weights
0	1	0	3	4.8333	20	7.8333	0.6	0.4	0	3	3	15.6	6
0.1	0.9	0	3	4.8333	20	7.8333	0.7	0.3	0	3	3	15.6	6
0.2	0.8	0	3	4.8333	20	7.8333	0.8	0.2	0	3	3	15.6	6
0.3	0.7	0	3	3	15.6	6	0.9	0.1	0	3	3	15.6	6
0.4	0.6	0	3	3	15.6	6	1	0	0	3	3	15.6	6
0.5	0.5	0	3	3	15.6	6							

`fmincon()` can be used to solve the new problems directly. The solutions under different weights can be obtained as shown in Table 7.2. The unbiased compromise solution is obtained at $w_1 = w_2 = 0.5$, with the solution of $\boldsymbol{x}_0 = [0.7708, 0]$. The solutions refer to the curves in Figure 7.3, with the cases $w_1 = 0$ and $w_1 = 1$ regarded as extreme solutions.

```
>> f=@(x)(x(1)+2*x(2))*sin(x(1)+x(2))*exp(-x(1)^2-x(2)^2)+5*x(2);
   g=@(x)exp(-x(2)^2-4*x(2))*cos(4*x(1)+x(2));
   Aeq=[]; Beq=[]; xm=[0;0]; C=[]; xM=[pi/2; pi/2]; ww1=[0:0.1:1];
   for w1=ww1, w2=1-w1;
      F=@(x)w1*f(x)+w2*g(x); x0=xm;
      x=fmincon(F,x0,A,B,Aeq,Beq,xm,xM);
      C=[C; w1 w2 x' f(x) g(x)];
   end
```

Table 7.2: Optimum solutions with different weights.

w_1	w_2	x_1	x_2	f	g	w_1	w_2	x_1	x_2	f	g
0	1	0.7854	0	0.2997	−1	0.1	0.9	0.7839	0	0.2994	−1
0.2	0.8	0.7820	0	0.2984	−0.9999	0.3	0.7	0.7795	0	0.2984	−0.9997
0.4	0.6	0.7760	0	0.2977	−0.9993	0.5	0.5	0.7708	0	0.2965	−0.9983
0.6	0.4	0.7623	0	0.2944	−0.9958	0.7	0.3	0.7458	0	0.2902	−0.9875
0.8	0.2	0.6996	0	0.2761	−0.9417	0.9	0.1	0.0032	0	0	0.9999
1	0	1.5708	0	0	1						

7.2.3 Best compromise solution of linear programs

For multiobjective linear programming problems, a special solution, known as the best compromise solution, is defined. Based on the mathematical formulas, a MATLAB solver can be written to find the solutions. Examples are used to show how to find the best compromise optimum solutions.

Consider a class of special linear programming problems:

$$J = \max_{\boldsymbol{x} \text{ s.t. } \begin{cases} \boldsymbol{A}\boldsymbol{x} \leqslant \boldsymbol{B} \\ \boldsymbol{A}_{\mathrm{eq}}\boldsymbol{x} = \boldsymbol{B}_{\mathrm{eq}} \\ \boldsymbol{x}_{\mathrm{m}} \leqslant \boldsymbol{x} \leqslant \boldsymbol{x}_{\mathrm{M}} \end{cases}} \boldsymbol{C}\boldsymbol{x}. \tag{7.2.3}$$

The model is different from the standard linear programming problem, since the variable $\boldsymbol{C}$ in the objective function is a matrix, rather than a vector. Therefore, each objective function $f_i(\boldsymbol{x}) = \boldsymbol{c}_i\boldsymbol{x}$, $i = 1, 2, \dots, p$ can be considered as the benefit for the ith party. Therefore, the optimization problem can be regarded as the balance of the benefit of all the parties. Of course, the benefits of parties are somehow conflicting, so it is not possible to let every party gain its maximum benefit. Some compromise must be made by all parties. The best compromise solution can be found uniquely.

The best compromise solution can be found using the following procedure:

(1) Solve each individual single objective function problem, and find the optimum objective function value of f_k, for all $k = 1, 2, \dots, p$.
(2) Through normalization method, construct a new objective function as

$$\boldsymbol{f}\boldsymbol{x} = -\frac{1}{f_1}\boldsymbol{c}_1\boldsymbol{x} - \frac{1}{f_2}\boldsymbol{c}_2\boldsymbol{x} - \cdots - \frac{1}{f_p}\boldsymbol{c}_p\boldsymbol{x}. \tag{7.2.4}$$

(3) The best compromise solution problem can be found by solving the following linear programming problem:

$$J = \min_{\boldsymbol{x} \text{ s.t. } \begin{cases} \boldsymbol{A}\boldsymbol{x} \leqslant \boldsymbol{B} \\ \boldsymbol{A}_{\mathrm{eq}}\boldsymbol{x} = \boldsymbol{B}_{\mathrm{eq}} \\ \boldsymbol{x}_{\mathrm{m}} \leqslant \boldsymbol{x} \leqslant \boldsymbol{x}_{\mathrm{M}} \end{cases}} \boldsymbol{f}\boldsymbol{x}. \tag{7.2.5}$$

Based on the above algorithm, the following MATLAB function can be written to find the best compromise solution. A maximization problem is solved in this function.

```
function [x,f0,flag,out]=linprog_comp(C,varargin)
[p,m]=size(C); f=0; % initialization setting
for k=1:p              % solve for each linear program
   [x,f0]=linprog(C(k,:),varargin{:}); f=f-C(k,:)/f0;
end
[x,f0,flag,out]=linprog(f,varargin{:}); f0=C*x; % solve the problem
```

Example 7.7. Find the best compromise solution of the problem in Example 7.1.

Solutions. If the objective function is changed into a maximization problem, the following commands can be used to compute the best compromise solution. The solution found is $\boldsymbol{x} = [0, 3, 4.8333]^{\mathrm{T}}$, and the total money spent is \$20, with total candy weight of 7.8333 kg.

```
>> C=[-4 -2.8 -2.4; 1 1 1]; A=[4 2.8 2.4; -1 -1 -1; -1 -1 0];
   B=[20; -6; -3]; Aeq=[]; Beq=[]; xm=[0;0;0]; xM=[]; %input
   [x f0]=linprog_comp(C,A,B,Aeq,Beq,xm,xM)  %best compromise solution
```

Example 7.8. Find the best compromise solutions to the multiobjective linear programming problem:

$$\min_{\boldsymbol{x} \text{ s.t. } \begin{cases} 2x_1+4x_2+x_4 \leqslant 110 \\ 5x_3+3x_4 \geqslant 180 \\ x_1+2x_2+6x_3+5x_4 \leqslant 250 \\ x_1,x_2,x_3,x_4 \geqslant 0 \end{cases}} \begin{bmatrix} 3x_1 + x_2 + 6x_4 \\ 10x_2 + 7x_4 \\ 2x_1 + x_2 + 8x_3 \\ x_1 + x_2 + 3x_3 + 2x_4 \end{bmatrix}.$$

Solutions. From the given multiobjective programming problem, the matrix $\boldsymbol{C}$ can be created. Note that matrix $\boldsymbol{C}$ given above should alter signs, and other constraints can be expressed in matrix form as

$$\boldsymbol{C} = -\begin{bmatrix} 3 & 1 & 0 & 6 \\ 0 & 10 & 0 & 7 \\ 2 & 1 & 8 & 0 \\ 1 & 1 & 3 & 2 \end{bmatrix}, \quad \boldsymbol{A} = \begin{bmatrix} 2 & 4 & 0 & 1 \\ 0 & 0 & -5 & -3 \\ 1 & 2 & 6 & 5 \end{bmatrix}, \quad \boldsymbol{B} = \begin{bmatrix} 110 \\ -180 \\ 250 \end{bmatrix}.$$

The function `linprog_comp()` written earlier can be used to compute the best compromise solution of the problem.

```
>> C=-[3,1,0,6; 0,10,0,7; 2,1,8,0; 1,1,3,2];
   A=[2,4,0,1; 0,0,-5,-3; 1,2,6,5]; B=[110; -180; 250];
   Aeq=[]; Beq=[]; xm=[0;0;0;0]; xM=[];        %input matrix
   [x,f0]=linprog_comp(C,A,B,Aeq,Beq,xm,xM) %best compromise solution
```

The best compromise solution and the objective functions are

$$\boldsymbol{x} = [0, 0, 21.4286, 24.2857]^{\mathrm{T}}, \quad \boldsymbol{f}_0 = [-145.7143, -170, -171.4286, -112.8571]^{\mathrm{T}}.$$

7.2.4 Least-squares linear programming

A multiobjective linear programming problem can be solved directly with the least-squares method. The linear least-squares solution definition is proposed, and a specific solution method is presented next.

Definition 7.3. The definition of a least-squares multiobjective linear programming problem is

$$\min_{\boldsymbol{x} \text{ s.t. } \begin{cases} \boldsymbol{A}\boldsymbol{x} \leqslant \boldsymbol{B} \\ \boldsymbol{A}_{\rm eq}\boldsymbol{x} = \boldsymbol{B}_{\rm eq} \\ \boldsymbol{x}_{\rm m} \leqslant \boldsymbol{x} \leqslant \boldsymbol{x}_{\rm M} \end{cases}} \frac{1}{2}\|\boldsymbol{C}\boldsymbol{x} - \boldsymbol{d}\|^2. \tag{7.2.6}$$

A linear programming problem least-squares solution can be evaluated directly with function `lsqlin()`, with the syntaxes

$\boldsymbol{x}$=`lsqlin(problem)`

$\boldsymbol{x}$=`lsqlin(`$\boldsymbol{C},\boldsymbol{d},\boldsymbol{A},\boldsymbol{B}$`)`

$\boldsymbol{x}$=`lsqlin(`$\boldsymbol{C},\boldsymbol{d},\boldsymbol{A},\boldsymbol{B},\boldsymbol{A}_{\rm eq},\boldsymbol{B}_{\rm eq},\boldsymbol{x}_{\rm m},\boldsymbol{x}_{\rm M}$`)`

$\boldsymbol{x}$=`lsqlin(`$\boldsymbol{C},\boldsymbol{d},\boldsymbol{A},\boldsymbol{B},\boldsymbol{A}_{\rm eq},\boldsymbol{B}_{\rm eq},\boldsymbol{x}_{\rm m},\boldsymbol{x}_{\rm M},\boldsymbol{x}_0$`)`

$\boldsymbol{x}$=`lsqlin(`$\boldsymbol{C},\boldsymbol{d},\boldsymbol{A},\boldsymbol{B},\boldsymbol{A}_{\rm eq},\boldsymbol{B}_{\rm eq},\boldsymbol{x}_{\rm m},\boldsymbol{x}_{\rm M},\boldsymbol{x}_0$`,options)`

Since the original problem is convex, the initial value vector $\boldsymbol{x}_0$ is not important, and it can be neglected. If a structured variable is used to describe the whole problem, the members in this variable are `C`, `d`, `Aineq`, `bineq`, `Aeq`, `beq`, `lb`, `ub`, `x0`, and `options`. The `solver` member should be set to `'lsqlin'`.

Example 7.9. Consider the multiobjective linear programming problem in Example 7.8. Find its least-squares solution.

Solutions. From the given multiobjective linear programming problem, it can be seen that the matrix $\boldsymbol{C}$ can be formed, and the other constraints are found. The following commands can be used, with function `lsqlin()`, so that the least-squares solution can be found as $\boldsymbol{x} = [0, 0, 28.4456, 12.5907]^{\rm T}$. The values of the objective functions are $[75.544, 88.1347, 227.5648, 110.5181]^{\rm T}$. Note that, for the same problem, since the solution methods and objective functions are selected different, the final optimal results are different. For this example, the best compromise solution is obviously different from the least-squares solution.

```
>> C=[3,1,0,6; 0,10,0,7; 2,1,8,0; 1,1,3,2]; d=zeros(4,1);
   A=[2,4,0,1; 0,0,-5,-3; 1,1,6,5]; B=[110; -180; 250];
   Aeq=[]; Beq=[]; xm=[0;0;0;0]; xM=[];
   x=lsqlin(C,d,A,B,Aeq,Beq,xm,xM), C*x %direct solution
```

7.3 Pareto optimal solutions

Since the solutions to the multiobjective programming problem are not unique, it is a good choice to provide the decision makers less options. In multiobjective programming problems, Pareto frontier is often used to describe optimal solutions. In this section, the concept and finding of Pareto frontiers are presented.

7.3.1 Nonuniqueness of multiobjective programming

It can be seen from the previous analysis that, in normal multiobjective programming problems, the solution is always nonunique. The solution may be different according to the preference of the decision makers. Now let us consider again the original multiobjective programming problem, assuming that a component of one objective function takes some discrete values, the number of the objective functions is decreased by 1. This will lead to new results for the optimization problem. Examples are used to demonstrate the analysis methods.

Example 7.10. With the discrete point method, solve again the multiobjective programming problem in Example 7.1.

Solutions. For the original problem, the total money spent can be expressed by a set of discrete points $m_i \in (15, 20)$, and then the original problem can be rewritten as a standard linear programming problem:

$$\min_{\boldsymbol{x} \text{ s.t.} \begin{cases} 4x_1+2.8x_2+2.4x_3=m_i \\ -x_1-x_2-x_3\leqslant-6 \\ -x_1-x_2\leqslant-3 \\ x_1,x_2,x_3\geqslant0 \end{cases}} -[1,1,1]\boldsymbol{x}.$$

The interpretation of the problem is that, if the cost is m_i, under the constraints, we wonder how much candy can be maximally bought. It is obvious that the maximum weight of candy then is n_i. For each m_i, different values of n_i can be drawn as shown in Figure 7.5. Not all the points on the curves are the solutions of the original problem, since the optimization of m_i was not considered.

```
>> f2=[-1,-1,-1]; Aeq=[4 2.8 2.4]; xm=[0;0;0]; %input matrix
   A=[-1 -1 -1; -1 -1 0]; B=[-6;-3]; mi=15:0.3:20; ni=[];
   for m=mi, Beq=m;
      x=linprog(f2,A,B,Aeq,Beq,xm);
      if length(x)==0, ni=[ni, NaN]; else, ni=[ni,-f2*x]; end
   end
   plot(mi,ni,'*-') %draw the two objective functions
```

7.3.2 Dominant solutions and Pareto frontiers

Since the solutions of a multiobjective programming problem are not unique, two important concepts are introduced here – dominant solutions and Pareto frontiers. Graphical illustrations of these concepts are given.

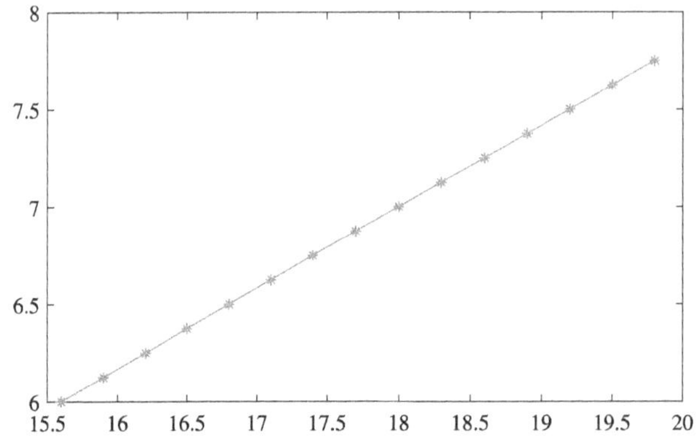

Figure 7.5: The relationship between two objective functions.

Definition 7.4. For the two solutions $\boldsymbol{x}_1$ and $\boldsymbol{x}_2$ of a multiobjective programming problem, if two conditions are satisfied: (1) for all objective functions, $\boldsymbol{x}_1$ is at least as good as $\boldsymbol{x}_2$; (2) for at least one objective function, $\boldsymbol{x}_1$ is significantly better than $\boldsymbol{x}_2$, solution $\boldsymbol{x}_1$ dominates $\boldsymbol{x}_2$, or $\boldsymbol{x}_2$ is dominated by $\boldsymbol{x}_1$.

Consider a problem with two objective functions. Some feasible scattered samples are obtained as shown in Figure 7.6.

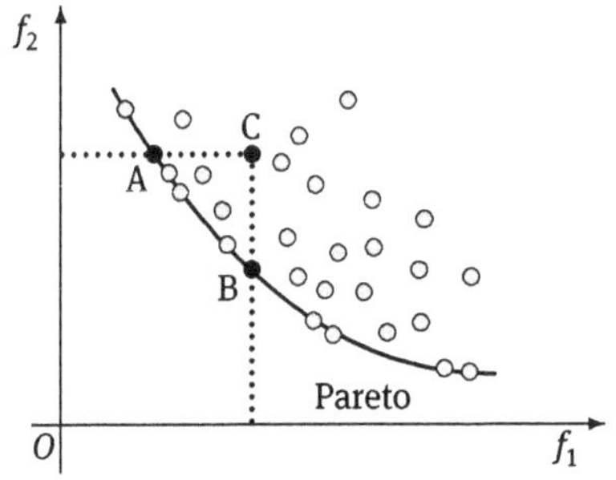

Figure 7.6: Pareto set illustration.

Since in the original problem the minimum solutions are required for the axes f_1 and f_2, it can be seen from Definition 7.4 that, although for an inner point C, the solutions at points A and B are dominating, the point C is not a good point and can be removed from the decision list. With some suitable regulations, all the dominated solutions can be removed. Therefore, a curve can be obtained and shown in the figure, from the remaining dominant solutions.

Definition 7.5. The points along the curve in Figure 7.6 are the solutions of the original problem. They are referred to as Pareto frontier, also written as Pareto front. In some references, they are also known as a nondominated set.

Pareto frontier is named after Italian engineer and economist Vilfredo Federico Damaso Pareto (1848–1923). With the concept of Pareto frontier, the solutions of the multiobjective programming problem are restricted to a curve.

7.3.3 Computations of Pareto frontier

It is easier to extract Pareto frontier for a problem with two objective functions. For problems with more objective functions, the solution is not thus simple. A tool is needed to carry out the task. There is no such a tool provided in MATLAB. In MathWorks' File Exchange website, there are several such tools. For instance, in [9, 44] Simone wrote a solver named Pareto filtering,[44] the programming style is concise and neat. The listing is provided here with modifications. Since minimization problems are expected in this book, the `ge` command is replaced by `le`.

```
function [p,idxs]=paretoFront(p)
[i,dim]=size(p); idxs=[1:i]';
while i>=1
   old_size=size(p,1);
   indices=sum(bsxfun(@le,p(i,:),p),2)==dim;
   indices(i)=false; p(indices,:)=[]; idxs(indices)=[];
   i=i-1-(old_size-size(p,1))+sum(indices(i:end));
end
```

The syntax of the function is `[`$\boldsymbol{p}$`,idxs]=paretoFront(`$\boldsymbol{P}$`)`, where $\boldsymbol{P}$ is an $n\times m$ matrix, while n is the number of points in the feasible solutions. The argument m is the number of objective functions. The returned $\boldsymbol{p}$ is an $n_1 \times m$ matrix, including all the points on the Pareto frontier. The argument `idxs` is a position vector describing the points in matrix $\boldsymbol{P}$.

Example 7.11. Extract the Pareto frontier for Example 7.1.

Solutions. Similar to the enumeration method, mesh grids for x_1, x_2 and x_3 variables are generated. The points not satisfying the constraints can be removed, and the feasible solutions are retained. Calling `paretoFront()` function to extract and label Pareto frontiers, as shown in Figure 7.7, it can be seen that some of the dominated points were not removed.

```
>> [x1,x2,x3]=meshgrid(0:0.1:4); % generate combinations
   ii=find(4*x1+2.8*x2+2.4*x3<=20&x1+x2+x3>=6&x1+x2>=3); % feasible
```

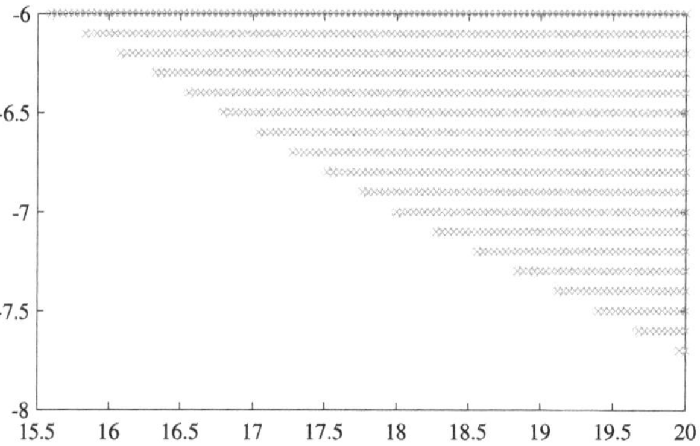

Figure 7.7: Pareto frontier of Example 7.1.

```
xx1=x1(ii); xx2=x2(ii); xx3=x3(ii);
f1=4*xx1+2.8*xx2+2.4*xx3; f2=-(xx1+xx2+xx3);
P=[f1(:), f2(:)]; p=paretoFront(P);  % extract Pareto frontier
plot(f1,f2,'x',p(:,1),p(:,2),'o') % draw Pareto frontier
```

7.4 Minimax problems

Minimax problems are a category of important problems in multiobjective programming. Assume that there is a set of p objective functions $f_i(\boldsymbol{x}), i = 1, 2, \ldots, p$. Each objective function can be understood as an n-dimensional surface. The maximum values of the objective functions at each $\boldsymbol{x}$ can be extracted, to form the new objective function $f(\boldsymbol{x})$. The new objective function can be regarded as an envelope surface of the p objective functions:

$$f(\boldsymbol{x}) = \max_i \,[f_1(\boldsymbol{x}), f_2(\boldsymbol{x}), \ldots, f_n(\boldsymbol{x})]^{\mathrm{T}}. \tag{7.4.1}$$

The envelope surface $f(\boldsymbol{x})$ of the objective function vector can be used as the scalar objective function, such that a new constrained optimization problem can be constructed. The minimax problem can be formulated:

$$J = \min_{\boldsymbol{x} \text{ s.t. } \boldsymbol{G}(\boldsymbol{x}) \leqslant \boldsymbol{0}} \max_i \boldsymbol{F}(\boldsymbol{x}) = \min_{\boldsymbol{x} \text{ s.t. } \boldsymbol{G}(\boldsymbol{x}) \leqslant \boldsymbol{0}} f(\boldsymbol{x}). \tag{7.4.2}$$

In other words, the minimax problem is targeting at finding the best solution in the worst case. Of course, the minimax problem here is sequential. The envelope surface of all the objective functions is extracted first, then the minimization problem is solved. The definition and solution methods of minimax problems are presented next.

Definition 7.6. Considering various constraints, a minimax problem is written as

$$J=\min_{\boldsymbol{x} \text{ s.t. } \begin{cases}\boldsymbol{A}\boldsymbol{x}\leqslant\boldsymbol{B}\\ \boldsymbol{A}_{\rm eq}\boldsymbol{x}=\boldsymbol{B}_{\rm eq}\\ \boldsymbol{x}_{\rm m}\leqslant\boldsymbol{x}\leqslant\boldsymbol{x}_{\rm M}\\ \boldsymbol{C}(\boldsymbol{x})\leqslant\boldsymbol{0}\\ \boldsymbol{C}_{\rm eq}(\boldsymbol{x})=\boldsymbol{0}\end{cases}} \max_i \boldsymbol{F}(\boldsymbol{x}). \tag{7.4.3}$$

A function named `fminimax()` is provided in the Optimization Toolbox in MATLAB. It can be used in solving minimax problems, with the syntaxes

[$\boldsymbol{x}$,$\boldsymbol{f}_{\rm opt}$,f_0,`flag,out]=fminimax(problem)`

[$\boldsymbol{x}$,$\boldsymbol{f}_{\rm opt}$,f_0,`flag,out]=fminimax(F,`$\boldsymbol{x}_0$)

[$\boldsymbol{x}$,$\boldsymbol{f}_{\rm opt}$,f_0,`flag,out]=fminimax(F,`$\boldsymbol{x}_0$,$\boldsymbol{A}$,$\boldsymbol{B}$,$\boldsymbol{A}_{\rm eq}$,$\boldsymbol{B}_{\rm eq}$,$\boldsymbol{x}_{\rm m}$,$\boldsymbol{x}_{\rm M}$)

[$\boldsymbol{x}$,$\boldsymbol{f}_{\rm opt}$,f_0,`flag,out]=fminimax(F,`$\boldsymbol{x}_0$,$\boldsymbol{A}$,$\boldsymbol{B}$,$\boldsymbol{A}_{\rm eq}$,$\boldsymbol{B}_{\rm eq}$,$\boldsymbol{x}_{\rm m}$,$\boldsymbol{x}_{\rm M}$,`CF,opt)`

The syntaxes of the function are very close to those presented for `fmincon()`. The difference is that the objective function is expressed as a vector. Of course, an anonymous or M function can be used to represent the objective function vector. Besides, in the returned arguments, apart from the objective function vector $\boldsymbol{f}_{\rm opt}$, the minimax value of f_0 is also returned.

If a structured variable is used to describe the problem, the member `solver` should be set to `'fminimax'`. The other members are the same as in the `fmincon()` function.

Example 7.12. To illustrate the physical interpretation to the minimax model, a multiobjective programming problem with two objective functions is constructed. Solve the problem and explain the results:

$$\min_{0.16\leqslant x\leqslant 1.2}\max_i \begin{bmatrix} \dfrac{1}{(x-0.2)^2+0.01}+\dfrac{1}{(x-0.9)^2+0.05}+1\\ \dfrac{1}{(x-0.2)^2+0.08}+\dfrac{1}{(x-1.1)^2+0.02}+1\end{bmatrix}.$$

Solutions. The two objective functions can be respectively expressed in two anonymous functions and the curves can be drawn. Meanwhile, the upper envelope of the two objective functions, $f_{\max}(x)$, can be evaluated and superimposed on the curve by a thick line, as shown in Figure 7.8.

```
>> f1=@(x)1./((x-0.2).^2+0.01)+1./((x-0.9).^2+0.05)+1;
   f2=@(x)1./((x-0.2).^2+0.08)+1./((x-1.1).^2+0.02)+1;
   fmax=@(x)max(f1(x),f2(x)); xm=0.16; xM=1.2;
   fplot(f1,[xm,xM]), hold on, fplot(f2,[xm,xM])
   fplot(fmax,[xm,xM],'LineWidth',2), hold off
```

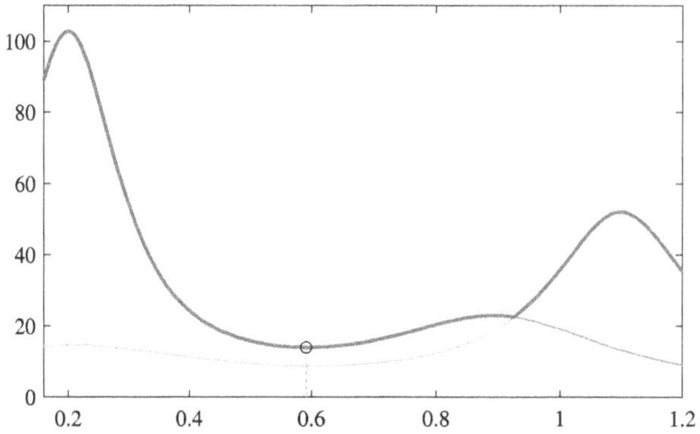

Figure 7.8: The two objective functions and their envelope.

With the upper envelope curve, if the minimum solution is expected in the interval of interest, it is immediately seen that the minimum of the upper envelope is located at a position near $x = 0.6$. This solution is the minimax solution of the original problem.

From the two anonymous functions, the objective function vector of the minimax problem can be described directly by another anonymous function F. Calling `fminimax()` function, the solution of the problem can be found directly at $x_1 = 0.5918$. It is close to estimated the solution in Figure 7.8. The objective function vector is $\boldsymbol{f}_x = [14.0131, 8.8762]^{\mathrm{T}}$. The minimax value is $f_0 = 14.0131$, the envelope function value.

```
>> F=@(x)[f1(x); f2(x)];
   A=[]; B=[]; Aeq=[]; Beq=[]; x0=rand(1);
   [x1,fx,f0,flag]=fminimax(F,x0,A,B,Aeq,Beq,xm,xM)
```

Example 7.13. Solve the following minimax problem:

$$\min_{0\leqslant x_1,x_2\leqslant 2} \max_i \begin{bmatrix} x_1^2\sin x_2 + x_2 - 3x_1x_2\cos x_1 \\ -x_1^2\mathrm{e}^{-x_2} - x_2^2\mathrm{e}^{-x_1} + x_1x_2\cos x_1x_2 \\ x_1^2 + x_2^2 - 2x_1x_2 + x_1 - x_2 \\ -x_1^2 - x_2^2\cos x_1x_2 \end{bmatrix}.$$

Solutions. The following statements can be used to solve the problem directly. If a random point is used in the solution process, the solution of the objective function is $\boldsymbol{x} = [0.7862, 1.2862]$. At this point, the objective function vector is $\boldsymbol{f}_0 = [-0.2640, -0.3876, -0.2500, -1.4963]$, and the minimax objective function value is $f_1 = -0.25$. The value of the returned argument `flag` is 4, not the expected 1. It means that the searching region around the optimum solution is flat, and different points near it also lead to different solutions, whose objective function value is the same.

```
>> F=@(x)[x(1)^2*sin(x(2))+x(2)-3*x(1)*x(2)*cos(x(1));
     -x(1)^2*exp(-x(2))-x(2)^2*exp(-x(1))+x(1)*x(2)*cos(x(1)*x(2));
```

```
    x(1)^2+x(2)^2-2*x(1)*x(2)+x(1)-x(2);
    -x(1)^2-x(2)^2*cos(x(1)*x(2))]; %multiobjective functions
  A=[]; B=[]; Aeq=[]; Beq=[]; xm=[0;0]; xM=[2;2];
  [x,f0,f1,flag]=fminimax(F,rand(2,1),A,B,Aeq,Beq,xm,xM)
```

For the surfaces of the four objective functions, the following commands can be used to draw them in the specified region, as shown in Figure 7.9.

```
>> [x1,x2]=meshgrid(0:0.02:2);
   F1=x1.^2.*sin(x2)+x2-3*x1.*x2.*cos(x1);
   F2=-x1.^2.*exp(-x2)-x2.^2.*exp(-x1)+x1.*x2.*cos(x1.*x2);
   F3=x1.^2+x2.^2-2*x1.*x2+x1-x2;
   F4=-x1.^2-x2.^2.*cos(x1.*x2);
   subplot(221), surf(x1,x2,F1), shading flat
   subplot(222), surf(x1,x2,F2), shading flat
   subplot(223), surf(x1,x2,F3), shading flat
   subplot(224), surf(x1,x2,F4), shading flat
```

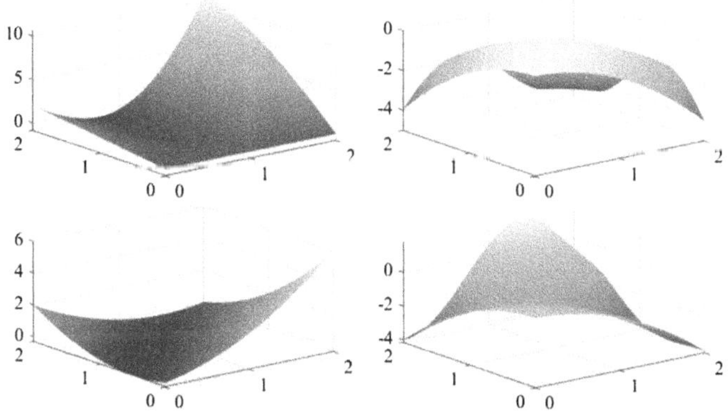

Figure 7.9: The surfaces of the four objective functions.

From the four objective functions, the envelope surface can be extracted and the surface of the new objective function $f(\mathbf{x})$ can be obtained as shown in Figure 7.10. The minimum value of the new objective function is needed in the minimax problem.

```
>> Fx=cat(3,F1,F2,F3,F4); Fx=max(Fx,[],3); %maximum values
   surf(x1,x2,Fx)                         %envelope surface
```

It can be seen from the upper envelop surface that its minimum region is rather flat, and the details cannot be visualized easily. The following commands can be used:

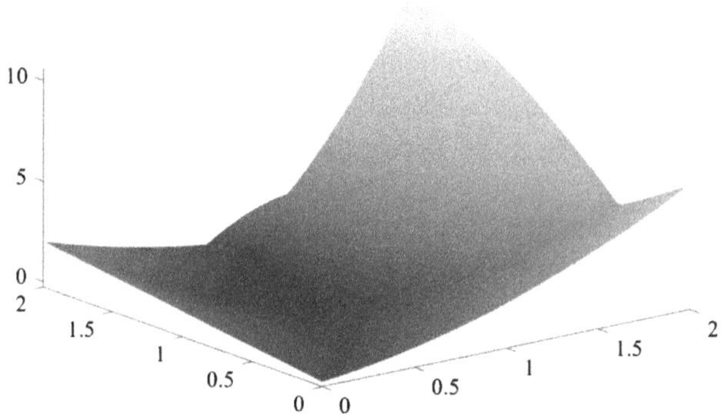

Figure 7.10: The maximum function surface of the four objective functions.

```
>> contour(x1,x2,Fx,300)
   hold on, plot(0.7862,1.2862,'o'), hold off
```

to draw the contour of the envelope objective function $f(\boldsymbol{x})$, and label the optimum solution, as shown in Figure 7.11. In fact, the optimum solution found with `fminimax()` function is only one of the optimum solutions. If the mesh grid data are considered, the optimum solution obtained is $x_1 = 0.66$, $x_2 = 1.16$, the value of the objective function is also −0.25. The point is marked in Figure 7.11 with ×. It is also located in the solution region. Many of the points in the bounded by the contour region have the same objective function value of −0.25.

```
>> [Fxm,i]=min(Fx(:)); Fxm, x1(i), x2(i)
```

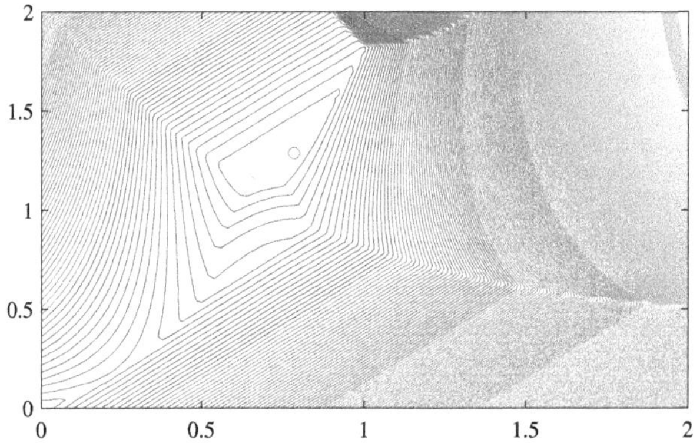

Figure 7.11: The contours of the envelope function surface.

Example 7.14. Solve the constrained minimax problem

$$\min_{\boldsymbol{x} \text{ s.t. } \begin{cases} x_1^2+x_2^2\geqslant 1 \\ (x_1+0.5)^2-(x_2+0.5)^2\geqslant 0.7 \end{cases}} \max_i \begin{bmatrix} x_1^2\sin x_2 + x_2 - 3x_1x_2\cos x_1 \\ -x_1^2\mathrm{e}^{-x_2} - x_2^2\mathrm{e}^{-x_1} + x_1x_2\cos x_1x_2 \\ x_1^2 + x_2^2 - 2x_1x_2 + x_1 - x_2 \\ -x_1^2 - x_2^2\cos x_1x_2 \end{bmatrix}.$$

Solutions. Mesh grids can be generated in the square region $(-2,2)\times(-2,2)$. The four objective function matrices can be computed directly. Then, the points which do not satisfy the constraints can be removed. Taking maximum values to the four objective functions, the data of the scalar objective function can be obtained. The contours of the envelope function surface can be obtained, as shown in Figure 7.12.

```
>> [x1,x2]=meshgrid(-2:0.005:2);
   F1=x1.^2.*sin(x2)+x2-3*x1.*x2.*cos(x1);
   F2=-x1.^2.*exp(-x2)-x2.^2.*exp(-x1)+x1.*x2.*cos(x1.*x2);
   F3=x1.^2+x2.^2-2*x1.*x2+x1-x2; F4=-x1.^2-x2.^2.*cos(x1.*x2);
   ii=find(x1.^2+x2.^2<1 & (x1+0.5).^2-(x2+0.5).^2<0.7);
   F1(ii)=NaN; F2(ii)=NaN; F3(ii)=NaN; F4(ii)=NaN;
   F0=cat(3,F1,F2,F3,F4); F0=max(F0,[],3);
   surface(x1,x2,F0); shading flat, view(0,90)
```

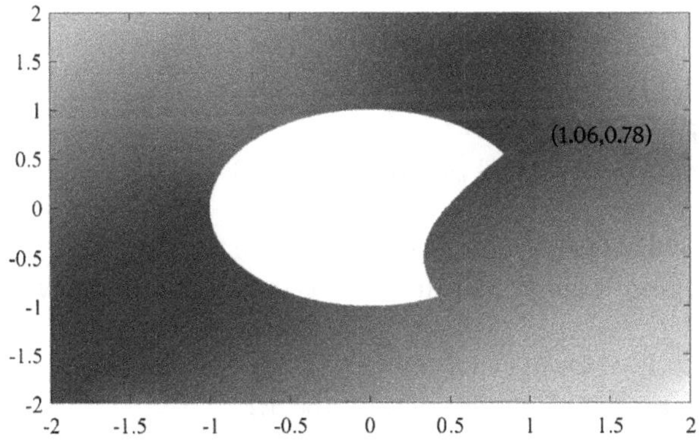

Figure 7.12: Contours of the envelope surface.

Now consider the problem solution with `fminimax()` function. The constraints can be expressed with the following MATLAB function. Since there are no equality constraints, the returned argument for it can still be expressed as an empty matrix.

```
function [c,ceq]=c7mminm(x);
ceq=[];
c=[-x(1)^2-x(2)^2+1; -(x(1)+0.5)^2+(x(2)+0.5)^2+0.7];
```

In fact, with function `fminimax()`, the minimax problem can be directly solved. The result obtained is superimposed on the surface in Figure 7.12. The coordinates of the point are $\boldsymbol{x} = [1.0614, 0.7829]^{\mathrm{T}}$, and the function vector values are $\boldsymbol{f}_x = [0.3084, -0.1512, 0.3084, -1.5059]^{\mathrm{T}}$. The minimax objective function value is $f_0 = 0.3084$.

```
>> F=@(x)[x(1)^2*sin(x(2))+x(2)-3*x(1)*x(2)*cos(x(1));
      -x(1)^2*exp(-x(2))-x(2)^2*exp(-x(1))+x(1)*x(2)*cos(x(1)*x(2));
      x(1)^2+x(2)^2-2*x(1)*x(2)+x(1)-x(2);
      -x(1)^2-x(2)^2*cos(x(1)*x(2))]; %express the objective function vector
   clear P; P.solver='fminimax'; P.options=optimset;
   P.Objective=F; P.nonlcon=@c7mminm; P.x0=rand(2,1);
   [x0,fx,f0,flag]=fminimax(P)
```

Example 7.15. Consider the multiobjective linear programming problem in Example 7.8. Solve the problem as a minimax problem.

Solutions. The following commands can be used to input directly the original problem into MATLAB environment. With the minimax solver, the solution of the problem can be found as $\boldsymbol{x}_1 = [9.5745, 1.4894, 26.8085, 15.3191]^{\mathrm{T}}$, $\boldsymbol{f}_x = [-122.1277, -122.1277, -235.1064, -122.1277]^{\mathrm{T}}$, and $f_0 = -122.1277$. The result obtained is different from that obtained in Example 7.8, since the viewpoint of the decision maker is different.

```
>> C=-[3,1,0,6; 0,10,0,7; 2,1,8,0; 1,1,3,2];
   A=[2,4,0,1; 0,0,-5,-3; 1,2,6,5]; B=[110; -180; 250];
   Aeq=[]; Beq=[]; xm=[0;0;0;0]; f=@(x)C*x(:);
   [x1,fx,f0,flag]=fminimax(f,rand(4,1),A,B,Aeq,Beq,xm)
```

A structured variable can be used to describe the original problem, and the following statements can be written. The results are exactly the same as those obtained above.

```
>> clear P; P.Objective=f; P.options=optimset;
   P.solver='fminimax'; P.Aineq=A; P.bineq=B; P.lb=xm;
   P.x0=rand(4,1); [x1,fx,f0,flag]=fminimax(P)
```

In fact, with the `fminimax()` function, other variations of similar problems can also be solved. For instance, the minimin problem

$$J = \min_{\boldsymbol{x} \text{ s.t. } \boldsymbol{G}(\boldsymbol{x})\leqslant \boldsymbol{0}} \min_i \boldsymbol{F}(\boldsymbol{x}). \tag{7.4.4}$$

The problem can be converted directly into the minimax problem

$$J = \min_{\boldsymbol{x} \text{ s.t. } \boldsymbol{G}(\boldsymbol{x}) \leqslant \boldsymbol{0}} \max_{i}(-\boldsymbol{F}(\boldsymbol{x})). \tag{7.4.5}$$

7.5 Exercises

7.1 Solve the following multiobjective programming problem:

(1) $\min_{\boldsymbol{x}} \left[\dfrac{x_1^2}{2} + x_2^2 - 10x_1 - 100,\ x_1^2 + \dfrac{x_2^2}{2} - 10x_2 - 100\right]^{\mathrm{T}}$,

(2) $\min_{\boldsymbol{x} \text{ s.t. } \begin{cases} x_1^2+x_2^2-1-0.1\cos\left(16\operatorname{atan}(x_1/x_2)\right)\geqslant 0 \\ (x_1-0.5)^2+(x_2-0.5)^2\leqslant 0.5 \\ 0\leqslant x_1,x_2\leqslant \pi \end{cases}} [x_1, x_2]^{\mathrm{T}}$,

(3) $\min_{\boldsymbol{x} \text{ s.t. } \begin{cases} x_1^2-4x_1+x_2+5 \\ x_1,x_2\geqslant 0 \end{cases}} \left[\sqrt{1+x_1^2},\ x_1^2 - 4x_1 + x_2 + 5\right]^{\mathrm{T}}$,

(4) $\min_{\boldsymbol{x} \text{ s.t. } \begin{cases} -\cos x_1 - e^{-x_2} + x_3 \leqslant 0 \\ 0\leqslant x_1\leqslant\pi,\ x_2\geqslant 0,\ x_3\geqslant 1.2 \end{cases}} [-x_1, -x_2, -x_3]^{\mathrm{T}}$.

7.2 Find the best compromise solution of the following multiobjective linear programming problem:

$$\begin{aligned} \max\ & z_1 = 100x_1 + 90x_2 + 80x_3 + 70x_4 \\ \min\ & z_2 = 3x_2 + 2x_4. \end{aligned}$$
$$\boldsymbol{x} \text{ s.t. } \begin{cases} x_1+x_2\geqslant 30 \\ x_3+x_4\geqslant 30 \\ 3x_1+2x_2\leqslant 120 \\ 3x_2+2x_4\leqslant 48 \\ x_1,x_2,x_3,x_4\geqslant 0 \end{cases}$$

7.3 Find the best compromise solution of the following multiobjective linear programming problem:

$$\max_{\boldsymbol{x} \text{ s.t. } \begin{cases} 2x_1+5x_2+10x_3\leqslant 100 \\ x_1+6x_2+8x_4\leqslant 250 \\ 5x_1+8x_2+7x_3+10x_4\leqslant 350 \\ x_1,x_2,x_3,x_4\geqslant 0 \end{cases}} \begin{bmatrix} 50x_1 + 20x_2 + 100x_3 + 60x_4 \\ 20x_1 + 70x_2 + 5x_3 \\ 3x_2 + 5x_4 \\ 2x_1 + 20x_3 + 2x_4 \end{bmatrix}.$$

7.4 Solve the following multiobjective programming problem:[14]

$$\max \quad [5x_1 - 2x_2,\ -x_1 + 4x_2]^{\mathrm{T}}.$$
$$\boldsymbol{x} \text{ s.t. } \begin{cases} -x_1+x_2+x_3=3 \\ x_1+x_2+x_4=8 \\ x_1+x_5=6 \\ x_2+x_6=4 \\ x_i\geqslant 0,\ i=1,2,\ldots,6 \end{cases}$$

7.5 Solve the following multiobjective programming problem and get its Pareto frontier:

$$\min \quad \begin{bmatrix} x_1^2 + x_2^2 \\ (x_1 - 5)^2 + (x_2 - 5)^2 \end{bmatrix}.$$
$$\boldsymbol{x} \text{ s.t. } -5\leqslant x_1,x_2\leqslant 10$$

7.6 Consider again the multiobjective programming problem in Example 7.2. Find the minimax solution.

8 Dynamic programming and shortest paths

The mathematical programming problems can be regarded as static problems since the objective functions and constraints are fixed before the problem solution process. In practice, there are problems whose objective functions are piecewise and periodic. When a plan is made in a manufacturing industry, the objective function of one year may depend upon the actually completed previous years plans. Therefore, the optimization problem is not static, and dynamic programming should be introduced.

Dynamic programming was pioneered by American scholar Richard Ernest Bellman (1920–1980) in 1953.[5] A new field in optimization was introduced. Dynamic programming is regarded as one of the three theoretical foundations of the so-called modern control theory. The theory finds its important applications in multistage decision making and in network optimization. For instance, [31] listed 47 dynamic programming applications. In this chapter, an introduction is presented. Its application in path planning problems of oriented graphs, known as the shortest path problem, is presented with examples.

In Section 8.1, a general description of dynamic programming is presented. Examples are used to show how to use the dynamic programming scheme to solve a linear programming problem. In Section 8.2, an application of dynamic programming, the optimal path planning, is presented. The concepts of oriented graph and its representation are put forward. Then, MATLAB-based solutions are illustrated. A solver based on the general-purpose Dijkstra algorithm is presented, which can be used to solve these problems. In Section 8.3, an introduction is proposed for the path planning problem of an undirected graph.

8.1 An introduction to dynamic programming

Dynamic programming is often described in multistage decision making problems. In this section, the fundamental concept of a multistage decision making problem is provided, and then the basic definition of a dynamic programming problem is introduced. The standard mathematical model of a dynamic programming problem is given, and the method to convert it into a multistage decision making problem is presented.

8.1.1 Concept and mathematical models in dynamic programming

Definition 8.1. Assume that a decision making process can be represented by some interrelated stages, where decisions in each stage should be made. Each of such decisions may affect the decisions of other stages. Therefore, the problem is known as a multistage decision making problem.

https://doi.org/10.1515/9783110667011-008

Definition 8.2. Assume that the kth stage can be expressed by the state space variable x_k, and the decision is s_k, then the state space equation $x_{k+1} = f_k(x_k, s_k)$ can be established. Such a state equation is used to describe the state transition law from the kth to the $(k+1)$th stage.

Definition 8.3. The mathematical model of a dynamic programming problem can be established for the state variable as

$$r_{k-1}(x_{k-1}, s_{k-1}) = \max_{\{x_k\}} \quad d_k + r_k(x_k, s_k), \tag{8.1.1}$$

where x_k is the state variable in the kth stage, and s_k is the decision at the kth stage. The variable d_k is the increment of the objective function from the kth to the $(k-1)$th stage. Notation $\{x_k\}$ is used to represent the kth stage state variable satisfying the constraints.

An illustration is provided in [17] to describe a dynamic programming problem, as shown in Figure 8.1. If the state x_n at $k = n$ is known, an optimization method can be used to solve for the nth step decision s_n, and the state x_{n-1} at the previous stage. With such a method, through backward formulation, the solution of the problem can be found.

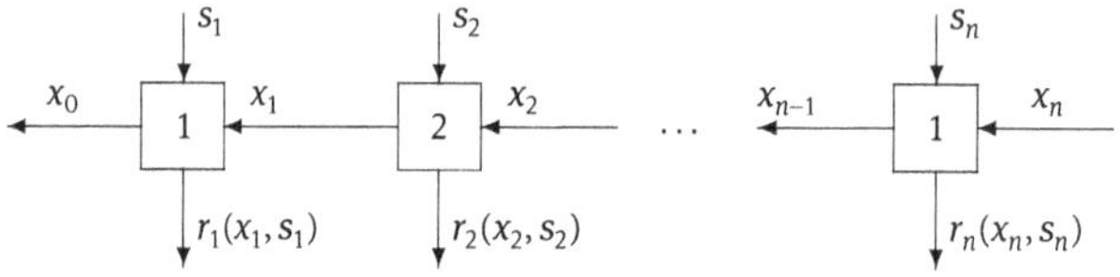

Figure 8.1: Illustration of a dynamic programming problem.

8.1.2 Dynamic programming solutions of linear programming problems

Many optimization problems considered so far can be converted into dynamic programming problems. Here a simple example of a linear programming problem is used to describe a multistage decision making process, and explore its solution process with the dynamic programming solution method.

Example 8.1. Solve the following simple linear programming problem with the dynamic programming method:

$$\max_{\boldsymbol{x} \text{ s.t. } \begin{cases} 5x_1+8x_2\leqslant 10 \\ x_1,x_2\geqslant 0 \end{cases}} \quad 4x_1 + 6x_2.$$

Solutions. Of course, direct use of a linear programming solver yields the optimum solution of $x_1 = 2$ and $x_2 = 0$, and the maximum value of the objective function is $f_0 = 8$.

```
>> P=optimproblem('ObjectiveSense','max'); %maximization problem
   x=optimvar('x',2,1,'LowerBound',0);      %decision variables
   P.Objective=4*x(1)+6*x(2); P.Constraints.c=5*x(1)+8*x(2)<=10;
   sols=solve(P); x0=sols.x; x1=x0(1), x2=x0(2), f0=[4,6]*x0
```

Now let us consider how we can solve the problem with the dynamic programming method. Dynamic programming solution converts the problem into a sequential optimization problem, and solves for one variable at a time. With the method demonstrated in [17], the original problem can be converted into the two-stage decision making problem shown in Figure 8.2.

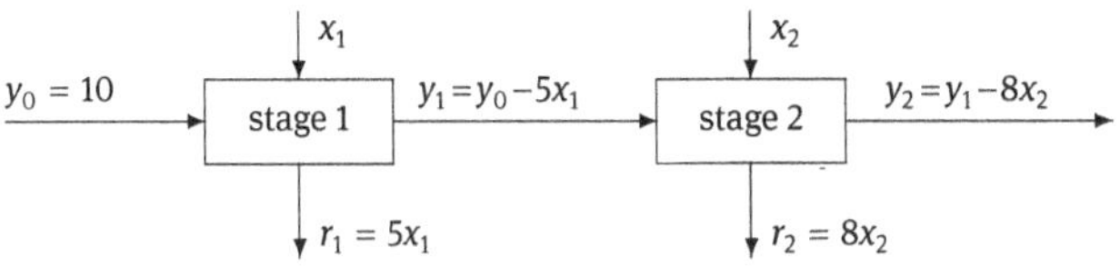

Figure 8.2: Converted dynamic programming problem.

The two state space equations $y_1 = y_0 - 5x_1$ and $y_2 = y_1 - 8x_2$ can be formulated. Now $x_1, x_2, y_1, y_2 \geqslant 0$. It is known that $y_0 = 10$. Therefore, it is found that $10 - 5x_1 \geqslant 0$ and $y_1 - 8x_2 \geqslant 0$, from which it is found that $0 \leqslant x_1 \leqslant 2$ and $0 \leqslant x_2 \leqslant y_1/8$. It is not hard to find that

$$f_{\max} = \max_{x_1}\left(4x_1 + \max_{x_2} 6x_2\right) = \max_{x_1}\left(4x_1 + 6\frac{10-5x_1}{8}\right) = \max_{x_1}\frac{x_1+30}{4},$$

where, to maximize $6x_2$, the value of x_2 should be taken as $x_2 = y_1/8$. From the above formula, it can be seen that the objective function achieves its maximum when x_1 takes its maximum value. Since the upper bound of x_1 is 2, therefore, $x_1 = 2$. It is found in turn that $y_1 = 0$, so that $x_2 = 0$. The maximum value of the objective function is 8, which is the same as that obtained with the linear programming solver.

The solution method illustrated above is just a simple attempt. For practical linear programming problems, the method shown in Chapter 4 should be used to solve them directly. It is not necessary to convert the problems manually into dynamic programming framework first, since the solution processes are even more complicated. It is not worth the efforts.

8.2 Shortest path problems in oriented graphs

Oriented graphs and shortest path problems are often encountered problems in many fields. Dynamic programming was first introduced in dealing with such problems.

With the dynamic programming theory, one needs to derive the shortest path from the terminal back to the starting point, to find the shortest path. In this section, fundamental definitions and description methods for oriented graphs are presented. An example is given to derive the optimal path manually. Then, the tools in Bioinformatics Toolbox[48] in MATLAB are demonstrated when solving the optimal path problems directly. Finally, a practical Dijkstra algorithm-based MATLAB implementation is provided.

8.2.1 Examples of oriented graphs

Oriented graphs are a class of special graphs. In such a graph, there are several nodes, and in-between the nodes, there are links, known as the edges. The edges are oriented, and they have their start and end nodes. On the edges, numbers are written, which are referred to as weights. The physical interpretation of the weights can be different, e. g., the actual distance between two nodes, or the cost to travel from one node to another. Based on the rules in the graph, the shortest path problem to be solved is to find a route, from the start point to the end point in the graph, such that the total sum of weights on the route is minimized.

An example on oriented graph is demonstrated next.

Example 8.2. Assume that there are 9 cities, as shown in Figure 8.3.[33] These cited are regarded as the nodes in the graph. In-between the cities, there are oriented edges, and traveling costs are labeled. The problem to be solved is how to travel from node ① to node ⑨ such that the total travel cost is minimized.

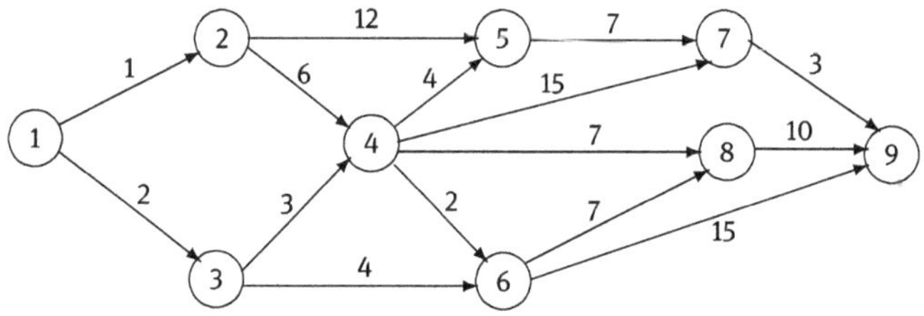

Figure 8.3: An oriented graph.

Definition 8.4. The three key factors in the oriented graph are nodes, edges, and weights.

The computer representation of an oriented graph is presented later. If in the entire graph, the number of oriented edges from any node A to any node B is no more than 1, this matrix is a good way to represent it. The information about nodes, edges, and weights can be described uniquely with a matrix.

Manual solution of the shortest path problem is presented in this section first, then a dynamic programming-based solution is given. Finally, matrix-based graph representation and MATLAB tool-based solution of the problem are provided.

8.2.2 Manual solutions of shortest path problem

Here an example is given to show the dynamic programming-based solution of the oriented graph problem. In many references, the iterative method for solving this problem is applied backwards. The method is presented in this section. Also the forward iteration method is provided.

Example 8.3. Consider the graph shown in Example 8.2. Find the shortest path from node ① to node ⑨ manually.

Solutions. Consider first the end node, that is, node ⑨. The initial distance 0 is labeled as (0). The next step is to find the shortest paths of the upper level nodes, and the nodes are ⑥, ⑦, ⑧. Since there is only one edge connected to node ⑨, the distances are labeled as (15), (3), and (10), respectively. From node ⑤ to ⑦, there is only one edge, thus node ⑤ should be labeled (10), which is the sum of labeled in ⑦ plus the weight of the edge. Next let us analyze the label on node ④. There are paths to travel from node ④ to nodes ⑤, ⑥, ⑦, and ⑧. Comparing the labels on the nodes, plus the distance to node ④, it is seen that the path from ④ to ⑤ yields the smallest value of 14, while paths to the nodes ⑥, ⑦, and ⑧ produce the values of 17, 18, and 17, respectively. Therefore, node ④ should be labeled (14). The labels on nodes ② and ③ should be the weights plus node labels. It can be seen that the smallest distances from nodes ② and ③ to node ④ are 20 and 17, respectively, therefore, they should be labeled as (20) and (17). The shortest path labeled on node ① should be labeled (19). It can be concluded that the optimum route should be ①→③→④→⑤→⑦→⑨, as shown in Figure 8.4.

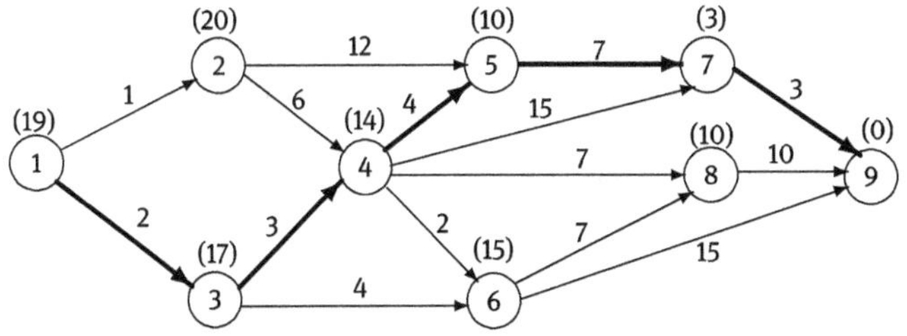

Figure 8.4: Manual solutions of shortest path problem.

Example 8.4. The forward iteration can also be made to solve the optimal path problem in Example 8.2.

Solutions. In fact, the forward search is also possible to solve the same problem. The reference at the start point can be labeled [0], other nodes can also be labeled accordingly, as shown in Figure 8.5. The optimal path found in this way is exactly the same as that found in the previous example.

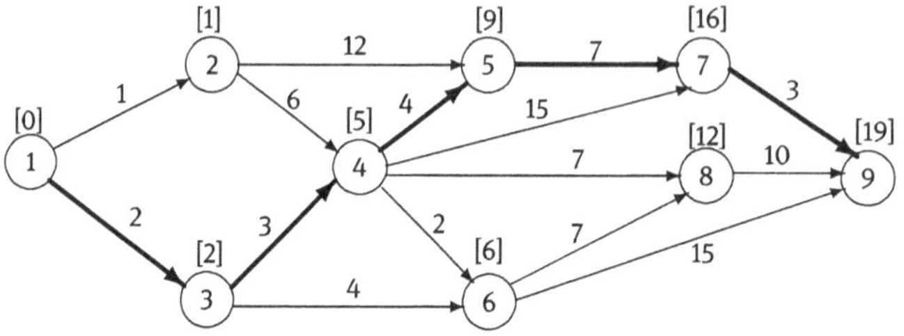

Figure 8.5: Forward iteration solution.

It can be seen from the previous example that the solution process is simple and straightforward. While for large-scale problems, the process may be very complicated, and may lead to errors, if one is not careful. A better and reliable method should be introduced.

8.2.3 Solution with dynamic programming formulation

The solution of the optimal path problem is demonstrated in the previous section. The method may be too tedious to use in complicated graphs. It may only be applied to small-scale problems. In real applications, in order to solve large-scale problems, a direct method based on dynamic programming is employed to find solutions. Better and straightforward solutions will be presented later in this section.

Example 8.5. Solve the shortest path problem in Example 8.2 with the dynamic programming method.

Solutions. Let $f_i(x_i)$ represent the optimal path to node x_i in the ith stage decision. The shortest path in the ith stage decision can be written as

$$f_{i-1}(x_{i-1}, s_{i-1}) = \min_{\substack{\text{all } (x_{i-1}, x_i) \\ \text{feasible paths}}} d(x_{i-1} - x_i) + f_i(x_i, s_i), \quad i = n, n-1, \ldots$$

The following steps can be taken to solve in turn the shortest path problem:

(1) Before the decision process, the terminal time is set to $f_4(x_4) = 0$, where $x_4 = 9$, which is node ⑨.

(2) When $i = 3$, there are two paths to select, x_3 has two selections, $x_3 = 7$ or $x_3 = 8$. It can be seen that $x_3 = 7$ (i. e., node ⑦) is a point on the optimum path and $f_3(x_3) = 3$. Indeed,
$f_3(x_3 = 7) = 3, f_3(x_3 = 8) = 10$, and then $f_3(x_3) = \min(3, 10) = 3$.
(3) It can be found that $f_2(x_2) = 10$ so that $x_2 = 5$, i. e., node ⑤. Indeed,
$f_2(x_2 = 5) = 3 + 7 = 10, f_2(x_2 = 6) = 15$,
$f_2(x_2 = 4) = \min(10 + 4, 3 + 15 + 10 + 7, 15 + 2) = 14$,
$f_2(x_2) = \min(10, 15, 14) = 10$.
(4) With the formula it is found that $f_1(x_1 = 3)$, that is, node ④, with $f_1(x_1) = 17$ since
$f_1(x_1 = 2) = \min(10 + 12, 14 + 6) = 20$,
$f_1(x_1 = 3) = \min(14 + 3, 15 + 4) = 17, f_1(x_1) = \min(20, 17) = 17$.
(5) Finally, it is found that we need node ③, with $f_0(x_0) = 19$, because
$f_0(x_0 = 1) = \min(20 + 1, 17 + 2) = 19$.

It is found backwards that ①→③→④→⑤→⑦→⑨.

The shortest distance is 19, the same as obtained earlier.

8.2.4 Matrix representation of graphs

Before presenting graphs, fundamental concepts regarding graphs are provided. Then, we concentrate on how to represent graphs with MATLAB. It has been discussed that graphs are composed of nodes, edges, and weights. If the edges are oriented, it is an oriented graph, otherwise it is an undirected graph.

There are many methods to represent graphs. The most suitable for computer manipulation is the matrix method. In this section, the concept of the incidence matrix is presented, followed by the methods in MATLAB descriptions.

Definition 8.5. Assuming that in a graph there are n nodes, an $n \times n$ matrix $\boldsymbol{R}$ can be used to represent the graph. Assuming that from node i to node j, there is an edge, and the weight is k, the matrix term can be written as $\boldsymbol{R}(i,j) = k$. Such a matrix is referred to as an incidence matrix.

If from node i to node j, there is no edge, one may set $\boldsymbol{R}(i,j) = 0$. In some algorithms, it is required to set $\boldsymbol{R}(i,j) = \infty$. Details will be demonstrated later.

Sparse matrix representation of an incidence matrix is supported in MATLAB. Assume that a graph is composed of n nodes, and there are m edges. The weight from node a_i to node b_i is w_i, $i = 1, 2, \dots, m$. Therefore, three vectors can be created. With them, the incidence matrix can be established:

```
a=[a1,a2,...,am,n]; b=[b1,b2,...,bm,n]; %start and end vectors
w=[w1,w2,...,wm,0]; %vector of weights
R=sparse(a,b,w);, %incidence matrix
```

Note that in each vector, the last entry is used to ensure that the incidence matrix is a square matrix, which is required by many solution methods. Of course, if the last entry is not provided, a rectangular incidence matrix is created, then the command `R(n,n)=0` can be used to convert it into a square matrix. A sparse matrix can be converted into a regular matrix with command `full()`, while a regular one can be converted with `sparse()` to a sparse matrix.

8.2.5 Finding the shortest path

Some functions are provided in the Bioinformatics Toolbox for handling oriented graphs and solving shortest path problems. The dedicated functions and syntaxes are:

(1) An oriented graph can be expressed uniquely by an incidence matrix. If a graphical display is needed, the dedicated function `biograph()` can be used to set up the graph object, and the overload function `view()` can be used to display the graph:

```
P=biograph(R,nodes,'ShowWeights','on'), %create P object
view(P), %display the graph P
```

where `nodes` are the strings of the nodes. They can be set to empty matrices for default display format.

(2) Call function `graphshortestpath()` to solve the shortest path problem directly. The syntax of the function is

```
[d,p]=graphshortestpath(R,n1,n2), %solve the problem
```

where $\boldsymbol{R}$ is the incidence matrix. It can be a regular or sparse matrix. The syntax of the function will be demonstrated later.

Other arguments are also allowed in function `biograph()`. For the oriented graph in Figure 8.3, $\boldsymbol{R}(i,j)$ represents the weight of an edge from node i to node j. When the object P is created, `graphshortestpath()` function can be used to solve the shortest path problem from node n_1 to n_2 directly. The returned argument d is the shortest distance, while $\boldsymbol{p}$ is a vector containing the nodes on the optimal path.

(3) Other functions can be called to decorate the displayed oriented graph. For instance, the color of nodes and edges can be set. These functions will be demonstrated further through examples.

Example 8.6. Solve again the problem in Example 8.2 with the facilities provided in the Bioinformatics Toolbox.

Solutions. It can be seen that from Figure 8.3 that all the start nodes, end notes, and weights of all the edges are summarized manually as shown in Table 8.1. With the following statements, the incidence matrix can be expressed in a sparse matrix.

Table 8.1: Information of the edges.

start	end	weights	start	end	weights	start	end	weights
1	2	1	1	3	2	2	5	12
2	4	6	3	4	3	3	6	4
4	5	7	4	7	15	4	8	7
4	6	2	5	7	7	6	8	7
6	9	15	7	9	3	8	9	10

Be careful, when constructing the incidence matrix $\boldsymbol{R}$, and make sure it is a square matrix:

```
>> ab=[1 1 2 2 3 3 4 4 4 4 5 6 6 7 8];
   bb=[2 3 5 4 4 6 5 7 8 6 7 8 9 9 9];
   w=[1 2 12 6 3 4 4 15 7 2 7 7 15 3 10];
   R=sparse(ab,bb,w); R(9,9)=0; full(R) %display incidence matrix
```

The following incidence matrix can be constructed in this way:

$$\boldsymbol{R}=\begin{bmatrix} 0 & 1 & 2 & 0 & 0 & 0 & 0 & 0 & 0\\ 0 & 0 & 0 & 6 & 12 & 0 & 0 & 0 & 0\\ 0 & 0 & 0 & 3 & 0 & 4 & 0 & 0 & 0\\ 0 & 0 & 0 & 0 & 4 & 2 & 15 & 7 & 0\\ 0 & 0 & 0 & 0 & 0 & 0 & 7 & 0 & 0\\ 0 & 0 & 0 & 0 & 0 & 0 & 0 & 7 & 15\\ 0 & 0 & 0 & 0 & 0 & 0 & 0 & 0 & 3\\ 0 & 0 & 0 & 0 & 0 & 0 & 0 & 0 & 10\\ 0 & 0 & 0 & 0 & 0 & 0 & 0 & 0 & 0 \end{bmatrix}.$$

With the incidence matrix $\boldsymbol{R}$, function `biograph()` in the Bioinformatics Toolbox can be used to set up the oriented graph object. Function `view()` can then be used to display the oriented graph, as shown in Figure 8.6(a).

```
>> h=view(biograph(R,[],'ShowWeights','on')) %assign the graph to h
```

With the oriented graph object $\boldsymbol{R}$, function `graphshortestpath()` can be called to solve the shortest path problem directly, and have it displayed. Meanwhile, the nodes and edges on the optimal path are set to red. The result obtained is displayed in Figure 8.6(b). It can be seen that the result obtained is the same as obtained manually.

```
>> [d,p]=graphshortestpath(R,1,9)   %shortest path from ① to ⑨
   set(h.Nodes(p),'Color',[1 0 0]) %coloring the nodes in red
```

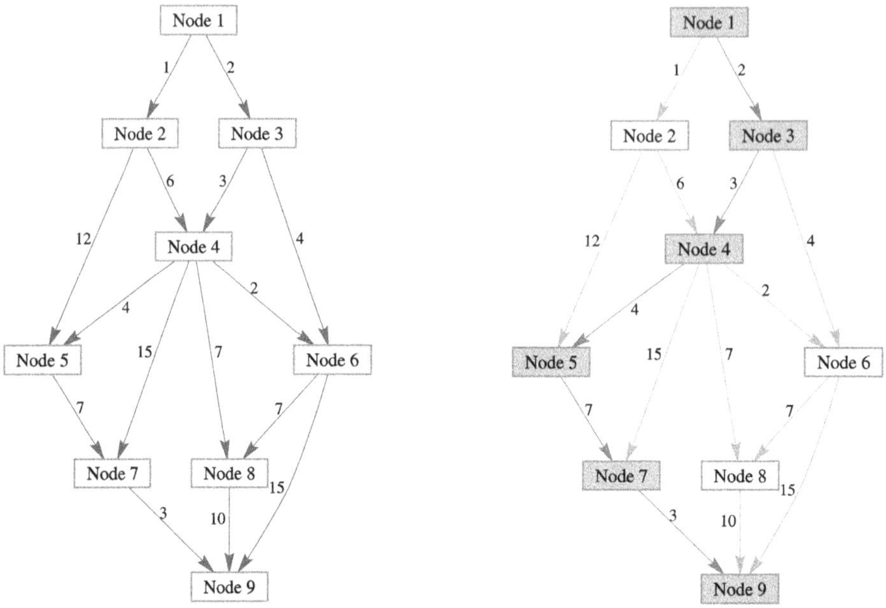

(a) automatic sketch of the graph **(b)** shortest path

Figure 8.6: Display of an oriented graph.

```
edges=getedgesbynodeid(h,get(h.Nodes(p),'ID')); % get edge handles
set(edges,'LineColor',[1 0 0])  % coloring the best path in red
```

Example 8.7. Assume that a factory is importing machines from a manufacturer abroad. Three export harbors can be selected by the manufacturer, and three import harbors can also be selected. Then, one of the two cities can be selected. Finally, the machines will be sent to the factory. The routes and transportation costs are displayed in Figure 8.7. Find a route which has the smallest transportation cost.

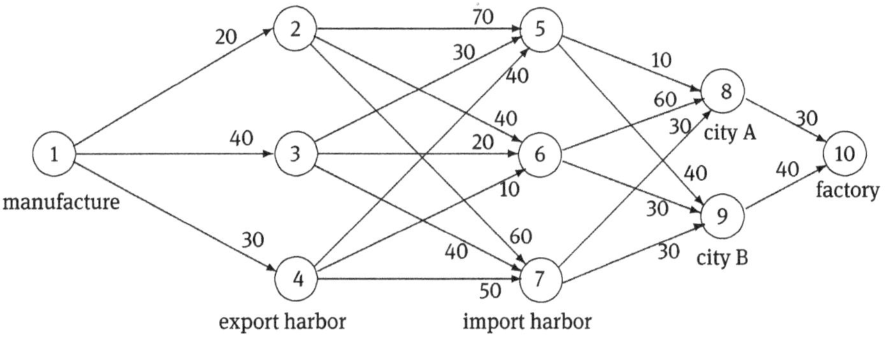

Figure 8.7: Transportation routes and cost.

Solutions. This problem is also a shortest path problem in an oriented graph. The above method can be used to create a table to describe the edges. Of course, if the table is not to be created, one may also set up the start nodes, end nodes, and weights vectors. Note that, in order to avoid errors in inputting the information, the user may start writing the vectors from node ①. All the edges from ① can be described. After that, the edges from node ② can be listed. All the information of the edges can be described in this way. The oriented graph is established as shown in Figure 8.8(a). Note that the automatically generated graph has some weights overlapped. Manual adjustment of the weights is made:

```
>> aa=[1 1 1 2 2 2 3 3 3 4 4 4 5 5 6 6 7 7 8 9];
   bb=[2 3 4 5 6 7 5 6 7 5 6 7 8 9 8 9 8 9 10 10];
   w=[20 40 30 70 40 60 30 20 40 40 10 50 10 40 60 30 30 30 30 40];
   R=sparse(aa,bb,w); R(10,10)=0; full(R) %square incidence matrix
   h=view(biograph(R,[],'ShowWeights','on'))
```

The incidence matrix can be obtained as shown below

$$R=\begin{bmatrix}0&20&40&30&0&0&0&0&0&0\\0&0&0&0&70&40&60&0&0&0\\0&0&0&0&30&20&40&0&0&0\\0&0&0&0&40&10&50&0&0&0\\0&0&0&0&0&0&0&10&40&0\\0&0&0&0&0&0&0&60&30&0\\0&0&0&0&0&0&0&30&30&0\\0&0&0&0&0&0&0&0&0&30\\0&0&0&0&0&0&0&0&0&40\\0&0&0&0&0&0&0&0&0&0\end{bmatrix}.$$

When the oriented graph is described by an incidence matrix, the problem can be solved directly, and the colored optimal path can be obtained as shown in Figure 8.8(b). It can be seen from the optimal path that the manufacture sends the machine to export harbor 4, and sends it to harbor 6. Then, via city B, the machine can be transported to the factory. The lowest cost is $d = 110$.

```
>> [d,p]=graphshortestpath(R,1,10) %optimal path computation
   set(h.Nodes(p),'Color',[1 0 0]) %set nodes on the path to red
   edges=getedgesbynodeid(h,get(h.Nodes(p),'ID')); %get the handles
   set(edges,'LineColor',[1 0 0])  %set the path to red color
```

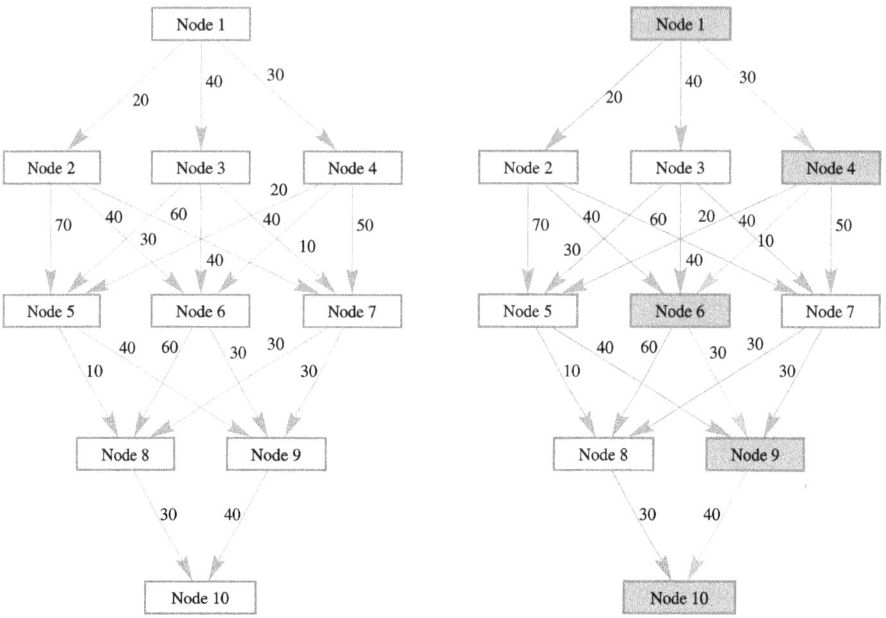

(a) automatic sketch **(b)** shortest path

Figure 8.8: Solution of a shortest path problem.

8.2.6 Dijkstra algorithm implementation

The shortest path between two nodes can also be found by Dijkstra algorithm.[20] In fact, if the start node is specified, the shortest path to all the other nodes can be found at the same time, without affecting the speed of the search method. Among the searching algorithms, Dijkstra's is one of the most effective. Assuming that the number of nodes is n, and the start node is s, the following procedures can be used:

(1) Initialization. Three vectors can be used to store the status of the nodes, where `visited` is used to indicate whether the node is updated, with initial value of 0; `dist` stores the shortest distance from the start node, with initial value of ∞; `parent` stores the parent node number, with default of 0. At the start node, `dist`(s)=0.

(2) Solution loop. Execute the loop $n-1$ times using counting variable i. The three vectors can be updated, where the unvisited node is checked with all its neighbors. The loop is continued until all the unvisited nodes are handled.

(3) Extract the shortest path to the end node t. With the `parent` vector, the optimal path can eventually be extracted.

Based on the Dijkstra algorithm, the following MATLAB function can be written. The function is different from the function `graphshortestpath()`, since attempts are made to find the shortest path from the start node to all other nodes at the same time.

```
function [d,path]=dijkstra(W,s,t)
[n,m]=size(W); ix=(W==0); W(ix)=Inf;   % set the weights to infinity
if n~=m, m0=max(n,m); W(m0,m0)=0; end  % set square matrix
visited(1:n)=0; dist(1:n)=Inf; dist(s)=0; d=Inf; w=[1:n]';
for i=1:(n-1)        % find the relationship of the node with the starting one
   ix=(visited==0); vec(1:n)=Inf; vec(ix)=dist(ix);
   [a,u]=min(vec); visited(u)=1;
   for v=1:n
      if (W(u,v)+dist(u)<dist(v))
         dist(v)=dist(u)+W(u,v); parent(v,i)=u;
end; end; end
u=parent(:,1)==s; p0(u,1)=s; p0(u,2)=w(u); w(u)=0;
for k=1:n, vec=parent(k,:); vec=vec(vec~=0);
   if length(vec)>0, vec=vec(end);
        if w(vec)==0
           v1=p0(vec,:); v1=v1(v1~=0); aa=[v1, k]; w(k)=0;
           for j=1:length(aa), p0(k,j)=aa(j); end
end, end, end
p0=p0(t,:); d=dist(t);
```

The syntax of the function is [***d***,***p***]=dijkstra(***W***,*s*,***t***), where ***W*** is the incidence matrix, while *s* and ***t*** are the start and end nodes, respectively. The end node can be expressed as a vector, such that the optimal paths to all the end nodes are obtained. The returned argument ***d*** contains the distances of the optimal paths to all the end nodes, while each row in ***p*** lists the nodes from the start node. Note that in this function, the 0 weights in the ***W*** matrix are set to ∞ automatically, such that Dijkstra algorithm can be executed normally.

Example 8.8. Solve again the problem in Example 8.2 with Dijkstra algorithm. Find the optimal paths and distances from node ① to all other nodes.

Solutions. The following commands can be used to find the optimal path from node ① to node ⑨. It can be seen that the result is exactly the same as obtained above:

```
>> ab=[1 1 2 2 3 3 4 4 4 4 5 6 6 7 8];
   bb=[2 3 5 4 4 6 5 7 8 6 7 8 9 9 9];
   w=[1 2 12 6 3 4 4 15 7 2 7 7 15 3 10];
   R=sparse(ab,bb,w); R(9,9)=0;           % incidence matrix
   W=ones(9); [d,p]=dijkstra(R.*W,1,9)  % find the optimal path
```

With the following statements, the optimal path information from node ① to all the other nodes is obtained, and the end node is expressed as a vector:

```
>> [d,p]=dijkstra(R.*W,1,2:9) % find simultaneously all the paths
   Tab=[1*ones(8,1), (2:9)', p, d']
```

The above statements find the optimal distances from node ① to each other node in matrix form. Also the optimal path information can better be explained in the form in Table 8.2:

$$d=\begin{bmatrix}1\\2\\5\\9\\6\\16\\12\\19\end{bmatrix},\quad p=\begin{bmatrix}1&2&0&0&0&0\\1&3&0&0&0&0\\1&3&4&0&0&0\\1&3&4&5&0&0\\1&3&6&0&0&0\\1&3&4&5&7&0\\1&3&4&8&0&0\\1&3&4&5&7&9\end{bmatrix}.$$

Table 8.2: The optimal path information.

start node	end node	optimum path p_1	p_2	p_3	p_4	optimal distance	start node	end node	optimum path p_1	p_2	p_3	p_4	p_5	p_6	optimal distance
1	2	1	2			1	1	3	1	3					2
1	4	1	3	4		5	1	5	1	3	4	5			9
1	6	1	3	6		6	1	7	1	3	4	5	7		16
1	8	1	3	4	8	12	1	9	1	3	4	5	7	9	19

8.3 Optimal paths for undigraphs

In the previously given oriented graphs, if there exists an return edge between nodes A and B, the oriented graph can be converted into an undigraph. In this section, a matrix representation of undigraphs is presented, then optimal path problems are discussed and solved.

8.3.1 Matrix description

In real applications, for instance, when considering city road maps, the graphs are usually undigraphs, since a road between nodes A and B is typically bidirectional. The manipulation of undigraphs is simple. If there are no circled roads, that is, where the start and end nodes are the same, an incidence matrix $\boldsymbol{R}$ can be created. Then, the incidence matrix $\boldsymbol{R}_1$ of the undigraph can be constructed as $\boldsymbol{R}_1=\boldsymbol{R}+\boldsymbol{R}^\mathrm{T}$.

Definition 8.6. If in a graph, the incidence matrix $\boldsymbol{R}$ is symmetric, the graph is referred to as symmetric, otherwise it is asymmetric.

In real applications, the undigraph is usually asymmetric. For instance, in a one-way road in the city, the edge is oriented. If a symmetric matrix $\boldsymbol{R}_1$ is given, one may manually modify the elements in it. For instance, if the road from i to j is one-way, one can manually set $\boldsymbol{R}_1(j,i) = 0$. In other examples, if the road from i to j is an uphill one, it is natural that $\boldsymbol{R}_1(j,i)$ and $\boldsymbol{R}_1(i,j)$ are different. Manual manipulation of the incidence matrix can be made. Later such problems will be demonstrated with examples.

Example 8.9. Convert the incidence matrix of Example 8.2 into an undigraph one.

Solutions. The following commands can be used to input the oriented graph into MATLAB environment and construct the incidence matrix $\boldsymbol{R}$. If the diagonal elements are all zeros, command $\boldsymbol{R} + \boldsymbol{R}^{\mathrm{T}}$ can be used to construct the incidence matrix:

```
>> ab=[1 1 2 2 3 3 4 4 4 4 5 6 6 7 8];
   bb=[2 3 5 4 4 6 5 7 8 6 7 8 9 9 9];
   w=[1 2 12 6 3 4 4 15 7 2 7 7 15 3 10];
   R=sparse(ab,bb,w); R(9,9)=0; C=full(R+R')
```

The incidence matrix of the undigraph can be obtained as shown

$$\boldsymbol{C} = \begin{bmatrix} 0 & 1 & 2 & 0 & 0 & 0 & 0 & 0 & 0 \\ 1 & 0 & 0 & 6 & 12 & 0 & 0 & 0 & 0 \\ 2 & 0 & 0 & 3 & 0 & 4 & 0 & 0 & 0 \\ 0 & 6 & 3 & 0 & 4 & 2 & 15 & 7 & 0 \\ 0 & 12 & 0 & 4 & 0 & 0 & 7 & 0 & 0 \\ 0 & 0 & 4 & 2 & 0 & 0 & 0 & 7 & 15 \\ 0 & 0 & 0 & 15 & 7 & 0 & 0 & 0 & 3 \\ 0 & 0 & 0 & 7 & 0 & 7 & 0 & 0 & 10 \\ 0 & 0 & 0 & 0 & 0 & 15 & 3 & 10 & 0 \end{bmatrix}.$$

Normally, for an undigraph, if the weights from nodes i to j and from j to i are different, for instance, in city traffic where one-way or uphill roads are involved, manual modifications to matrix $\boldsymbol{R}_1$ should be made. For instance, the matrix editing interface `openvar` is provided in MATLAB. The following command can be used to open the editor, as shown in Figure 8.9. The user may edit the elements in the editor in a visual way.

```
>> openvar C
```

C

9x9 double

	1	2	3	4	5	6	7	8	9
1	0	1	2	0	0	0	0	0	0
2	1	0	0	6	12	0	0	0	0
3	2	0	0	3	0	4	0	0	0
4	0	6	3	0	4	2	15	7	0
5	0	12	0	4	0	0	7	0	0
6	0	0	4	2	0	0	0	7	15
7	0	0	0	15	7	0	0	0	3
8	0	0	0	7	0	7	0	0	10
9	0	0	0	0	0	15	3	10	0

Figure 8.9: MATLAB matrix editor interface.

8.3.2 Route planning for cities with absolute coordinates

If the nodes are allocated at the absolute coordinates (x_i, y_i) and the link between the cities are given, also the weights can be calculated using the Euclidian distance between the cities. The optimal path problem can then be solved. Next, an example is given to show the solution methods.

Example 8.10. Assume that there are 11 cities, located at the coordinates $(4, 49)$, $(9, 30)$, $(21, 56)$, $(26, 26)$, $(47, 19)$, $(57, 38)$, $(62, 11)$, $(70, 30)$, $(76, 59)$, $(76, 4)$, and $(96, 4)$. The roads between them are shown in Figure 8.10. Find the shortest path from city A to city B. If the roads between cities 6 and 8 are not usable, search for the shortest path again.

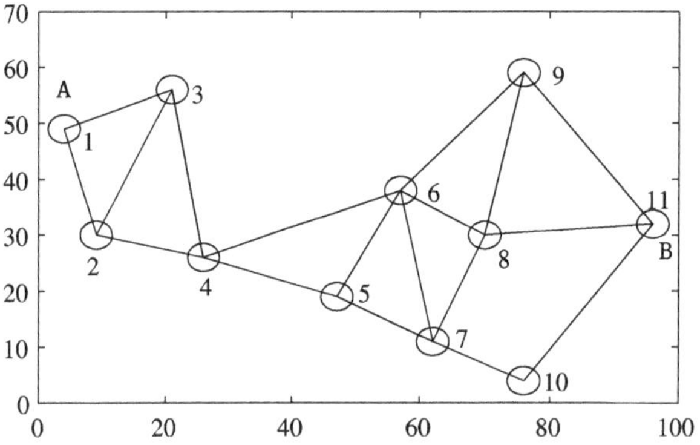

Figure 8.10: City locations and route map.

Solutions. Input the oriented graph in a sparse incidence matrix, and set the weights to 1. The incidence matrix of an undigraph can be generated. The actual Euclidian distance between the cities can be computed from

$$d_{ij} = \sqrt{(x_i - x_j)^2 + (y_i - y_j)^2}.$$

Therefore, with the following statements, by dot multiplying the weighting matrix by the weights, the optimum solution can be found as $1 \to 2 \to 4 \to 6 \to 8 \to 11$; the shortest distance is 111.6938.

```
>> x=[4,9,21,26,47,57,62,70,76,76,96];
   y=[49,30,56,26,19,38,11,30,59,4,32]; n=11;
   for i=1:n, for j=1:n
     D(i,j)=sqrt((x(i)-x(j))^2+(y(i)-y(j))^2); %Euclidian distance
   end, end
   n1=[1 1 2 2 3 4 4 5 5 6 6 6 7 8 7 10 8 9];
   n2=[2 3 3 4 4 5 6 6 7 7 8 9 8 9 10 11 11 11]; %incidence matrix
   R=sparse(n1,n2,1); R(11,11)=0; R=R+R';
   [d,p]=dijkstra(R.*D,1,11) %solve problem
```

If the roads between nodes 6 and 8 are broken, one may manually set $\boldsymbol{R}(8,6) = \boldsymbol{R}(6,8) = \infty$. Now the following commands can be used to solve the optimal path problem. The best route is $1 \to 2 \to 4 \to 5 \to 7 \to 8 \to 11$, with the shortest path having length 122.9394.

```
>> R(6,8)=Inf; R(8,6)=Inf;
   [d,p]=dijkstra(R.*D,1,11) %optimization with broken roads
```

8.4 Exercises

8.1 Solve the following simple linear programming problem as a dynamic programming problem, and use graphical and numerical methods to validate the results:

$$\max_{\boldsymbol{x} \text{ s.t. } \begin{cases} 8x_1+12x_2 \leqslant 30 \\ x_1, x_2 \geqslant 0 \end{cases}} 14x_1 + 4x_2.$$

8.2 Find the shortest path in the graphs from node A to node B, shown in Figure 8.11, (a) and (b).

8.3 If one wants to travel from city A to city B, he/she will pass three islands. On island 1, there are three sights C, D, and E; on island 2, there are three sights, F, G, and H; while on island 3, there are two sights, I and J. Assume that one may stay in each island once to visit one of the sights. Then, he/she may travel to city B. If the

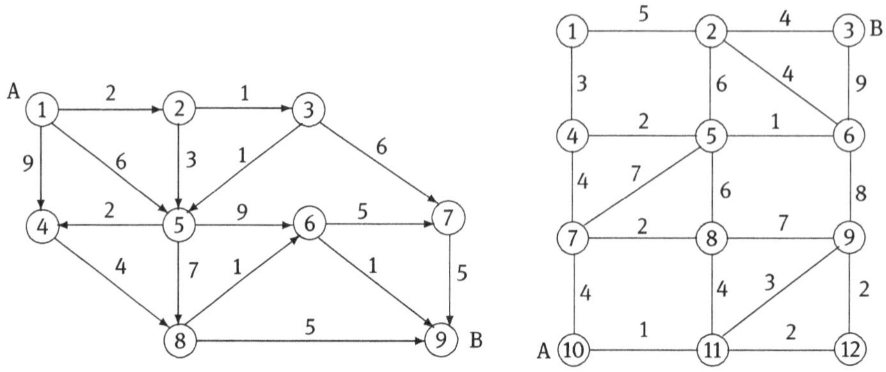

(a) oriented graph problem

(b) undigraph problem

Figure 8.11: Shortest path problem.

traveling costs between different locations are shown in Table 8.3, find a traveling route such that the total cost is minimized.[31]

Table 8.3: The traveling costs between any two nodes.

start	end	cost	start	end	cost	start	end	cost	start	end	cost
A	C	550	A	D	900	A	E	770	C	F	680
C	G	790	C	H	1 050	D	F	580	D	G	760
D	H	660	E	F	510	E	G	700	E	H	830
F	I	610	F	J	790	G	I	540	G	J	940
H	I	790	H	J	270	I	B	1 030	J	B	1 390

8.4 A person lives in city C_1. He needs to travel to other cities $C_2, \dots, C_8$. The traveling cost is provided in matrix $\boldsymbol{R}$, where $R_{i,j}$ represent the cost from city C_i to city C_j. Design the cheapest traveling map for him to go from city C_1 to other cities if

$$\boldsymbol{R} = \begin{bmatrix} 0 & 364 & 314 & 334 & 330 & \infty & 253 & 287 \\ 364 & 0 & 396 & 366 & 351 & 267 & 454 & 581 \\ 314 & 396 & 0 & 232 & 332 & 247 & 159 & 250 \\ 334 & 300 & 232 & 0 & 470 & 50 & 57 & \infty \\ 330 & 351 & 332 & 470 & 0 & 252 & 273 & 156 \\ \infty & 267 & 247 & 50 & 252 & 0 & \infty & 198 \\ 253 & 454 & 159 & 57 & 273 & \infty & 0 & 48 \\ 260 & 581 & 220 & \infty & 156 & 198 & 48 & 0 \end{bmatrix}.$$

8.5 Solve the shortest path problem in Figure 8.12.

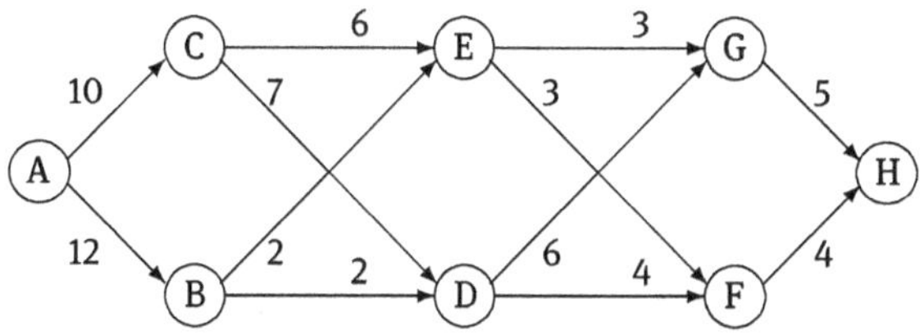

Figure 8.12: Shortest path problem of an oriented graph.

9 Introduction to intelligent optimization methods

In the previous chapters, various programming problems and solutions were presented. The major method used was to select an initial point, from which a search method could be used to find the solutions. These search methods are referred to as conventional optimization methods.

In real applications, the conventional methods are complained about, since for nonconvex problems, different initial points may lead to different solutions, such that local minimum solutions are usually found, not the global ones. To solve such problems, various algorithms are introduced. Particularly attractive are intelligent algorithms, such as genetic algorithms, particle swarm optimization methods, ant colony algorithms, and so on, where genetic and particle swarm optimization algorithms are the most widely used intelligent algorithms. In Section 9.1, some of the commonly used intelligent optimization methods are briefly introduced. As in many other previous chapters, only simple concepts are presented, and technical details are not covered, since the focus of this series is to show how tools are used.

In Section 9.2, Global Optimization Toolbox in MATLAB is illustrated, and some of the direct solvers, including genetic algorithms, particle swarm optimization methods, pattern search algorithm and simulated annealing algorithms, are briefly explained. The use and characteristics of the solvers are also presented.

In fact, in Chapter 5, an attempt has been made to find global optimum solutions of ordinary nonlinear programming problems. Some other global solvers are also written for other problems. They are also compared in Section 9.3 with intelligent optimization methods through various examples to see which methods are more suitable for finding global optimum solutions.

In Chinese, there is an old saying "Try to show yourself a horse, not a mule". Through fair comparisons of genuine problem solutions, the users may have their own views on intelligent optimization methods.

9.1 Intelligent optimization algorithms

The so-called intelligent optimization algorithms are those inspired by natural phenomena and human intelligence. Their representatives are genetic algorithms, particle swarm optimization algorithms, ant colony algorithms, immunization methods, and so on. In this section, the genetic algorithms and particle swarm optimization are briefly introduced.

9.1.1 Genetic algorithms

Genetic algorithms (GAs) are a class of nature-inspired evolutionary computing methods following the law of "survival of the fittest".[43] The first genetic algorithm

https://doi.org/10.1515/9783110667011-009

method was invented in 1975 by Professor John Holland (1929–2015) at Michigan University.

The main idea of the method is to search from a population of randomly distributed individuals. They are encoded in a certain way, regarded as genes with chromosomes. The population evolves generation by generation through processes such as reproduction, crossover, and mutation, until the individuals with the best values of a fitness function are found.

The procedures of a simple genetic algorithm are as follows:

(1) Create an initial population $\boldsymbol{P}_0$ with N individuals. Evaluate the fitness function for each individual. The initial population $\boldsymbol{P}_0$ is normally created randomly.
(2) Set the generation number to $i = 1$, meaning the first generation.
(3) Compute the values of selective functions, that is, select some individuals using probabilistic rules from the current population.
(4) With the processes of reproduction, crossover, and mutation, form the population of the next generation $\boldsymbol{P}_{i+1}$.
(5) Set $i = i + 1$. If the stop conditions are not satisfied, go back to step (3) to continue the evolution process.

In the genetic algorithms' research field, the objective functions in conventional optimization problems are referred to as fitness functions. Traditional simple genetic algorithms can only be used in solving unconstrained problems. Through over 40 years of evolution itself, some of the advanced genetic algorithms may be able to solve constrained optimization problems.

Compared with conventional optimization methods, the genetic algorithms have the following major differences:[13]

(1) In searching for the optimum points, genetic algorithms allow searching from many initial points simultaneously in a parallel manner. Therefore, it is expected that GAs are more likely to find global optimum points than the traditional methods, which initiate searches from a single point.
(2) The finite decision variable bounds must be provided, and one cannot solve problems with infinite bounds like those in the Optimization Toolbox.
(3) Genetic algorithms do not depend upon the gradient information of the objective functions. Only the fitness functions, i. e., objective functions, are needed in the search process.
(4) Genetic algorithms evaluate and select the fitness function according to a probabilistic rule rather than a deterministic one. Therefore, there may exist slight differences in each run.

The populations and individuals in the genetic algorithms used in the book are encoded by real numbers, which may be different from those in traditional genetic algorithms where binary encoding is used. For the users it is not necessary to consider the encoding and decoding processes. The decision variables can be regarded as normal variables in other optimization solvers. For integer programming problems, the internal encoding is a binary encoding method.

9.1.2 Particle swarm optimization methods

Particle swarm optimization (PSO) is another category of nature-inspired evolutionary computing optimization algorithms. Particle swarm optimization was initially proposed in [29], which is motivated by the phenomenon when birds are seeking food. It supposes to be a useful algorithm for finding global solutions when solving optimization problems. Assume that within a certain region, there is a piece of food which can be considered as the global optimum point, and there is a flock of randomly distributed birds (or particles). Each bird has its personal best value $p_{i,b}$, and the flock has its best value g_b so far. The position and speed of each bird can be updated with the following formulas:

$$\begin{cases} v_i(k+1) = \phi(k)v_i(k) + \alpha_1\gamma_{1i}(k)[p_{i,\mathrm{b}} - x_i(k)] + \alpha_2\gamma_{2i}(k)[g_{\mathrm{b}} - x_i(k)], \\ x_i(k+1) = x_i(k) + v_i(k+1), \end{cases} \tag{9.1.1}$$

where γ_{1i}, γ_{2i} are the random numbers uniformly distributed in the interval $[0, 1]$. The $\phi(k)$ variable is the momentum function, while the variables α_1 and α_2 are acceleration constants.

It seems that in certain considerations, the particle swarm algorithm is similar to a genetic algorithm. The particles are constructed in the same format as the individuals in a GA. The information of the particles is updated, which is similar to the evolution by generations, although the internal formulas are different. After several round of updating, the global optimum solutions may be obtained.

9.2 MATLAB Global Optimization Toolbox

In earlier versions of MATLAB, there was a toolbox called “Genetic Algorithm and Direct Search Toolbox”, which later evolved into the Global Optimization Toolbox. Apart from that, there are many free toolboxes written in MATLAB, including the influential ones such as the Genetic Algorithm Optimization Toolbox, developed by Christopher Houck, Jeffery Joines, and Michael Kay from North Carolina State University and the Genetic Algorithm Toolbox, developed by Peter Fleming and Andrew Chipperfield at the University of Sheffield.

It can be seen from the existing Global Optimization Toolbox and other related toolboxes that most of the existing solver are only capable of solving unconstrained optimization problems. Very few claim to be able to solve constrained nonlinear programming problems. Apart from genetic algorithms, where binary encoding may be adopted and they can be used in integer programming problems, most of the other intelligent algorithms may find it difficult to handle mixed-integer programming problems, unless specially designed.

In the current Global Optimization Toolbox in MATLAB, four solvers are provided. They are the genetic algorithm-based solver `ga()`, particle swarm optimization solver `particleswarm()`, pattern search solver `patternsearch()`, and simulated annealing solver `simulannealbnd()`. These four solvers can all be used to solve unconstrained optimization problems with decision variable bounds. Besides, the genetic algorithm and pattern search solvers claim that they are able to handle constrained nonlinear programming problems. Genetic algorithm solver can also be used in handling mixed-integer programming problems.

(1) Genetic algorithm solver

The syntaxes of `ga()` are listed below:

$[\boldsymbol{x}, f_0, \texttt{flag}, \texttt{out}] = \texttt{ga}(f, n)$

$[\boldsymbol{x}, f_0, \texttt{flag}, \texttt{out}] = \texttt{ga}(f, n, \boldsymbol{A}, \boldsymbol{B}, \boldsymbol{A}_{\rm eq}, \boldsymbol{B}_{\rm eq}, \boldsymbol{x}_{\rm m}, \boldsymbol{x}_{\rm M})$

$[\boldsymbol{x}, f_0, \texttt{flag}, \texttt{out}] = \texttt{ga}(f, n, \boldsymbol{A}, \boldsymbol{B}, \boldsymbol{A}_{\rm eq}, \boldsymbol{B}_{\rm eq}, \boldsymbol{x}_{\rm m}, \boldsymbol{x}_{\rm M}, \texttt{nfun})$

$[\boldsymbol{x}, f_0, \texttt{flag}, \texttt{out}] = \texttt{ga}(f, n, \boldsymbol{A}, \boldsymbol{B}, \boldsymbol{A}_{\rm eq}, \boldsymbol{B}_{\rm eq}, \boldsymbol{x}_{\rm m}, \boldsymbol{x}_{\rm M}, \texttt{nfun}, \texttt{intcon})$

$[\boldsymbol{x}, f_0, \texttt{flag}, \texttt{out}] = \texttt{ga}(\texttt{problem})$

It can be seen that the syntaxes of the function are rather close to those of function `fmincon()`. The argument `intcon` can be used in describing mixed-integer programs. The difference is that the number of decision variables, n, must be provided by the user. The initial value $\boldsymbol{x}_0$ need not be provided. If the structured variable is used, the members used are listed in Table 9.1.

Table 9.1: `ga()` function members.

member names	member explanations
`fitnessfcn`	the fitness function should be described by a function handle. For optimization problems, it is used to describe the objective function
`nvars`	number of decision variables, n
`options`	control parameters which can be set initially by `optimset()` or `optimoptions()`
`solver`	should be set to `'ga'`. The above four members must be provided
`Aineq` and others	the members include `bineq`, `Aeq`, `beq`, `lb` and `ub`. Besides, `nonlcon` and `intcon` can also be provided, and they are the same as those in other solvers

(2) Pattern search method

The syntaxes for the pattern search function `patternsearch()` are:

$[x, f_0, \texttt{flag}, \texttt{out}]=\texttt{patternsearch}(f, x_0)$

$[x, f_0, \texttt{flag}, \texttt{out}]=\texttt{patternsearch}(f, x_0, A, B, A_{eq}, B_{eq}, x_m, x_M)$

$[x, f_0, \texttt{flag}, \texttt{out}]=\texttt{patternsearch}(f, x_0, A, B, A_{eq}, B_{eq}, x_m, x_M, \texttt{nfun})$

$[x, f_0, \texttt{flag}, \texttt{out}]=\texttt{patternsearch}(\texttt{problem})$

Compared with `ga()` function, `patternsearch()` does not support the `intcon` member, that is, the mixed-integer programming is not supported. Besides, an initial reference vector x_0 should be provided, and the number of decision variables, n, need not be used. If a structured variable `problem` is used, the member `fitnessfcn` cannot be used, and the commonly used `objective` member should be used to describe the objective functions.

(3) Particle swarm optimization method

The syntaxes of particle swarm optimization solver `particleswarm()` are:

$[x, f_0, \texttt{flag}, \texttt{out}]=\texttt{particleswarm}(f, n)$

$[x, f_0, \texttt{flag}, \texttt{out}]=\texttt{particleswarm}(f, n, x_m, x_M)$

$[x, f_0, \texttt{flag}, \texttt{out}]=\texttt{particleswarm}(f, n, x_m, x_M, \texttt{options})$

$[x, f_0, \texttt{flag}, \texttt{out}]=\texttt{particleswarm}(\texttt{problem})$

It can be seen that `particleswarm()` can only be used to solve unconstrained optimization problems, with decision variable bounds. If a structured variable is used, the members `objective`, `nvars`, `lb`, and `ub` should be provided.

(4) Simulated annealing algorithm

The syntaxes of simulated annealing solver `simulannealbnd()` are:

$[x, f_0, \texttt{flag}, \texttt{out}]=\texttt{simulannealbnd}(f, x_0)$

$[x, f_0, \texttt{flag}, \texttt{out}]=\texttt{simulannealbnd}(f, x_0, x_m, x_M)$

$[x, f_0, \texttt{flag}, \texttt{out}]=\texttt{simulannealbnd}(f, x_0, x_m, x_M, \texttt{options})$

$[x, f_0, \texttt{flag}, \texttt{out}]=\texttt{simulannealbnd}(\texttt{problem})$

It can be seen that the syntaxes are similar to those of `particleswarm()`. The difference is that n is not needed, and the initial vector x_0 should be provided instead. This function can only be used in solving unconstrained optimization problems with decision variable bounds.

9.3 Examples and comparative studies of intelligent optimization methods

As mentioned previously, all four solvers in the Global Optimization Toolbox can be used in solving unconstrained optimization problems with decision variable bounds.

Besides, some solvers claim that they are able to solve constrained or even mixed-integer programming problems. In this section, several examples from the previous sections to chapters are used for comparative studies with the intelligent solvers, and the global optimum solvers provided in the book. The benefits and disadvantages of the algorithms are assessed with examples.

9.3.1 Unconstrained optimization problems

All the solvers in the Global Optimization Toolbox can be used to solve unconstrained optimization problems. Here, two test examples are used to compare the advantages and disadvantages of the global optimization solvers. This may provide some useful information when selecting solvers for practical problems.

Example 9.1. Consider the modified Rastrigin function in Example 3.22. Compare the four solvers in Global Optimization Toolbox, and also the `fminunc_global()` solver in Chapter 3, developed by the author. Each solver is executed 100 times. See whether the global optimum solutions are obtained. Measure the elapsed time and draw useful conclusions.

In order to have the materials in this chapter independent, the original problems are all presented again. For this example,

$$f(x_1,x_2) = 20 + \left(\frac{x_1}{30}-1\right)^2 + \left(\frac{x_2}{20}-1\right)^2 - 10\left[\cos 2\pi\left(\frac{x_1}{30}-1\right) + \cos 2\pi\left(\frac{x_2}{20}-1\right)\right],$$

where $-100 \leqslant x_1, x_2 \leqslant 100$.

Solutions. For the same problem, each solver is executed 100 times. The results can be monitored and the success rate for finding the global optimum solutions are recorded as shown in Table 9.2. The elapsed time indicates the time needed to solve the problem 100 times with one solver. In the execution for one solver, the user must comment off the other solvers:

```
>> f=@(x)20+(x(1)/30-1)^2+(x(2)/20-1)^2-... %objective function
      10*(cos(2*pi*(x(1)/30-1))+cos(2*pi*(x(2)/20-1)));
   A=[]; B=[]; Aeq=[]; Beq=[]; xm=-100*ones(2,1); xM=-xm;
   F1=[]; F2=[]; F3=[]; F4=[]; tic
   for i=1:100, x0=100*rand(2,1); %run each solver 100 times
      [x,f0]=ga(f,2,A,B,Aeq,Beq,xm,xM); F1=[F1; x,f0];
      [x,f0]=patternsearch(f,x0,A,B,Aeq,Beq,xm,xM); F2=[F2; x',f0];
      [x,f0]=particleswarm(f,2,xm,xM); F3=[F3; x,f0];
      [x,f0]=simulannealbnd(f,x0,xm,xM); F4=[F4; x',f0];
   end, toc
```

Table 9.2: Comparisons of different algorithms.

comparisons	`ga()`	`pattern-search()`	`particle-swarm()`	`simulan-nealbnd()`	`fminunc_global()`
success rate	20 %	7 %	90 %	7 %	83 %
elapsed time (seconds)	6.11	1.97	1.53	23.42	57.07

```
i1=length(find(F1(:,3)<1e-5)), i2=length(find(F2(:,3)<1e-5))
i3=length(find(F3(:,3)<1e-5)), i4=length(find(F4(:,3)<1e-2))
```

It can be seen that the best performance out of the five solvers showed the particle swarm optimization solver. The `fminunc_global()` function developed by the author ranked the second, the success rate was slightly lower than that of the PSO solver. The elapsed time is much larger, but it is usually acceptable, since the average of each run is only 0.57 seconds. The success rate of the genetic algorithm is rather low, normally, if it is run 5 to 6 times, the global optimum solution may be found. It may be considered as a useful global optimum solver.

The pattern search and simulated annealing solvers are initial value dependent, and the success rate is extremely low. The interval [0,100] for random numbers was deliberately chosen, and, unfortunately, the success rate for the global optimum solutions is rather low. If the interval is selected as $[0,1]$, the success rate is almost zero. It can be seen from the last test condition that for the simulated annealing solver the error bound is set to a large one of `1e-2`. If it is set to `1e-5` as in other solvers, the success rate is zero, indicating that the accuracy of the algorithm is very low. The two algorithms are not suitable for handling the problem studied here.

Example 9.2. The Griewangk benchmark problem is tested here for $n = 50$:

$$\min_{x}\left(1+\sum_{i=1}^{n}\frac{x_i^2}{4\,000}-\prod_{i=1}^{n}\cos\frac{x_i}{\sqrt{i}}\right),\quad \text{where } x_i\in[-600,600].$$

Compare the benefits of the intelligent optimization solvers.

Solutions. It is obvious that the global minimum solution is $x_i = 0$. Now we test each of the intelligent solvers and `fminunc_global()` function, and the results are obtained in Table 9.3. Since the scale of the problem is higher, the accuracy of using intelligent algorithms is usually very low. Therefore, in comparing the success rates, different error bounds are used.

```
>> n=50; % set the scale to 50 decision variables
   f=@(x)1+sum(x.^2/4000)-prod(cos(x(:)./[1:n]'));
   A=[]; B=[]; Aeq=[]; Beq=[]; xm=-600*ones(n,1); xM=-xm;
   F1=[]; F2=[]; F3=[]; F4=[]; F5=[]; tic
```

```
for i=1:100, i, x0=600*rand(n,1); % run the solvers 100 times
   [x,f0]=ga(f,n,A,B,Aeq,Beq,xm,xM); F1=[F1; f0];
   [x,f0]=patternsearch(f,x0,A,B,Aeq,Beq,xm,xM); F2=[F2; f0];
   [x,f0]=particleswarm(f,n,xm,xM); F3=[F3; f0];
   [x,f0]=simulannealbnd(f,x0,xm,xM); F4=[F4; f0];
   [x,f0]=fminunc_global(f,-600,600,n,20); F5=[F5; f0];
end, toc
min(F4), max(F5)
```

Table 9.3: Comparisons of success rates of the solvers.

error levels	ga()	pattern-search()	particle-swarm()	simulan-nealbnd()	fminunc_global()
$\epsilon = 10^{-4}$	44 %	3 %	17 %	0 %	100 %
$\epsilon = 10^{-3}$	97 %	3 %	17 %	0 %	100 %
$\epsilon = 10^{-2}$	98 %	6 %	29 %	0 %	100 %
elapsed time (seconds)	251.16	295.03	42.11	2 196.71	81.79

It is obvious that the reliable one is the solver `fminunc_global()` developed by the author. The success rate is 100 %. The maximum value of the objective function during the 100 runs is 8.5778×10^{-12}, about 10^{10} times smaller than for the rest of the solvers. The elapsed time is smaller than for most of the compared algorithms. The genetic algorithm is ranked second among all algorithms tested. When the error bound is selected as 10^{-2}, i. e., 0.01, the success rate may reach 98 %. Unfortunately, the precision of the method is much too low. The particle swarm optimization solver ranked third, with the success rate of only 29 %. The accuracies are much lower than of `fminunc_global()`. The pattern search and simulated annealing solvers are almost unusable in solving this problem, with the success rate below 10 %, or even 0 %. In the simulated annealing method, when 100 executions are made, the minimum objective function value is 9.8217, much greater than the theoretical 0.

Even though $n = 500$, with `fminunc_global()` solver we reached the objective function value of $f_0 = 4.3997 \times 10^{-11}$, and the elapsed time was 11.20 seconds in each run. These results cannot, of course, be achieved with the other intelligent solvers.

```
>> n=500; f=@(x)1+sum(x.^2/4000)-prod(cos(x(:)./[1:n]'));
   tic, [x,f0]=fminunc_global(f,-600,600,n,20), toc
```

Of course, this selected example is not fair, since the theoretical optimum value involved was zero. The solver `fminunc_global()` is a random initial point-based one, and it has natural superiority for this particular problem. The user may consider using a conversion method to shift the global minimum solution away from zero, and compare again. The shift can be made using a similar method as in constructing the modified Rastrigin function. This is left as an exercise.

The examples used in this chapter are typical, and can be solved well with other methods. They are not specially chosen to show the defects of intelligent optimization algorithms. Under the default setting of the parameters, the intelligent solvers provided in the Global Optimization Toolbox are not satisfactory in finding the global optimum solutions. Even after some of the options or control parameters are tuned, the global optimum solutions are unlikely to be found.

Since all the code in the solution process is reproducible, the users may find their own test problems to evaluate the behaviors of the intelligent solvers, and see what the possible problems are, and how to improve the algorithms. Of course, the analytical solutions of some problems are known, the results can be compared with global minimum solutions. In practice, the solutions are not known. The users have to validate the results by themselves. The validation process is simple, since the objective function values can be compared directly. For instance, if two solutions are found by different methods, that with a larger objective function value is, of course, a local optimum solution. Only those with smaller objective function values are candidates for global minimum solutions.

For constrained optimization problems, one should test whether all the constraints are satisfied. If they are, those with larger objective function values must be local ones. Those with smaller objective function values are candidates to be global ones, unless they are beaten by other results.

9.3.2 Constrained optimization problems

Two functions, `ga()` and `patternsearch()`, are provided in Global Optimization Toolbox in MATLAB, claiming that they are able to solve constrained nonlinear programming problems. In this section, several test problems are tried and the behaviors of the solvers are assessed. The conclusions may be useful to the readers.

Example 9.3. Solve the nonlinear programming problem in Example 5.11 with genetic algorithm and pattern search solvers, and compare their behaviors. The objective function is

$$\begin{aligned} f(\boldsymbol{x}) = {} & 0.7854x_1x_2^2(3.3333x_3^2 + 14.9334x_3 - 43.0934) - 1.508x_1(x_6^2 + x_7^2) \\ & + 7.477(x_6^3 + x_7^3) + 0.7854(x_4x_6^2 + x_5x_7^2), \end{aligned}$$

and the corresponding constraints are

$$\begin{aligned} & x_1x_2^2x_3 \geqslant 27, \quad x_1x_2^2x_3^2 \geqslant 397.5, \quad x_2x_3x_6^4/x_4^3 \geqslant 1.93, \quad x_2x_3x_7^4/x_5^3 \geqslant 1.93, \\ & A_1/B_1 \leqslant 1100, \quad \text{where } A_1 = \sqrt{(745x_4/(x_2x_3))^2 + 16.91 \times 10^6}, \quad B_1 = 0.1x_6^3, \end{aligned}$$

$$A_2/B_2 \leqslant 850, \quad \text{where } A_2 = \sqrt{(745x_5/(x_2x_3))^2 + 157.5 \times 10^6}, \quad B_2 = 0.1x_7^3,$$
$$x_2x_3 \leqslant 40, \quad 5 \leqslant x_1/x_2 \leqslant 12, \quad 1.5x_6 + 1.9 \leqslant x_4, \quad 1.1x_7 + 1.9 \leqslant x_5,$$
$$2.6 \leqslant x_1 \leqslant 3.6, \quad 0.7 \leqslant x_2 \leqslant 0.8, \quad 17 \leqslant x_3 \leqslant 28,$$
$$7.3 \leqslant x_4, x_5 \leqslant 8.3, \quad 2.9 \leqslant x_6 \leqslant 3.9, \quad 5 \leqslant x_7 \leqslant 5.5.$$

Solutions. The original problem can be described by the following statements, and the genetic algorithm solver can be used to solve the problem. The solutions obtained with the genetic algorithm are listed in Table 9.4, and the last row marked * is obtained with the solver `fmincon()`. The elapsed times are much higher than that needed in `fmincon()`, which is 0.036 seconds. The solutions by the genetic algorithm are quite similar for this example, and the results are closer to that obtained with the solver `fmincon()`.

```
>> f=@(x)0.7854*x(1)*x(2)^2*(3.3333*x(3)^2+14.9334*x(3)-43.0934)...
      -1.508*x(1)*(x(6)^2+x(7)^2)+7.477*(x(6)^3+x(7)^3)...
      +0.7854*(x(4)*x(6)^2+x(5)*x(7)^2);
   A=[]; B=[]; Aeq=[]; Beq=[]; xx=[];
   xm=[2.6,0.7,17,7.3,7.3,2.9,5]; xM=[3.6,0.8,28,8.3,8.3,3.9,5.5];
   for i=1:7
      tic, [x,f0,flag]=ga(f,7,A,B,Aeq,Beq,xm,xM,@c5mcpl), toc
      if flag==1, xx=[xx; x  f0]; end
   end
```

Table 9.4: The first 7 solutions of `ga()` function.

group	x_1	x_2	x_3	x_4	x_5	x_6	x_7	$f(x)$	elapsed time (seconds)
1	3.4993	0.7	17	7.3	7.7144	3.3506	5.2867	2 994.2	5.63
2	3.4993	0.7	17	7.3	7.7144	3.3506	5.2867	2 994.1	5.86
3	3.4994	0.70002	17	7.3	7.7144	3.3506	5.2867	2 994.3	5.32
4	3.4994	0.70001	17	7.3001	7.7144	3.3506	5.2867	2 994.2	5.82
5	3.4993	0.7	17	7.3	7.7144	3.3506	5.2867	2 994.2	7.30
6	3.4993	0.7	17	7.3	7.7144	3.3506	5.2867	2 994.2	5.52
7	3.4993	0.70001	17	7.3	7.7144	3.3506	5.2867	2 994.2	6.43
*	3.5	0.7	17	7.3	7.7153	3.3505	5.2867	2 994.4	0.036

It can be seen that the objective function values obtained are all slightly smaller than that obtained in Example 5.11. Substituting the solution back into the constraints function `c5mcpl()`, it is found that the 15 inequality constraints are $[-2.1492, -98.036, -1.9254, -0.058158, -18.3179, -0.02211, -28.1, 0.001, -7.001, -0.3741, 0.00097]$, two of them are larger than 0, violating the constraints, therefore, although the value of the objective function is smaller, the results are invalid.

For this particular example, the solution by the ordinary solver `fmincon()` is more accurate and more efficient. In fact, several runs of `fmincon()` solver are tested, and they all yield the same result, with no exceptions. Probably the original problem is a convex one.

The following commands can also be written to solve the problem with the pattern search solver. Ten runs are made and the results are given in Table 9.5. It can be seen that the success rate for finding the global optimum solution is about 60 %. Although the elapsed time of the algorithm is shorter than for the genetic algorithm solver, it is much longer than for the `fmincon()` solver.

Table 9.5: Ten solutions obtained with the solver `patternsearch()`.

group	x_1	x_2	x_3	x_4	x_5	x_6	x_7	$f(x)$	elapsed time (seconds)
1	3.5	0.7	18.426	7.3	7.7151	3.3495	5.2865	3249.6	0.67
2	3.5	0.7	17	7.3	7.7153	3.3505	5.2867	2994.4	0.15
3	3.5942	0.7188	26.011	7.3	7.7145	3.3465	5.2859	5199.3	0.46
4	3.5	0.7	17	7.3	7.7153	3.3505	5.2867	2994.4	0.12
5	3.5	0.7	17	7.3	7.7153	3.3505	5.2867	2994.4	0.11
6	3.5	0.7	17	7.3	7.7153	3.3505	5.2867	2994.4	0.61
7	3.5006	0.7001	17	7.3	7.7153	3.3505	5.2867	2995.2	1.57
8	3.5	0.7	17	7.3	7.7153	3.3505	5.2867	2994.4	0.74
9	3.5	0.7	17.503	7.3	7.7152	3.3502	5.2866	3082.3	0.26
10	3.5	0.7	17	7.3	7.7153	3.3505	5.2867	2994.4	0.35
*	3.5	0.7	17	7.3	7.7153	3.3505	5.2867	2994.4	0.036

```
>> xx=[];
   for i=1:10, tic, x0=10*rand(7,1);
      [x,f0,k]=patternsearch(f,x0,A,B,Aeq,Beq,xm,xM,@c5mcpl);
      if k==1, xx=[xx; x' f0]; end, toc
   end
```

Example 9.4. Solve again the constrained nonlinear programming problem in Example 5.10 with the intelligent solvers:

$$\min_{x \text{ s.t. } \begin{cases} x_3+9.625x_1x_4+16x_2x_4+16x_4^2+12-4x_1-x_2-78x_4=0 \\ 16x_1x_4+44-19x_1-8x_2-x_3-24x_4=0 \\ -0.25x_5-x_1\leqslant-2.25 \\ x_1-0.25x_5\leqslant2.25 \\ -0.5x_5-x_2\leqslant-1.5 \\ x_2-0.5x_5\leqslant1.5 \\ -1.5x_5-x_3\leqslant-1.5 \\ x_3-1.5x_5\leqslant1.5 \end{cases}} x_5.$$

Solutions. When the genetic algorithm solver ga() is called 10 times, different results are obtained, and the results are listed in Table 9.6. Unfortunately, none of the solutions are close to the global optimum solutions. The minimum objective function is at least more than 10 times larger than the best known solution, indicating that ga() function failed in solving such a problem. While in Example 5.10, function fmincon_global() was called 100 times and each time the global solution was found. Therefore, the success rate is 100 %.

Table 9.6: Run ga() solver 10 times.

group	x_1	x_2	x_3	x_4	x_5	$f(x)$
1	−0.5584	3.1305	4.8629	0.7501	13.7460	13.7460
2	−0.2379	2.9400	5.4545	0.7029	10.9320	10.9320
3	4.1276	1.8729	−3.5772	1.0901	8.5531	8.5531
4	3.4040	2.1638	−4.1593	1.1104	10.5300	10.5300
5	1.2786	3.3040	−6.1398	−0.1653	8.6973	8.6973
6	4.5058	1.3569	1.6581	1.1254	10.5950	10.5950
7	0.8231	3.3592	−1.8408	0.3073	9.4258	9.4258
8	2.6299	2.7754	−8.5554	1.0850	9.2296	9.2296
9	2.2530	2.6968	−6.4534	1.1560	9.7220	9.7220
10	2.3595	2.4852	−4.3134	1.1925	4.1094	4.1094
*	2.4544	1.9088	2.7263	1.3510	0.8175	0.8175

```
>> clear P; P.fitnessfcn=@(x)x(5);
   P.nonlcon=@c5mnls; P.solver='ga'; P.nvars=5;
   P.Aineq=[-1,0,0,0,-0.25; 1,0,0,0,-0.25; 0,-1,0,0,-0.5;
            0,1,0,0,-0.5; 0,0,-1,0,-1.5; 0,0,1,0,-1.5];
   P.Bineq=[-2.25; 2.25; -1.5; 1.5; -1.5; 1.5];
   P.options=optimoptions('ga'); xx=[];
   for i=1:10        % run 10 times the ga() solver
      [x,f0,flag]=ga(P), if flag==1, xx=[xx; x]; end
   end
```

The patternsearch() function can also be used in solving the same problem. It is executed 10 times, and the results are consistent. They are all $\boldsymbol{x} = [1.2075, 0, 2.3396, 4, 4.1698]^{\mathrm{T}}$, and the value of the objective function is the same $f_0 = 4.1698$. For this example, the pattern search method also failed in finding the global optimum solution.

```
>> P.Objective=@(x)x(5); P.solver='patternsearch';
   P.options=optimset; xx=[];
   for i=1:10        % run 10 times patternsearch() function
```

```
    P.x0=10*rand(5,1); [x,f0,flag]=patternsearch(P)
    if flag==1, xx=[xx; x' f0]; end
end
```

Even though fmincon_global() is not used, the plain solver fmincon() in MATLAB is called 100 times. When a random initial search point is selected, the global success rate is 39 %, and the maximum objective function value is only 1.1448, at least not ridiculously large as those obtained by the genetic algorithm and pattern search solvers.

```
>> P.Objective=@(x)x(5); P.solver='fmincon'; xx=[];
   for i=1:100        % run fmincon() 100 times and record the result
      P.x0=-10+20*rand(5,1); [x,f0,flag]=fmincon(P)
      if flag==1, xx=[xx; x' f0]; end
   end
```

Example 9.5. Use Global Optimization Toolbox to solve the nonconvex quadratic programming problem in Example 5.13:

$$\min \quad \boldsymbol{c}^{\mathrm{T}}\boldsymbol{x}+\boldsymbol{d}^{\mathrm{T}}\boldsymbol{y}-\frac{1}{2}\boldsymbol{x}^{\mathrm{T}}\boldsymbol{Q}\boldsymbol{x},$$

$$\boldsymbol{x}\ \text{s.t.}\begin{cases}2x_1+2x_2+y_6+y_7\leqslant 10\\2x_1+2x_3+y_6+y_8\leqslant 10\\2x_2+2x_3+y_7+y_8\leqslant 10\\-8x_1+y_6\leqslant 0\\-8x_2+y_7\leqslant 0\\-8x_3+y_8\leqslant 0\\-2x_4-y_1+y_6\leqslant 0\\-2y_2-y_3+y_7\leqslant 0\\-2y_4-y_5+y_8\leqslant 0\\0\leqslant x_i\leqslant 1,\ i=1,2,3,4\\0\leqslant y_i\leqslant 1,\ i=1,2,3,4,5,9\\y_i\geqslant 0,\ i=6,7,8\end{cases}$$

where $\boldsymbol{c}=[5,5,5,5]$, $\boldsymbol{d}=[-1,-1,-1,-1,-1,-1,-1,-1,-1]$, and $\boldsymbol{Q}=10\boldsymbol{I}$.

Solutions. With the method in Example 5.13, the problem-based method can be used to describe the nonconvex quadratic programming problem. The problem can then be converted into a structured variable p. Modifying the members for the solver ga(), the function ga() can be called to directly solve the problem:

```
>> P=optimproblem; c=5*ones(1,4); d=-1*ones(1,9); Q=10*eye(4);
   x=optimvar('x',4,1,'Lower',0,'Upper',1);
   y=optimvar('y',9,1,'Lower',0); % set two decision vectors
   cons1=[2*x(1)+2*x(2)+y(6)+y(7)<=10; 2*x(1)+2*x(3)+y(6)+y(8)<=10;
          2*x(2)+2*x(3)+y(7)+y(8)<=10;
          -8*x(1)+y(6)<=0; -8*x(2)+y(7)<=0; -8*x(3)+y(8)<=0;
```

```
        -2*x(4)-y(1)+y(6)<=0; -2*y(2)-y(3)+y(7)<=0;
        -2*y(4)-y(5)+y(8)<=0; y([1 2 3 4 5 9])<=1];
   P.Constraints.cons1=cons1; p=prob2struct(P);   % to structured var
   f=@(x)c*x(1:4)'+d*x(5:13)'-0.5*x(1:4)*Q*x(1:4)';   % objective function
   p.solver='ga'; ff=optimset; ff.TolX=eps; ff.TolFun=eps;
   p.options=ff; p.fitnessfcn=f; p.nvars=13;
   tic, [x0,f0,flag]=ga(p), toc
```

Note that, in the execution of function `ga()`, the decision variable $\boldsymbol{x}$ is provided as a row vector, which is not the same as for most of the optimization solvers. Therefore, one must be very careful in describing fitness functions, otherwise, an error message "dimension incompatible in matrix multiplications" will be displayed.

After 58.3 seconds of waiting, the optimum solution found is

$$\boldsymbol{x}_0 = [0.9991, 1, 1, 0.9999, 1.001, 1.001, 1.001, 1.001, 1.001, 2.9992, 3.001, 3, 0.998],$$

with the objective function value of −14.9992, slightly larger than −15 obtained in Example 5.13. Therefore, the `fmincon_global()` solver is slightly better in this example than the genetic algorithm solver, and the elapsed time is much less.

Now let us consider using the pattern search method to solve the problem again. Note that we need to redefine the objective function, since the decision vector in the solver `patternsearch()` should be specified in the column vector, not the row vector used earlier. The following commands can be used. Within 0.178 seconds, the result can be found. The elapsed time is clearly less than that of the `fmincon_global()` solver.

```
>> f=@(x)c*x(1:4)+d*x(5:13)-0.5*x(1:4)'*Q*x(1:4);   % objective function
   p.Objective=f; p.solver='patternsearch'; p.x0=100*rand(13,1);
   tic, [x0,f0]=patternsearch(p); toc
   norm(x0-round(x0)), e=max(p.Aineq*x0-p.bineq)
```

With the above statements, the global optimum solution can be found directly as $\boldsymbol{x} = [1, 1, 1, 1, 1, 1.001, 1.001, 1.0004, 1.0002, 2.9997, 2.9992, 3.0008, 1.001]$, the objective function value is −15.0033, slightly smaller than that obtained in Example 5.13, which was −15. If the maximum value of the constraints is evaluated, it is found that $e = 9.9972 \times 10^{-4} > 0$, which violates the original constraints. Therefore, the result obtained is not valid, but it is rather close to the global optimum solution. The result obtained in Example 5.13 is the global optimum solution of the original problem.

Example 9.6. A nature-inspired intelligent algorithm was provided in [11]. With the algorithm, a test example was used to validate it, as shown in Example 5.16. In fact, it has been pointed out in Example 5.16 that the results by the `fmincon_global()` solver are significantly better than all the results listed in the reference. Now compare the genetic and pattern search algorithms and see whether better solutions can be found. The objective function is

$$f(x) = l(x_1x_2 + x_3x_4 + x_5x_6 + x_7x_8 + x_9x_{10}),$$

and the constraints are

$$\frac{6Pl}{x_9x_{10}^2} - \sigma_{\max} \leqslant 0, \quad \frac{6P(2l)}{x_7x_8^2} - \sigma_{\max} \leqslant 0,$$

$$\frac{6P(3l)}{x_5x_6^2} - \sigma_{\max} \leqslant 0, \quad \frac{6P(4l)}{x_3x_4^2} - \sigma_{\max} \leqslant 0, \quad \frac{6P(5l)}{x_1x_2^2} - \sigma_{\max} \leqslant 0,$$

$$\frac{Pl^3}{E}\left(\frac{244}{x_1x_2^3} + \frac{148}{x_3x_4^3} + \frac{76}{x_5x_6^3} + \frac{28}{x_7x_8^3} + \frac{4}{x_9x_{10}^3}\right) - \delta_{\max} \leqslant 0,$$

$$\frac{x_2}{x_1} - 20 \leqslant 0, \quad \frac{x_4}{x_3} - 20 \leqslant 0, \quad \frac{x_6}{x_5} - 20 \leqslant 0, \quad \frac{x_8}{x_7} - 20 \leqslant 0, \quad \frac{x_{10}}{x_9} - 20 \leqslant 0,$$

with the upper and lower decision variable bounds:

$$1 \leqslant x_{1,7,9} \leqslant 5, \quad 30 \leqslant x_{2,8,10} \leqslant 65, \quad 2.4 \leqslant x_{3,5} \leqslant 3.1, \quad 45 \leqslant x_{4,6} \leqslant 60,$$

where $L = 100$, $l = 100$, $P = 50\,000$, $\delta_{\max} = 2.7$, $\sigma_{\max} = 14\,000$, and $E = 2 \times 10^7$.

Solutions. Seven sets of results can be obtained with the genetic algorithm as shown in Table 9.7. Although the results are much better than those provided in [11], the objective function values are significantly larger than that provided in Example 5.16, which may be the global optimum solution. The elapsed time is around 1.35~2.57 seconds, which is about 4~8 times that needed in Example 5.16. The results obtained by the pattern search algorithm are uncertain, sometimes even worse than those provided in [11].

Table 9.7: Seven sets of results with ga() function.

group	x_1	x_2	x_3	x_4	x_5	x_6	x_7	x_8	x_9	x_{10}	$f(x)$
1	2.992	59.84	2.78	55.55	2.528	50.43	2.21	44.03	1.777	34.73	61 987
2	2.994	59.82	2.82	55.18	2.534	50.37	2.22	43.98	1.791	34.59	62 146
3	2.992	59.84	2.83	55.00	2.549	50.22	2.22	43.96	1.752	34.97	62 166
4	2.992	59.84	2.79	55.40	2.532	50.38	2.20	44.09	1.753	34.97	61 985
5	2.992	59.84	2.78	55.54	2.536	50.35	2.21	44.07	1.751	34.99	61 954
6	2.998	59.78	2.79	55.42	2.546	50.25	2.21	44.03	1.753	34.96	62 045
7	2.992	59.84	2.85	54.79	2.548	50.23	2.20	44.09	1.751	34.99	62 190
*	2.992	59.84	2.78	55.55	2.524	50.47	2.20	44.09	1.750	35.00	61 915

```
>> l=100; ff=optimset; ff.TolX=eps; ff.TolFun=eps;
   f=@(x)l*(x(1)*x(2)+x(3)*x(4)+x(5)*x(6)+x(7)*x(8)+x(9)*x(10));
   xm=[1,30,2.4,45,2.4,45,1,30,1,30]'; A=[]; B=[]; Aeq=[]; Beq=[];
   xM=[5,65,3.1,60,3.1,60,5,65,5,65]'; xx=[];
   for i=1:7, tic
      [x,fv,flag]=ga(f,n,A,B,Aeq,Beq,xm,xM,@c5mglo1,ff)
      toc, if flag==1, xx=[xx; x,fv]; end
   end
   x0=100*rand(10,1);
   [x,fv,flag]=patternsearch(f,x0,A,B,Aeq,Beq,xm,xM,@c5mglo1,ff)
```

In fact, it is not difficult to find that, compared with the global solver in Chapter 5, most of the results obtained with the intelligent solvers from Optimization Toolbox are not as good. The reputation of the intelligent solvers is usually towards the negative side. For nonlinear constrained programming problems, it is recommended to use the global solver `fmincon_global()`. At least one should not expect too much when using intelligent optimization solvers.

In real applications, the eventual target is to find the solution satisfying all the constraints and such that the objective function is minimized. The solution process includes reasonable application of relevant tools. Practical optimum problem solution is not pure academic work. We should not use the algorithm only because it is new or looks nice mathematically. The actual goal is to find the solution with the best quality, that is, find a meaningful global optimum solution.

9.3.3 Mixed-integer programming

As it was presented earlier, only function `ga()` supports the use of the member `intcon`. It can be used in solving mixed-integer programming problems. The solver `patternsearch()` cannot be used for such problems. In other words, the only function in the Global Optimization Toolbox capable of solving mixed-integer programming is the `ga()` solver. Practical examples are used here to demonstrate the effectiveness of the solver, compared with other traditional solvers.

Example 9.7. Use the genetic algorithm solver to find solutions of 0–1 knapsack problem in Example 6.30.

Solutions. The following commands can be used in solving the original knapsack problems directly. Since integer encoding may be automatically used in the genetic algorithm solver, the global optimum solution to the original problem can be found directly as $\boldsymbol{x}_0 = [1, 0, 1, 0, 1, 1, 1, 0, 0, 0]$, and the objective function value is -324, the same as that obtained in Example 6.30.

```
>> w=[19,15,4.6,14.6,10,4.4,1.2,11.4,17.8,19.6]; W=40;
   p=[89,59,19,43,100,72,44,16,7,64]; n=10;
   P=optimproblem('ObjectiveSense','max');
   x=optimvar('x',n,1,'Type','integer','Lower',0,'Upper',1);
   P.Objective=p*x; P.Constraints.c=w*x<=W;
   p=prob2struct(P); p.nvars=n; p.solver='ga';
   f=[p.f]'; p.options=optimset;
   p.fitnessfcn=@(x)sum(f(:).*x(:));
   tic, [x0,f0,flag]=ga(p), toc
```

Example 9.8. Use the genetic algorithm to solve the mixed 0–1 programming problems studied in Example 6.25, namely

$$\min\quad 5x_4+6x_5+8x_6+10x_1-7x_3-18\ln(x_2+1)-19.2\ln(x_1-x_2+1)+10.$$
$$\boldsymbol{x}\ \text{s.t.}\begin{cases}-0.8\ln(x_2+1)-0.96\ln(x_1-x_2+1)+0.8x_3\leqslant 0\\ -\ln(x_2+1)-1.2\ln(x_1-x_2+1)+x_3+2x_6-2\leqslant 0\\ x_2-x_1\leqslant 0\\ x_2-2x_4\leqslant 0\\ x_1-x_2-2x_5\leqslant 0\\ x_4+x_5\leqslant 1\\ 0\leqslant \boldsymbol{x}\leqslant [2,2,1,1,1,1]^{\mathrm{T}}\\ x_4,x_5,x_6\in\{0,1\}\end{cases}$$

Solutions. The method in Example 6.25 can still be used in describing a mixed 0–1 integer programming problem. Then, the solver `ga()` can be directly used to solve it. The results by the genetic algorithm are given in Table 9.8. It can be seen that for this example, although the genetic algorithm may find a feasible solution each time, the global optimum solutions cannot be obtained.

Table 9.8: Five runs of the `ga()` solver.

group	x_1	x_2	x_3	y_1	y_2	y_3	objective function	elapsed time (seconds)
1	1.524	1.523	0.92792	1	0	0	7.0675	0.24
2	1.2036	1.2026	0.79207	1	0	0	7.2587	0.30
3	1.2958	1.2948	0.83309	1	0	0	7.1555	0.43
4	1.5044	1.5034	0.9198	1	0	1	15.069	0.28
5	1.9993	1.9983	1	1	0	0	8.2089	0.26
*	1.3010	0	1	0	1	0	6.0980	0.21

```
>> clear P; P.intcon=4:6; xx=[]; P.nonlcon=@c6mmibp;
   f=@(x)5*x(4)+6*x(5)+8*x(6)+10*x(1)-7*x(3) ...
         -18*log(x(2)+1)-19.2*log(x(1)-x(2)+1)+10; %object function
   P.ub=[2 2 1 1 1 1]'; P.lb=[0 0 0 0 0 0]'; P.Bineq=[0;0;0;1];
   P.Aineq=[-1 1 0 0 0 0; 0 1 0 -2 0 0; 1 -1 0 0 -2 0;
             0 0 0 1 1 0];
```

```
P.solver='ga'; P.nvars=6; P.options=optimset; P.fitnessfcn=f;
for i=1:5            % solution with genetic algorithm
   tic, [x,fm,flag]=ga(P), if flag==1, xx=[xx; x fm]; end; toc
end
```

Example 9.9. Solve the integer and mixed-integer linear programming problem in Example 6.8 with the genetic algorithm, namely

$$\min_{x \text{ s.t. } \begin{cases} 2x_2+x_3+4x_4+2x_5\leqslant 54 \\ 3x_1+4x_2+5x_3-x_4-x_5\leqslant 62 \\ x_1,x_2\geqslant 0,x_3\geqslant 3.32,x_4\geqslant 0.678,x_5\geqslant 2.57 \end{cases}} -2x_1 - x_2 - 4x_3 - 3x_4 - x_5.$$

Solutions. The following commands can be used to solve the integer linear programming problem directly, and the first 10 trials are shown in Table 9.9. It can be seen that each time the solution is different, which can be a coincidence. Once (in group 7) the global optimum solution is found. In this example, 10 trials are made, the global optimum solution is found only once.

```
>> clear P; P.solver='ga'; P.options=optimset;
   P.lb=[0; 0; 3.32; 0.678; 2.57]; P.nvars=5; xx=[];
   P.fitnessfcn=@(x)[-2 -1 -4 -3 -1]*x'; P.intcon=1:5;
   P.Aineq=[0 2 1 4 2; 3 4 5 -1 -1]; P.Bineq=[54; 62];
   for i=1:10
      [x0,f0,flag]=ga(P); if flag==1, xx=[xx; x0 f0]; end
   end
```

Table 9.9: Ten sets of solutions with solver ga().

group	x_1	x_2	x_3	x_4	x_5	fx	group	x_1	x_2	x_3	x_4	x_5	fx
1	6	0	11	9	3	-86	2	18	0	6	2	20	-86
3	12	0	8	8	7	-87	4	15	0	6	10	4	-88
5	5	0	12	8	5	-87	6	18	0	4	11	3	-88
7	19	0	4	10	5	-89	8	22	0	4	1	23	-86
9	10	0	10	4	14	-86	10	8	0	10	9	3	-86
*	19	0	4	10	5	-89							

If the solver is called once to solve the mixed-integer programming problem, the solution obtained is $\boldsymbol{x}_0 = [12, 0.000165, 8, 9, 5]^{\mathrm{T}}$, with the objective function value of -88.0003. It is immediately found that the search for the global optimum solution is not successful, since the solution is far away from the true one. If the code is run again, maybe another local minimum is found. Many times the commands were tried, yet none of the calls found the global optimum solution. It should be concluded that under the default setting, the genetic algorithm cannot be used in solving mixed-integer programming problems.

```
>> P.intcon=[1,4,5]; [x0,f0,flag]=ga(P)
```

It can be seen from the examples that since binary encoding can be used in genetic algorithm solvers, they can be used to solve certain integer programming problems. Unfortunately, they usually cannot get the global optimum solutions. Besides, in solving mixed-integer programming problems, the behavior of the genetic algorithm solver is not ideal.

For ordinary mixed-integer programming problems, the solvers in Chapter 6 are recommended. For instance, for mixed-integer linear and nonlinear programming problems, functions `intlinprog()` and `new_bnb20()` are recommended, respectively. For mixed-integer nonlinear programming problems, one may also add an extra outer loop, as that in `fmincon_global()` function, to ensure global minimum solutions.

9.3.4 Discrete programming problems with the genetic algorithm

In Section 6.3.5, methods are introduced to solve ordinary discrete programming problems. It can be seen from the example that, if the numbers of discrete decision variables and their combinations are small enough, loops can be used to solve the problem. If there are too many of them, loops are not suitable. Genetic algorithm solvers can be used to solve more complicated discrete programming problems. An example is given here to demonstrate the application of the genetic algorithm when looking for solutions of discrete programming problems.

Example 9.10. Solve the discrete programming problem in Example 6.18 with the genetic algorithm. The mathematical model is given in Example 9.6, and based on this model, more constraints are posed. Variables x_1 and x_2 are integers, x_3 and x_5 may only take the discrete values in $[2.4, 2.6, 2.8, 3.1]$, while x_4 and x_6 may only select values from $[45, 50, 55, 60]$.

Solutions. Compared with the method in Example 6.18, two discrete vectors can be created:

$$\boldsymbol{v}_1 = [2.4, 2.6, 2.8, 3.1], \quad \boldsymbol{v}_2 = [45, 50, 55, 60].$$

Let $x_3 \sim x_6$ are the subscripts created by vectors $\boldsymbol{v}_1$ and $\boldsymbol{v}_2$. They can be set to integers, with bounds of 1 to 4. The following commands are used to describe the objective function. In order not to destroy the mathematical model, the decision variable $\boldsymbol{x}$ in the mathematical model can be restored. The objective function can then be computed.

```
function f=c9mglo1(x)
l=100; x=x(:).'; v1=[2.4,2.6,2.8,3.1]; v2=[45,50,55,60];
x([3,5])=v1(x([3,5])); x([4,6])=v2(x([4,6]));
f=l*(x(1)*x(2)+x(3)*x(4)+x(5)*x(6)+x(7)*x(8)+x(9)*x(10));
```

Similarly, the same method can be used to describe the constraints

```
function [c,ceq]=c9mglo2(x)
x=x(:).'; v1=[2.4,2.6,2.8,3.1]; v2=[45,50,55,60];
x(3:6)=round(x(3:6)); x([3,5])=v1(x([3,5]));
x([4,6])=v2(x([4,6])); [c,ceq]=c5mglo1(x);
```

With the above defined objective function and constraints, the following commands can be used to call the genetic algorithm solver seven times, and the results are shown in Table 9.10.

```
>> xm=[1,30,1,1,1,1,1,30,1,30]'; A=[]; B=[]; Aeq=[]; Beq=[];
   xM=[5,65,4,4,4,4,5,65,5,65]'; xx=[]; n=10; intcon=1:6;
   v1=[2.4,2.6,2.8,3.1]; v2=[45,50,55,60];
   for i=1:7, tic
      [x,fv,flag]=ga(@c9mglo1,n,A,B,Aeq,Beq,xm,xM,@c9mglo2,intcon)
      if flag==1,
         x=x(:).'; x(3:6)=round(x(3:6));
         x([3,5])=v1(x([3,5])); x([4,6])=v2(x([4,6]));
         xx=[xx; x,fv];
      end, toc
   end
```

It can be seen from Table 9.10 that the discrete decision variables can be handled simply and straightforwardly. The integer and discrete parts obtained are satisfactory.

Table 9.10: Seven solutions obtained with ga() function.

group	x_1	x_2	x_3	x_4	x_5	x_6	x_7	x_8	x_9	x_{10}	$f(x)$
1	3	60	3.1	55	2.6	50	2.2492	43.6527	2.0220	32.5647	64 453.695
2	3	60	3.1	55	2.6	50	2.2860	43.2991	1.7933	34.5682	64 147.942
3	3	60	3.1	55	2.6	50	2.2865	43.2939	1.8062	34.4441	64 170.439
4	3	60	3.1	55	2.6	50	2.2055	44.0819	1.8444	34.0859	64 059.836
5	3	60	3.1	55	2.6	50	2.2045	44.0916	1.8989	33.5928	64 149.194
6	3	60	3.1	55	2.6	50	2.5184	41.2525	2.0492	32.3374	65 066.776
7	3	60	3.1	55	2.6	50	2.7857	39.2236	1.8054	34.4516	65 196.258
*	3	60	3.1	55	2.6	50	2.2046	44.0911	1.7498	34.9951	63 893.436

However, the noninteger solutions are not precisely evaluated, and the objective function values are not satisfactory, significantly larger than that obtained in Example 6.18.

The genetic algorithm is easy to use in describing discrete programming problems, which is more reasonable than the loop structures used in Example 6.18, and it can be easily extended to the cases where loop structures are not applicable. However, it is usually hard to find the global optimum solutions. If the loop structures in the discrete programming problems as in Example 6.18 cannot be used, the genetic algorithm is recommended.

9.4 Exercises

9.1 It is pointed out in Example 9.2 that the Griewangk benchmark problem is not fair when comparing the advantages of the algorithms. A modified version of the problem should be made. For instance, it can be modified as

$$\min_{x}\left(1+\sum_{i=1}^{n}\frac{(x_i/i-1)^2}{4\,000}-\prod_{i=1}^{n}\cos\frac{x_i/i-1}{\sqrt{i}}\right),\quad \text{where } x_i\in[-600,600].$$

Assess the behaviors of the solvers for the same objective function.

9.2 The facilities of Optimization Toolbox functions in MATLAB are powerful enough in solving convex problems, since if a solution is found, the result is the global optimum solution. Normally, there is no need to handle convex problems with intelligent optimization methods. Can you find any example or exercise where the intelligent optimal solvers beat the global nonlinear programming problem solver?

9.3 In the previous several chapters, many examples were used. Try the intelligent methods to solve them and see whether the results may beat any of those from the examples.

9.4 In the constrained optimization problems, if the genetic algorithm solvers are used under the default setting, the behavior of the solver may not be good. Some control options of the solver may be modified. For instance, the size of population, the mutation rate, and so on. The quality of the solver may be improved. However, this is too demanding for average users who are not experts in genetic algorithms. If you are familiar with genetic algorithms, try to modify the control options and see whether you are able to solve the problem in Example 9.4, and see whether you can find global optimum solutions.

9.5 In the previous chapters, we solved many optimization problems. Use intelligent solvers to solve them again, and see whether you are able to find better solutions than those provided in the book. If you are not able to beat any of the solutions, then later, when you are tackling nonlinear or integer programming problems, it is the time to say “NO” to the conventional optimization algorithms.

Bibliography

[1] Ackley D H. A Connectionist Machine for Genetic Hillclimbing. Boston, USA: Kluwer Academic Publishers, 1987.

[2] Avriel M, Williams A C. An extension of geometric programming with applications in engineering optimization. Journal of Engineering Mathematics, 1971, 5: 187–194.

[3] Bard J F. Practical Bilevel Optimization – Algorithms and Applications. Dordrecht: Springer Science + Business Media, 1998.

[4] Bazaraa M S, Sherali H D, Shetty C M. Nonlinear Programming – Theory and Algorithms. 3rd ed. New Jersey: Wiley-interscience, 2006.

[5] Bellman R. Dynamic Programming. Princeton: Princeton University Press, 1957.

[6] Bhatti M A. Practical Optimization Methods with Mathematica Applications. New York: Springer-Verlag, 2000.

[7] Bixby R E, Gregory J W, Lustig, I J, Marsten R E, Shanno D F. Very large-scale linear programming: a case study in combining interior point and simplex methods. Operations Research, 1992, 40: 885–897.

[8] Boyd S, Ghaoui L El, Feron E, Balakrishnan V. Linear Matrix Inequalities in Systems and Control Theory. Philadelphia: SIAM books, 1994.

[9] Cao Y. Pareto set. MATLAB Central File ID: # 15181, 2007.

[10] Cha J Z, Mayne R W. Optimization with discrete variables via recursive quadratic programming: Part 2 – algorithms and results. Transactions of the ASME, Journal of Mechanisms, Transmissions, and Automation in Design, 1994, 111: 130–136.

[11] Chakri A, Ragueb H, Yang X-S. Bat algorithm and directional bat algorithm with case studies. In: Yang X-S, ed., Nature-Inspired Algorithms and Applied Optimization, 189–216. Switzerland: Springer 2018.

[12] Chew S H, Zheng Q. Integral Global Optimization. Lecture Notes In Economics and Mathematical Systems, Vol. 298. Berlin: Springer-Verlag, 1988.

[13] Chipperfield A, Fleming P. Genetic Algorithm Toolbox User'S Guide. Department of Automatic Control and Systems Engineering, University of Sheffield, 1994.

[14] Cohon J L. Multiobjective Programming and Planning. New York: Academic Press Inc, 1978.

[15] Cook W. In Pursuit of the Travelling Salesman – Mathematics at the Limits of Computation. Princeton: Princeton University Press, 2012.

[16] Cottle R, Johnson E, Wets R. George B Dantzig (1914–2005). Notice of the American Mathematical Society, 2007, 54(3): 344–362.

[17] Danø S. Nonlinear and Dynamic Programming – An Introduction. Wein: Springer-Verlag, 1975.

[18] Dantzig G B. Linear Programming and Extensions – A Report Prepared for United States Air Force project Rand. Princeton: Princeton University Press, 1963.

[19] D'Errico J. Fminsearchbnd. MATLAB Central File ID: #8277, 2005.

[20] Dijkstra E W. A note on two problems in connexion with graphs. Numerische Mathematik, 1959, 1: 269–271.

[21] Dolan E D, Moré J J, Munson T S. Benchmarking optimization software with COPS 3.0. Technical Report ANL/MCS-TM-273, Argonne National Laboratory, 2004.

[22] Floudas C A, Pardalos P M. A Collection of Test Problems for Constrained Global Optimization Algorithms. Berlin: Springer-Verlag, 1990.

[23] Floudas C A, Pardalos P M, Adjiman C S, Esposito W R. Handbook of Test Problems in Local and Global Optimization. Dordrecht: Kluwer Scientific Publishers, 1999.

[24] Gao L F. Optimization Theory and Methods. Shenyang: Northeastern University Press, 2005 (in Chinese).

https://doi.org/10.1515/9783110667011-010

[25] Goldberg D E. Genetic Algorithms in Search, Optimization and Machine Learning. Reading, MA: Addison-Wesley, 1989.
[26] Henrion D. Global optimization toolbox for Maple. IEEE Control Systems Magazine, 2006, 26(5): 106–110.
[27] Hillier F S, Lieberman G J. Introduction to Operations Research. 10th ed. New York: McGraw-Hill Education, 2015.
[28] Hock W, Schittkowski K. Test Examples for Nonlinear Programming Code. Berlin: Springer-Verlag, 1990.
[29] Kennedy J, Eberhart R. Particle swarm optimization. Proceedings of IEEE International Conference on Neural Networks. Perth, Australia, 1995, 1942–1948.
[30] Kuipers K. BNB20 solves mixed integer nonlinear optimization problems, MATLAB Central File ID: #95, 2000.
[31] Lew A, Mauch H. Dynamic Programming – A Computational Tool. Berlin: Springer, 2007.
[32] Leyffer S. Deterministic methods for mixed integer nonlinear programming. Ph.D. thesis, Department of Mathematics & Computer Science, University of Dundee, U.K., 1993.
[33] Lin Y X. Dynamic Programming and Sequential Optimization. Zhengzhou: Henan University Press, 1997 (in Chinese).
[34] Löfberg J. YALMIP: a toolbox for modeling and optimization in MATLAB. Proceedings of IEEE International Symposium on Computer Aided Control Systems Design. Taipei, 2004, 284–289.
[35] Löfberg J. YALMIP [R/OL]. https://yalmip.github.io/.
[36] Lu D, Sun X L. Nonlinear Integer Programming. New York: Springer, 2006.
[37] MathWorks. Solve sudoku puzzles via integer programming: problem-based [R/OL]. http://www.mathworks.com/help/optim/examples/solve-sudoku-puzzles-via-integer-programming.html, 2017.
[38] Moler C. Experiment with MATLAB. Beijing: BUAA Press, 2014.
[39] Nelder J A, Mead R. A simplex method for function minimization. Computer Journal, 1965, 7: 308–313.
[40] Pan P-Q. Linear Programming Computation. Berlin: Springer, 2014.
[41] Price C J, Coope I D. Numerical experiments in semi-infinite programming. Computational Optimisation and Applications, 1996, 6(2): 169–189.
[42] Scherer C, Weiland S. Linear Matrix Inequalities in Control. Delft University of Technology, 2005.
[43] Shao J L, Zhang J, Wei C H. Foundations in Artificial Intelligence. Beijing: Electronic Industry Publishers, 2000.
[44] Simone. Pareto filtering. MATLAB Central File ID: # 50477, 2015.
[45] Sturmfels B. Solving Systems of Polynomial Equations. CBMS Conference on Solving Polynomial Equations, Held at Texas A & M University, American Mathematical Society, 2002.
[46] Taha H A. Operations Research – An Introduction. 10th ed. Harlow: Pearson, 2017.
[47] The MathWorks Inc. Robust control toolbox user's manual, 2007.
[48] The MathWorks Inc. Bioinformatics users manual, 2011.
[49] Vanderbei R J. Linear Programming – Foundations and Extensions. 4th ed. New York: Springer, 2014.
[50] Willems J C. Least squares stationary optimal control and the algebraic Riccati equation. IEEE Transactions on Automatic Control, 1971, 16(6): 621–634.
[51] Xu J P, Hu Z N, Wang W. Operations Research. 2nd ed. Beijing: Science Press, 2004.
[52] Xue D Y. Fractional-Order Control Systems – Fundamentals and Numerical Implementations. Berlin: de Gruyter, 2017.
[53] Xue D Y, Chen Y Q. Scientific Computing with MATLAB. 2nd ed. Boca Radon: CRC Press, 2016.

MATLAB function index

Bold page numbers indicate where to find the syntax explanation of the function. The function or model name marked by * are the ones developed by the authors. The items marked with ‡ are those downloadable freely from internet.

abs 42, 44, 46, 48, 49, 61
are 27–29, 31
assignin 42, 48
axis 44, 105

binvar 144, 231
biograph **284**, 285
blkdiag 247
break 21, 22, 42, 48, 91, 166, 241

c3fun3* 75
c3mdej1* 76
c3mdixon* 69, 70, 73, 79
c3mgrad* 76
c3mode* 94
c3mopt* 63, 64
c3myopt* 71, 72
c3myout* 71
c5exmcon* **165**, 166
c5me2o* 200, 201
c5mglo1* **179**, 312
c5mheat* 198, 199
c5mnls* 169
c5mpi* **186**
c5mpool* 195, 196
c5msinf* **190**, 191, **192**, **193**
c5mtest* **180**, 181
c5mtown* 164
c6exinl* 225
c6exnls* 169, 173
c6fun3* 75
c6mbp1* 218
c6mdisp* 226, 227
c6mmibp* 234, 235
c9mglo1* 316
ceil 224
class 81, 172
clear 76, 115, 136, 162, 163, 166, 167, 169, 171, 173, 220
collect 10, 11
conj 30, 42
continue 240
contour 72, 272
contour3 73
conv 12
cos 14, 17, 18, 22–24, 26, 40, 43
cputime 49

dec2mat 143
default_vals* 41, **42**, 48
diag 132, 187, 188
diff 22, 57, 188
dijkstra* **289**, 290, 293
double 47, 49, 145–147, 222

eps 21
error 42, 289
exp 14, 17, 18, 22–24, 26, 76
expm 46, 49
eye 30, 39, 136, 143, 144, 147, 174, 187, 188, 239, 241, 309
ezplot 50

fcontour 80
feasp **142**, 143
fill 105
fimplicit 13, 14, 18, 19, 44, 49, 50, 56, 97, 155
find 11, 44, 46, 49, 139, 155, 156, 165, 181, 210–215, 218, 225, 240, 246, 257, 267, 273, 303
find_subtours* 240, 241
fminbnd **84**
fminbnd_global* **84**, 85, 87
fmincon **160**, 161, 162, 164, 166–169, 171, 172, 175, 218, 259, 309
fmincon_global* **172**, 173–177, 179, 181, 187–189, 199
fminimax **269**, 270, 271, 274
fminsearch **62**, 64–66, 68–70, 75
fminsearchbnd‡ 84, **85**, 86, 87
fminunc 64, 66–70, **72**, 73–81, 90, 94, 97
fminunc_global* **81**, 82, 158, 303, 304
for 42, 48, 69, 74, 81, 82, 120, 128, 139, 172, 173, 181, 186, 217, 218, 239–241, 246, 260–262, 265, 289, 293, 302, 304, 306–309, 312, 314, 316

fplot 15–17, 23, 43, 56–58, 61, 105, 208, 256, 269
fseminf **189**, 191, 193
fsolve **25**, **26**, 27–30, 41, 42, 48, 50
fsurf 64, 79, 86
full 130, 138, 284, 285, 287, 291
funm 46, 49

ga **300**, 302–304, 306, 308, 310–316
get 286, 287
getedgesbynodeid 286, 287
getlmis **142**, 143
gevp **142**
graphshortestpath **284**, 285
griddata 77, 78
griddatan 78

hankel 144
hold 23, 43, 44, 49, 61, 71, 97, 105, 155, 255, 256, 269, 272

if 12, 21, 22, 41, 42, 48, 81, 84, 91, 120, 139, 157, 166, 172, 191, 217, 218, 240, 241, 246, 265, 289, 306–309, 312, 314, 316
imag 30, 41, 42, 48
Inf 81, 84, 114, 223, 239, 289, 293
int2str 48, 71
interp1 77
intlinprog **220**, 221, 230, 231
intvar 144
isnan 191

jacobian 24, 74, 76, 159, 167

length 11, 25, 41, 42, 48, 84, 181, 186, 215, 217, 218, 240, 241, 246, 265, 289, 303
line 15, 17, 57, 105, 241
linprog **110**, **111**, 112–118, 120, 123, 124, 217, 221, 260, 262, 265
linprog_comp* **262**, 263
linspace 190–193, 241
lmiterm **142**, 143
lmivar **142**, 143
load 92, 93
log 218, 234, 235, 313
lsqcurvefit **89**, 90, 91, 93
lsqlin **264**
lsqnonlin **258**, 259
lyap2lmi* **139**, **140**

max 156, 179, 190–192, 269, 271, 273, 304, 310
meshgrid 72, 73, 154, 156, 165, 210, 213, 216, 218, 225, 267, 271, 273
min 190, 191, 272, 289
mincx **142**
more_sols* **41**, 43–47, 50
more_vpasols* **48**, 49
mpsread 123, 124
myout* 72

NaN 155, 156, 165, 257, 265, 273
nargin 21, 22, 120, 246
nargout 42
ndgrid 192, 193, 211, 212, 214, 215
new_bnb20* **223**, 224, 225, 227, 228, 235
norm 12, 13, 21, 22, 25–27, 29–31, 34–36, 38, 40, 42, 46–49, 68–70, 76, 79, 90, 94, 112, 134, 135, 145, 146, 175, 177, 181, 231, 310
nr_sols* **21**, 23, 25

ode45 93–95
ones 27, 29, 30, 65, 68, 84, 88, 89, 116, 136, 147, 174, 187, 188, 196, 224, 225, 230, 231, 289, 290
opt_con1*/opt_con2* 161, 162
opt_fun2* 167
optimize **144**, 145, **146**, 147, 148, 195, 197, 222, 231
optimoptions 50, **66**, 67, **78**, 79, 169, 220, 308
optimproblem **124**, 125–129, 133–137, 174, 176, 177, 221, 231, 232, 236, 237, 239, 241, 243, 246, 279, 309, 313
optimset **28**, 29, 30, 41, 43, 65, 66, 68, 70, 72–76, 86, 112–114, 124, 125, 161, 163, 166, 313
optimvar **124**, 125–129, 133–137, 174, 176, 177, 221, 231, 232, 236, 237, 239, 241, 243, 246, 279, 309, 313
otherwise 12

paretoFront‡ **267**, 268
particleswarm **301**, 302–304
patternsearch 300, **301**, 302–304, 307, 309, 310, 312
pi 18, 19
plot 23, 43, 44, 49, 61, 71, 89, 90, 92, 94, 188, 189, 191, 257, 265, 268, 272
polyval 12

prob2struct **129**, 130, 134, 135, 138, 175–177, 310, 313

quadprog **131**, 132, 134

rand 42, 48, 68–70, 73, 76, 78, 79, 81, 84, 87, 94, 97, 169, 171–173, 175, 177, 180, 188, 190–193, 196, 201, 218, 225, 233, 234, 270, 271, 274, 302, 304, 307, 309, 310, 312
real 30, 41, 48
reshape 69, 120, 177
roots **11**, 12, 13, 47
roots1* **11**, 12, 13
round 120, 175, 177, **220**, 221, 231, 232, 236, 238, 239, 241, 243, 246, 310
rref 107–109

sdpsettings 146–148
sdpvar **144**, 145–148, 184, 185, 194, 197, 222
sec_sols* **22**, 23, 25
set 285, 287
setlmis **142**, 143
shading 155, 156, 271, 273
showconstr **129**
showproblem **129**
show_sudoku* 246, 247
simplify 76
simulannealbnd 300, **301**, 302–304
sin 23, 24, 26, 40, 43
size 23, 36, 42, 44, 48, 49, 88, 89, 120, 124, 139, 214, 216, 240, 246, 262, 267, 289
solve **32**, 33, 34, **125**, 126–129, 133–137, 221, 231, 232, 236, 238, 239, 241, 243, 246, 279
solvebilevel **184**, 185
solvesdp 222
solve_sudoku* **246**, 247
sort 49, 210–215, 217, 218, 225
sparse 239, 241, 246, **283**, 285, 289, 291, 293
sqrt 12, 15, 47, 170, 198, 239, 240, 246, 293
strcmp 81, 172
struct2cell 48
subs 10, 11, 34–36, 40, 47, 49, 57, 159
sum 69, 121, 128, 133–135, 137, 146, 148, 157, 176, 177, 187, 188, 192, 193, 197, 222, 236–239, 241, 246, 267, 303, 304, 313
surfc 155, 156, 165
switch/case 11, 71
sym 12, 37–40, 45, 49, 107, 113, 135, 139
sym2poly 47
syms 10, 11, 14, 15, 18, 19, 22, 24, 32–36, 40, 44, 47, 49, 57, 64, 74, 76, 105, 155, 159, 167, 208

tic/toc 37–40, 42, 45, 48, 69, 79, 82, 124, 171, 173, 179, 181, 190–193, 214, 217, 218, 221, 222, 224, 225, 227, 228, 230, 231
transport_linprog* **120**, 121, 222
tsp_solve* **240**, 241

value **144**, 148, 184, 185, 195, 197, 222, 231
varargin 42, 84, 172, 262
varargout 42
view 155, 273, 284, 285, 287
vpa 34, 35
vpasolve 35–39, **40**, 45, 48, 57, 159

while 21, 22, 42, 48, 61, 68, 70, 91, 166, 240, 241, 267
writeproblem 129

ylim 59

zeros 23, 42–50, 112, 117, 120, 132, 139, 187, 188, 231, 233, 240, 264
zlim 73

Index

0–1 knapsack problem 243, 312
0–1 linear programming 229–233, 245, 248
0–1 programming problem 4, 207, 214–217, 235–237, 245, 313

Abel–Ruffini theorem 8, 13, 37
absolute error tolerance 28
acceleration constant 299
Ackley benchmark problem 97
additional parameter 63, 69, 70, 73, 94, 160, 161, 217, 223
affine matrix function 140
algebraic Riccati equation 26–30, 37, 38, 141
analytical solution 3, 7, 13, 15, 16, 24, 31, 33, 34, 37, 55–57, 65, 69, 98, 99, 134, 159, 205, 305
anonymous function 14, 16, 22, 24–26, 30, 42, 44, 56, 61, 63–65, 70, 73, 75, 77, 78, 80, 81, 83, 86, 90, 91, 93, 94, 97, 160, 161, 170, 175, 187, 190, 223, 259, 269, 270
ant colony algorithm 297
assignment problem 4, 207, 236–238
asymmetrical 134

backward difference method 61, 95
banana function 74
Beale test function 98
benchmark problem 55, 65, 73, 79, 97, 122, 134, 174, 303, 317
best compromise solution 261–264, 275
best known solution 205, 308
bilevel linear programming 182–184
bilevel programming 144, 153, 181–185
bilevel quadratic programming 182–184
binary encoding 298, 299, 315
binary programming 214
Bioinformatics Toolbox 280, 284, 285
block diagonal 142
boundary value problem 87, 93–95, 101
bounded knapsack problem 242
branch and bound method 219, 223
brutal force method 209

Cardano's formula 10
chick–rabbit cage problem 8, 32, 36, 37
chromosome 298
closed-form solution 10, 13
clue 135, 244, 245
colored 287
completing the square 9
complex coefficient 12, 13, 29, 33
complex conjugate 19, 29–31, 38, 41
complex root 35
complex solution 29, 31, 40, 41, 43, 45, 47, 52–54
compromise solution 254, 256, 257, 261
computational complexity 48, 207–209, 212
computational load 78, 187, 190, 192, 209, 211, 229
concave-cost 137, 171, 176–178
constrained optimization 55, 103, 130, 157, 194, 197, 200, 204, 298, 305, 317
constraint 103–105, 107, 110, 111, 114–116, 119, 124–131, 133, 134, 136–138, 141, 144, 145, 147, 148, 153–158, 160–163, 165–167, 169, 170, 172, 173, 175–182, 184–186, 188–193, 195, 197, 199, 200, 204, 205, 209, 210, 213–215, 217, 224–229, 234, 236, 238–240, 242, 243, 245, 254, 256, 260, 263–265, 267, 269, 273, 277, 278, 305, 306, 310–312, 315, 316
contour plot 71, 74, 80, 155, 164, 272
control option 14, 26, 28, 32, 43, 63, 65, 66, 70, 72, 73, 78, 111, 125, 148, 160, 165, 167, 259, 317
Control System Toolbox 27, 31
convex problem 105, 131, 138, 153, 174, 193, 317
crossover 298
cubic equation 2, 9–11
curve fitting 87, 89–93

De Jong benchmark problem 65
decision variable 56, 59, 60, 65, 67, 69, 72, 81, 83–85, 92–95, 104, 105, 107, 111, 114, 116–118, 120–124, 127, 129, 136, 141, 142, 144, 153, 154, 156, 157, 160–163, 168, 169, 171–173, 178, 184–186, 194, 195, 197, 199, 200, 204, 207–209, 211–217, 220–224, 226, 227, 229–231, 236, 255–257, 259, 299, 310, 315, 316
decision variable bound 55, 83–85, 172, 212, 298, 300, 301, 311
deformed Riccati equation 53
differential equation 1, 7, 27, 93–95, 101

Dijkstra algorithm 4, 277, 288–290
direct transpose 30
discrete programming 207, 213, 226–229, 315, 317
Dixon problem 69
domain of definition 59, 60
dominant solution 265, 266
dominated solution 266, 267
dot operation 14, 22, 86, 88
dot product 293
double precision 12, 31, 40, 46
double subscripts 117, 118, 120, 127, 136–138, 176
dual simplex 3, 111
dynamic programming 3, 4, 277–293

efficiency 3, 23, 25, 28, 55, 62–64, 69, 75, 77–79, 97, 101, 104, 158, 204, 219, 221, 222, 224, 233–236
Eggholder function 99
eigenvalue 11
empty matrix 111–114, 160, 161, 163, 165, 169, 172, 273, 284
end node 241, 280, 281, 287–290
enumerate method 4, 207, 209–221, 223, 225, 230, 234, 238, 247, 248, 267
envelop 269, 270
envelop surface 268, 271, 273
equality constraint 104, 111, 113, 114, 117, 124, 158, 160, 170, 181, 198, 232, 233, 273
error message 29, 31, 66, 83, 166, 232, 310
error norm 13, 25, 40, 46, 68, 93, 135
error tolerance 21, 22, 26, 28, 65–68, 142, 177
Euclidian distance 239, 292, 293
evolution 297–299
extraneous root 47

feasible solution 103, 106, 131, 136, 138, 141–143, 153, 207–211, 215, 217, 225, 229, 267, 313
Ferrari's formula 11
finite vector 172
Finkbeiner–Kall quadratic programming 150
first-order derivative 22, 56, 57, 59, 62, 159
fitness function 298, 300, 310
fitting 89, 259
fitting quality 100
Freudenstein–Roth function 98
function call 25, 28, 32, 64–70, 84, 87, 122, 160, 165, 166, 171, 172, 210, 220
function handle 22, 25, 26, 42, 63, 70, 72, 78, 81, 84, 86, 90, 160, 175, 259, 300

generalized eigenvalue 141, 142
generalized Riccati equation 38, 39, 44, 45
genetic algorithm 4, 81, 83, 297–300, 303–307, 309–317
Global Optimization Toolbox 297, 299–302, 305, 309, 312
global optimum solution 3, 4, 55, 59, 60, 79, 81–87, 105, 106, 111, 131, 132, 135, 136, 147, 148, 153, 155, 156, 158, 159, 164, 169, 171–179, 181, 195, 199, 202, 203, 205, 208–218, 225, 227, 228, 234, 235, 250, 258, 297, 299, 302, 303, 305–308, 310–315, 317
gradient 55, 62, 63, 66, 67, 73–77, 159, 166–168, 298
graphical method 13–20, 31, 35, 52, 55–57, 105, 107, 149, 154, 156, 164, 208, 253, 255, 256
Griewangk benchmark problem 97

$\mathcal{H}_\infty$ norm 145
Hankel matrix 144
Hartmann function 99
heat exchange network 4, 196–199
heating system 196
Hermitian matrix 138
Hermitian transpose 30
Hessian matrix 131, 134, 137
high-precision 7, 31, 41, 48, 49

identity 2
identity matrix 30
implicit equation 1, 13, 14, 18, 19, 50, 96
incident matrix 283–285, 287, 289–291, 293
indexed matrix 245
individual 298, 299
inequality constraint 104, 105, 109–113, 158, 160, 161, 180, 182, 187, 190, 199, 217, 232, 233, 235, 306
infinite loop 41
initial research point 4, 22, 25, 26, 29, 40, 41, 58, 59, 68, 71, 111, 131, 160, 165, 166, 169, 171, 177, 200, 259, 309
inner constraint 184
inner objective function 184

inner polygon 185–188
integer linear programming 120, 124, 207, 219–222, 236, 243, 314
intelligent algorithm 4, 81, 229, 297, 300, 302–305, 307, 310, 312, 317
intermediate result 22, 23, 25, 26, 61, 63, 66, 67, 71, 77, 162, 163, 257
intermediate variable 94, 198
interpolation 77, 78, 100
intersection 15, 17–20, 41, 43, 56, 236
irrational degree 47
irrational number 34, 47, 48
isolated solution 20, 41, 50
iteration 23, 28, 60, 61, 64–67, 71, 76, 113, 115, 161, 162, 165, 166, 168, 171, 239, 281

Jacobian matrix 21, 22, 24–26

Karush–Kuhn–Tucker algorithm 182
kinetic equation 53
knapsack problem 4, 207, 235, 242, 243, 312
Kursawe benchmark problem 98

Lagrange multiplier 158, 159
large-scale problem 66, 78, 192, 219, 220, 242, 282
least squares 7, 87–90, 92, 100, 258, 259, 263, 264
Levenberg–Marquardt algorithm 50
linear combination 88
linear constraint 154, 163, 169, 217, 231
linear matrix inequality 3, 103, 138–143
linear programming 2, 3, 103–132, 138, 141, 145, 182, 183, 260, 262, 263, 277–279, 293
linear regression 88–90
LMI 138–146
local optimum 4, 59, 84, 158, 169, 178, 195, 199, 305
loop structure 28, 41, 61, 68, 70, 91, 93, 128, 153, 171, 216, 260, 317

Marsha test function 156, 157, 162
mathematical programming 1–3, 103, 110, 122, 253, 277
MATLAB workspace 30, 42, 64, 70, 73, 91, 92, 115, 123, 142, 146, 237, 246
matrix of ones 27, 28, 30
matrix partition 142
maximum allowed iterations 28, 65, 165
maximum area 186
maximum value 2, 99, 100, 155, 179, 190, 192, 193, 268, 273, 278, 279, 304, 310
medium-scale problem 219
member 28, 65, 72, 73, 84, 111, 113, 115, 124, 125, 130, 132, 160, 161, 165, 167, 175, 177, 220, 223, 259, 264, 269, 300, 301, 309, 312
mesh grid 154, 156, 164, 192, 210, 211, 244, 245, 267, 272
Miele–Cantrell function 99
minimax problem 4, 253, 268–276
mixed 0–1 programming problem 229–235, 248
mixed-integer linear programming 207, 216, 219–221, 223, 224, 314, 315
mixed-integer programming 4, 207, 216, 218, 219, 221–223, 227, 235, 300, 301, 312, 314, 315
modified Rastrigin function 302, 305
momentum function 299
MPS file 122–124
multiobjective linear programming 261–264, 274
multiobjective programming 4, 253–275
multiple decision variable 116, 168
multiple-valley 87
multistage decision making 277, 278
multiterm linear algebraic equation 7
multiterm Sylvester equation 7
multivariate equation 21, 22
multivariate function 21, 78, 90
mutation 298, 317

nature-inspired 297, 299, 310
negative-definite matrix 138
Newton–Raphson method 7, 20–24, 26, 61
node 239–241, 280–285, 287–293
nonconvex 4, 106, 132, 136–138, 146, 147, 150, 153, 171, 174, 175, 178, 201, 202, 205, 297, 309
nondominant set 267
nonlinear constraint 153, 160, 161, 163, 165, 168, 170, 178, 180, 186, 195, 200, 217, 225, 226, 234
nonlinear function 21
nonlinear programming 153–200, 207, 217, 219, 297, 299, 305, 307
nonuniqueness 10, 95, 128, 157, 185, 253, 257, 264, 265

norm 25–28, 31, 34, 36, 38, 49, 69, 70, 76, 112, 146, 175, 178, 181, 231, 259
NP-complete problem 209
NP-hard problem 209, 238
number of iterations 25, 26, 28, 63, 190

objective function 55–82, 84–86, 90, 93, 94, 96, 97, 99, 103–109, 111–121, 123–134, 141, 142, 144, 146–148, 153–173, 175–182, 184, 185, 187–193, 195, 197, 198, 204, 205, 207–218, 222, 223, 226–232, 234–236, 239, 253–260, 262–274, 277–279, 298, 300, 301, 304–306, 308–317
Optimization Toolbox 4, 55, 62, 64, 66, 70, 72, 73, 78, 83, 90, 110, 123, 129, 131, 146, 153, 159, 171, 182, 189, 195, 204, 219, 223, 258, 269, 297, 298, 312, 317
oriented graph 4, 277, 279–293
outer constraint 184
outer objective function 182, 184
overdetermined equation 50, 87

P problem 209
parallel computing 55, 62, 78, 79, 83, 298
parametric equation 1
parent node 288
Pareto frontier 253, 257, 264, 265, 267
partial derivative 168
particle swarm optimization 4, 297, 299–301, 303, 304
partitioned matrix 140, 142, 143
pattern search algorithm 4, 297, 300, 303–305, 307–311
periodic function 277
physical interpretation 56, 103, 189, 269, 280
piecewise function 179, 180, 277
pivot 108, 109
planform 273
polygon 185, 188
polynomial equation 1–3, 7–9, 11–13, 32–34, 36, 37
polynomial matrix equation 27, 46
pooling and blending 4, 193–195
population 298, 317
positive-definite 27, 131, 139, 143
precision 12, 22, 35, 41, 48, 70, 90, 112, 148, 304
probabilistic rule 298
profit maximization 121, 122, 194
prototype function 88–92
pseudopolynomial 7, 46–48
punish function 157

quadratic constraint 147, 148, 152
quadratic form 37, 130, 134, 146
quadratic function 60
quadratic programming 103, 124, 131–138, 145, 146, 150, 153, 161, 171, 174–177, 309
quadratic transportation 136
quartic equation 2, 10–13, 33, 34
quasi-Newton algorithm 63, 67, 77
quasianalytical 3, 7, 35–41, 45, 48, 49, 51, 159

random initial value 41, 81, 134, 169, 172
Rastrigin function 79, 80, 82
Raydan problem 98
real root 13
recursive formula 61
reduced row echelon form 107, 108
regular matrix 130, 284
relative error tolerance 28
relax variable 109, 110, 112
repeated root 11, 12
reproduction 298, 305
Riccati equation 26, 27, 29, 37, 39, 40
Riccati inequality 141, 143
Riemann sheet 47
Robust Control Toolbox 138, 142–144, 151
Rosenbrock function 73–75, 98

samples 77, 78, 88, 90, 189, 192, 227
scalar 46, 124, 200, 253, 254
scalar function 56, 154, 253, 254
scalar objective function 258, 260, 268, 273
scattered samples 77, 78, 100, 266
Schur complement 140, 141, 143
Schwefel function 99
secant method 22, 23, 25
second-order derivative 56, 57, 61, 158
semiinfinite programming 4, 153, 189–193
shooting method 93
shortest path 4, 277, 279–282, 284–288, 292, 293
simplex method 2, 3, 63, 66, 103, 104, 106–110, 112
simulated annealing algorithm 4, 297, 300, 301, 303, 304

small-scale problem 4, 130, 209, 211, 214, 223, 234, 282
sorting 210, 215
sparse matrix 130, 239, 283, 284, 293
special matrix 142, 144
spline interpolation 77
square matrix 140, 284
square root 9, 15, 33
start node 241, 284, 287–289
state variable 278
structured variable 28, 32, 65, 70, 72, 73, 75, 76, 81, 84, 111, 113, 123, 124, 129, 132, 134, 135, 137, 160–163, 166, 167, 169, 171–174, 176, 177, 180, 219–221, 223, 224, 235, 259, 269, 274, 300, 301, 309
submatrix 92
suboptimal 211–214
subtour 239–241
subtour detecting 240
successful rate 82, 83, 179, 181
sudoku 4, 235, 244–247, 250
switch structure 71, 222
Sylvester equation 7
symbolic computation 2, 7, 33
symbolic expression 14–16, 18, 19, 24, 32, 33, 35–37, 39, 40, 49, 56
Symbolic Math Toolbox 31, 74
symmetrical matrix 139, 142, 143, 291

tangent line 20
Taylor series expansion 60
three-dimensional plot 73
time elapse 38, 49, 69, 79, 83, 124, 173, 179, 181, 190–193, 214, 217, 218, 221, 222, 224–226, 228, 230–235, 302–304, 306, 307, 310, 311, 313
Townsend function 163
traveling salesman problem 207, 235, 238–242, 250
trust region algorithm 63, 67, 76

unbounded knapsack problem 242
unconstrained optimization 3, 55–97, 103, 146, 154, 157, 299–302
uncontrollable 29
underdetermined equation 49, 50, 96
undetermined coefficient 88–93, 100
undigraph 4, 290, 291, 293, 294
uniform distribution 78, 82, 299
unique solution 7, 245, 262

valley 5, 64, 79, 80, 97, 100, 164
variable substitution 10, 11, 35, 36, 63, 168, 226
vertex 105–107
Viéte theorem 8, 10

warning message 28, 31, 39, 50, 65, 68, 87, 137, 146, 166
weight 242, 250, 260, 261
weighted sum 258
weighting 131, 260, 261, 293

YALMIP Toolbox 138, 144–148, 152, 182, 184, 185, 194, 195, 197, 219, 222, 231

zoom 14–17, 19, 31

www.ingramcontent.com/pod-product-compliance
Lightning Source LLC
Chambersburg PA
CBHW080917220326
41598CB00034B/5599